CHALLENGING THE P[illegible] ORDER

Europe and the International Order

Series Editor: Joel Krieger

Published

R. J. Dalton and M. Kuechler (eds), *Challenging the Political Order*
A. de Swaan, *In Care of the State*
P. Hall, *Governing the Economy*
J. A. Hellman, *Journeys among Women*
L. Holmes, *The End of Communist Power*
J. Jenson, E. Hagen and C. Reddy (eds), *Feminization of the Labour Force*
J. Krieger, *Reagan, Thatcher and the Politics of Decline*
A. Lipietz, *Towards a New Economic Order*
A. S. Markovits and P. S. Gorski, *The German Left*
F. Fox Piven (ed.), *Labor Parties in Postindustrial Societies*
G. Ross, S. Hoffmann and S. Malzacher, *The Mitterrand Experiment*

Forthcoming

J. Keeler, *The Scope and Limits of Democratic Reform*
C. Murphy, *Global Governance and Industrial Change*
G. Ross, *The Construction of Europe*

Challenging the Political Order

New Social and Political Movements in Western Democracies

Edited by Russell J. Dalton and Manfred Kuechler

Oxford University Press · New York
1990

Oxford University Press

Oxford New York Toronto
Delhi Bombay Calcutta Madras Karachi
Petaling Jaya Singapore Hong Kong Tokyo
Nairobi Dar es Salaam Cape Town
Melbourne Auckland
and associated companies in
Berlin Ibadan

First published 1990 by Polity Press in association with Blackwell Publishers.

First published 1990 in North America by Oxford University Press, Inc., 200 Madison Avenue, New York, New York 10016.

Library of Congress Cataloging-in-Publication Data

Challenging the political order : new social and political movements in western democracies / edited by Russell J. Dalton and Manfred Kuechler.
p. cm. — (Europe and the international order)
Includes bibliographical references.
ISBN 0–19–520832–3. — ISBN 0–19–520833–1 (pbk.)
1. Social movements—Europe. 2. Radicalism—Europe. 3. Europe––Politics and government—1945– 4. Europe—Social conditions—20th century. I. Dalton, Russell J. II. Kuechler, Manfred.
III. Series: Europe and the international order (New York, N.Y.)
HN373.5.C47 1990
303.48'4'094—dc20 90–6772
CIP

Typeset in 10 on 11 pt Baskerville
by Photo·graphics, Honiton, Devon
Printed in Great Britain by
T. J. Press, Padstow, Cornwall

Contents

Part III: Networks of Action

Part IV: New Movements and Political Parties

Part V: New Social Movements in Perspective

Notes on Contributors

Karl-Werner Brand is Lecturer in Political Theory and Political Sociology at the Technical University of Munich, West Germany. His research interests include social movements and the history of political thought. He is the author of *Neue Soziale Bewegungen* (1982), a co-author of *Aufbruch in eine andere Gesellschaft* (1983), and the editor of *Neue Soziale Bewegungen in Westeuropa und den USA* (1985).

Wilhelm Bürklin is Lecturer (*Akademischer Rat*) in Political Science at the University of Kiel, West Germany. His major research interests are party systems, voting behavior, theory of democracy, and policy implementation in Southeast Asia. He is the author of *Grüne Politik* (1984) and *Wählerverhalten und Wertewandel* (1988).

Russell J. Dalton is Professor of Political Science at the University of California, Irvine. His research interests include comparative political behavior, political parties, and political change in advanced industrial societies. He is co-author of *Germany Transformed* (1981), co-editor of *Electoral Change in Advanced Industrial Democracies* (1984), and author of *Citizen Politics in Western Democracies* (1988) and *Politics in West Germany* (1989). He is now writing a book-length study of the environmental movement in Western Europe.

Joyce Gelb is Director of the Rosenberg/Humphrey Program in Public Policy at City College of New York, and on the Political Science faculty of the City College and Graduate Center, CUNY. She is author of the just published *Feminism and Politics: A Comparative Perspective* (1989), as well as *Women and Public Policies* (1986) and numerous articles on women in comparative politics and issues related to ethnicity and urban policy.

Ronald Inglehart is Professor of Political Science and Program Director at the Institute for Social Research at the University of Michigan. His research focuses on social and political change; he has helped design and analyze the Eurobarometer surveys since their inception. His theory of value change, first proposed in the early 1970s, has subsequently received confirmation from research conducted in dozens of countries. Author of nearly a hundred publications, his most recent book is *Culture Shift in Advanced Industrial Society* (1989).

Max Kaase is Professor of Political Science and Comparative Social Science Research at the University of Mannheim, West Germany. He has published more than a hundred articles in scholarly journals and edited volumes in the areas of political sociology, democratic political systems, and mass communication. He is a principal co-editor (with Samuel Barnes) of *Political Action: Mass Participation in Five Western Democracies* (1979).

Herbert P. Kitschelt is Assistant Professor of Political Science at Duke University. He is the author of five books, including *The Logic of Party Formation* (1989) and *Beyond the European Left* (1990), and numerous articles in American and European journals on such topics as energy policy-making, new social movements, and political parties. He is currently completing a book about theories of technology policy-making.

P. Bert Klandermans is Senior Lecturer in Social Psychology at the Free University, Amsterdam. His major research interests are mobilization and participation in social movements, persuasive communication, and collective action. He is editor of the JAI book series, *International Social Movements Research*, and has edited two volumes on social movements: *From Structure to Action: Comparing Movement Participation Across Cultures* (1988, with Hanspeter Kriesi and Sidney Tarrow), and *Organizing for Change: Social Movement Organizations in Europe and the United States* (1989).

Manfred Kuechler is Professor of Sociology at Hunter College and the Graduate Center of the City University of New York (CUNY). Before he took residence in the United States in 1985, he was a professor at the University of Frankfurt, West Germany. His numerous articles in scholarly journals and contributions to edited volumes are in the areas of voting behavior, social movements, and research methodology.

Ferdinand Müller-Rommel is Lecturer (*Akademischer Rat*) in Political Science at the University of Lüneburg, West Germany. He works in the areas of comparative politics, political parties, and government in Western Europe. He is the author of *New Politics in Western Europe* (1989), *Cabinets in Western Europe* (1988), and *Small Parties in Western Europe* (forthcoming).

Claus Offe is Professor for Political Science and Sociology at the Center for Social Policy Research of the University of Bremen, West Germany. His research is in the area of political institutions, labor market, and social policies. Among his publications are *Contradictions of the Welfare State* (1984) and *Disorganized Capitalism* (1985).

Thomas Rochon is an Associate Professor in the Center for Politics and Policy at Claremont Graduate School. His research interests include political movements, elite politics, and social policy. He has recently published *Mobilizing for Peace* (1988), an analysis of the European peace movement, and is now working on a study of retrenchment in West European welfare states.

Dieter Rucht is Senior Research Associate at the Wissenschaftszentrum Berlin für Sozialforschung (WZB), West Germany. He works in the areas of social movements in modern welfare states, environmental policies and politics, and political participation. He is the author of *Von Wyhl nach Gorleben* (1980) and *Planung und Partizipation* (1982), a co-author of *Aufbruch in eine andere Gesellschaft* (1983), and co-editor of *Neue Soziale Bewegungen in der Bundesrepublik Deutschland* (1986).

Sidney Tarrow is Maxwell M. Upson Professor of Government at Cornell University and the author of *Peasant Communism in Southern Italy* (1967), *Between Center and Periphery, Democracy and Disorder* (1989), and co-editor of *Communism in Italy and France* (1975, with Donald Blackmer) and *From Structure to Action* (1988, with Bert Klandermans and Hanspeter Kriesi). He is currently working on a comparative study of protest cycles in modern democracies under the sponsorship of the National Endowment for the Humanities.

Frank L. Wilson is Professor and Head of Political Science at Purdue University. He is involved in research on social movements and political parties in France. His principal publications include *Interest Group Politics in France* (1987), *French Political Parties under the Fifth Republic* (1982), *European Politics Today: The Democratic Experience* (forthcoming), and numerous articles in scholarly journals.

Preface

The recent flowering of new movements concerned with environmentalism, women's rights, peace, consumerism, and the other pressing issues of advanced industrial societies has affected the political order of most Western democracies. This development stimulated an extensive series of historical and descriptive essays on these movements, which often made dramatic claims about the societal or political importance of these movements. To their supporters, these new social and political forces are seen as the vanguard of a new society; to their critics, new social movements represent a fundamental threat to the social and political order. The goal of this book is to examine these claims and determine what is "new" about these new movements.

In recent years, a new round of systematic, often comparative, theory-testing research has begun. Recognizing the potential theoretical and political importance of these new movements, private research foundations and government agencies on both sides of the Atlantic have sponsored detailed studies of these new social interests and their actions within the political process.

To consolidate this burgeoning literature on new social movements and assess the progress research has made, Russell Dalton and Wilhelm Bürklin jointly submitted a proposal to the National Science Foundation (INT 85–21364) and the *Deutsche Forschungsgemeinschaft* for a research seminar; Dalton and Bürklin developed the conference themes, secured separate national funding, and invited the conference participants. Approximately two dozen scholars who are actively engaged in relevant research projects met in Tallahassee in 1987 to compare findings and debate the significance of these movements. This volume presents a subset of the research reports presented at this conference, focusing on the impact of new social movements on the established political order of advanced industrial societies. Several additional chapters were solicited for the book to develop this topic more fully.

We intentionally brought together scholars with different methodologies and different perspectives on these movements; we wanted to provide a forum for comparing research findings, and this diversity of views is well represented in this collection. After considering rival theories of the origins of these new movements (presented as the first section of this volume), most of our discussions focused on the role of new social movements within the political process of Western democracies. We explored the widespread belief that new social movements are isolated from established interest groups and routine channels of interest representation, producing a propensity toward a potentially destabilizing pattern of unconventional political action. Another section of this study considers whether new social movements are weakening the structure of party government in Western democracies (partisan dealignment) or creating new political forces which will restructure party cleavages (partisan realignment). In addressing these questions, the book provides a firm vantage point for judging the implications of new social movements for the political order of Western democracies – and our view is more moderate and sanguine than most of the early theoretical and political literature.

Because of his growing involvement in a new study of the political and cultural basis of economic success in the newly industrializing countries of East Asia, Bürklin was unable to oversee the preparation of this volume. It was therefore agreed to shift the editorial responsibilities to Russell Dalton and Manfred Kuechler who were both at Florida State University at the time. Willi Bürklin is a contributor to the introductory chapter; his efforts in organizing the conference and his advice throughout the preparation of this manuscript deserve our recognition and appreciation.

In compiling this book we are also indebted to the participants of the conference. We would especially like to thank those who contributed their wisdom and insights to our discussion – Jim Fendrich, Scott Flanagan, Bill Gamson, Jost Halfmann, Larry Isaac, Roger Karapin, Ellis Krauss, Franz Lehner, Andrei Markovits, Margit Mayer, Lester Milbrath, Joyce Mushaben, Helmut Norpoth, Karl-Dieter Opp, Karlheinz Reuband, Dietrich Tränhardt, and Ed Walsh – but who are not represented in this volume. The financial support of the National Science Foundation, *Deutsche Forschungsgemeinschaft*, and the College of Social Sciences at Florida State University was essential to the success of this project. Russell Dalton received additional support for this work under a National Science Foundation grant (SES 85–10989). We would also like to thank Herr Grünhagen at the University of Kiel for his assistance with the German preparations for the conference, Robert Rohrschneider for his efforts in organizing the conference in Tallahassee, and Mary Schneider at Florida State University for her help in preparing the book manuscript. Joel Krieger, David Held, and Valerie Aubry provided the support and assistance that helped us through the publication process; the mark of a good publishing team. Finally, book editors are something like symphony conductors: we receive the applause for the work of others – thus we want to thank the artists in our orchestra for their contributions (and their tolerance of our advice for revisions), and remind the reader that the book is really their accomplishment.

RJD, Tallahassee
MK, New York City

Part I
Introduction

1

The Challenge of New Movements

RUSSELL J. DALTON, MANFRED KUECHLER, and WILHELM BÜRKLIN

The challenge that new movements pose to the political order in Western democracies springs from within. It is not a revolutionary attack against the system, but a call for democracies to change and adapt. The challenge comes from individuals and new social groups demanding that democracies open the political process to a more diverse and citizen-oriented set of interests. It manifests itself in the proliferation of citizen-interest lobbies and single-issue groups in America, civic associations in France, *Bürgerinitiativen* in West Germany, and citizen-action groups in other industrial democracies. The challenge also arises from the new issues and political demands these citizen groups are bringing into the political process.

At least briefly, we can provide an overview of the origins of this challenge, tracing the emergence of new movements to a variety of sources that have affected most Western industrial societies. One contributing factor dates back to the student movement of the 1960s. The halcyon days of the early 1960s quickly gave way to critical protests against the goals of Western society and the functioning of the democratic political process. Students at Berkeley embroiled the campus in turmoil over the issues of free speech and civil rights; a conflict over administrative regulations at a small university outside Paris escalated into the May Revolts and brought France to the edge of collapse; student protests in Berlin again gave the city the appearance of a war zone. Like the revolutionary fever of 1848, student rebellion spread among Western industrial democracies.

Although the student movement began to dissipate within a few years, it marked the beginning of a broader wave of social change that has affected virtually all advanced industrial democracies. The novelty of the student movement was not so much the issues it raised, but in broadening the discourse on these matters from elite debate to a mass movement – and with this broadening of participation the content of the debate often grew beyond the narrow ideological discourse of the students. The general public developed political interests beyond traditional economic and class issues to a range of new social, cultural, and quality of life issues that emerged from the modernizing process occurring in these societies (Inglehart, 1977, 1989). A new style of political action developed, as citizens shifted from traditional methods of interest representation to a more participatory political style (Barnes, Kaase et al., 1979; Jennings and van Deth, 1990).

This volume focuses on the new social and political movements that act as manifestations of these new political forces. A host of new environmental groups were formed in North America and Europe during the early 1970s; a new, more assertive women's movement joined the earlier women's groups; consumer and self-help groups multiplied throughout the 1970s; and these groups were joined, at least temporarily, by a revitalized peace movement in the early 1980s. These new groups are now important and contentious actors in the political process of many Western democracies. These organizations translate the public's changing values and issue interests into a potential political force; they channel the energies of the movement; and they decide on the political goals and strategies of the movement.

Although environmental, feminist, peace organizations and other new social movements take on a wide variety of forms, several analysts claim that at the core of these groups is a *qualitatively new* aspect of citizen politics in Western democracies (Brand et al., 1986; Melucci, 1980; Touraine, 1983; Van der Loo et al., 1984; Capra and Spretnak, 1984). Drawing upon the New Left ideology of the student movement and the unconventional political tactics of student protests, many of these new groups apparently represent a fundamental change from the prevailing social goals and political style of Western industrial democracies. No single group fully typifies this new type of interest representation, but its essential characteristics are visible among various groups within the environmental, women's, and peace movements. West German sociologists coined the term "Neue soziale Bewegungen" to describe this phenomenon (Brand, 1982; Brand et al., 1983), and the term "New Social Movements" (NSM) entered the English research vocabulary as an identifier for this new type of interest organization.

The NSM concept probably originated among West German social scientists because these movements marked a dramatic new development for West German society that captured the attention of political observers, and because the German movements accentuated the traits identified with the new social movement concept. The spread of citizen action-groups (*Bürgerinitiativen*) in the 1970s, the development of national umbrella groups such as the BBU (*Bundesverband Bürgerinitiativen Umweltschutz*), and eventually the formation of the Green Party in 1980 created an extensive network of new social groups. The political actions and rhetoric of the West German movements also loudly proclaimed that these were groups unlike other interest groups or previous social movements.

While the West German movements may provide the most visible examples of the new social movement phenomenon, similar developments exist in most other industrial democracies. A well-developed structure of environmental groups, self-help organizations, women's groups and other new social movements exists in the Netherlands. The Dutch environmental movement, for example, is one of the most vibrant in Europe, and the Dutch peace movement was one of the most effective national groups in opposing the stationing of new NATO nuclear missiles in the early 1980s. The Danish social movements display a rich diversity, including a strong core of student-oriented and unconventional political groups; and these movements have achieved notable success in mobilizing popular support and influencing governmental actions. In more muted form, the British, French, and Belgian movements display similar political orientations. In this instance, as in the case of the labor movement, the American social movements may be the deviant case. The broad ideological *Weltanschauung* that exists among activists in European social movements (old and new) is often less developed in America, although some American environmentalists, feminists, and peace groups have developed this new political orientation (Capra and Spretnak, 1984; Freeman, 1975).

It is claimed that new social movements challenge the contemporary political order on several fronts. On the ideological level, these movements advocate a new social paradigm which contrasts with the dominant goal structure of Western industrial societies (Dunlap and van Liere, 1978; Milbrath, 1984; Raschke, 1985). New social movements also illustrate a style of unconventional political action – based on direct action – that contrasts with the traditional neo-corporatist pattern of interest intermediation in many contemporary democracies. Even the organizational structures of these movements are supposedly unique, stressing participatory decision-making, a decentralized structure, and opposition to bureaucratic procedures. Thus, it is claimed that new social movements challenge the basic goals, structure, and organizational style of Western industrial democracies.

Both the supporters and critics of new social movements are quick to acknowledge this challenge. Leaders of environmental parties across Europe openly declare that they represent a party of a new type, qualitatively different from all other parties. The political theory and rhetoric of new social movements often proclaim their anti-establishment views with unbridled enthusiasm (e.g., Bahro, 1986; Kelly, 1984; Porritt, 1984). The political opponents of new social movements may be even more vocal in stressing the challenge these groups pose to Western industrial democracies. A frequent criticism, especially among conservative politicians, holds that new social movements are simply a front for revolutionary and anti-system political groups (Langguth, 1984; Fogt, 1987; Kaltefleiter and Pfaltzgraff, 1985). Even the more thoughtful academic critics of new social movements have often resorted to alarmist claims. Michel Crozier and his colleagues (1975) maintain that the social forces represented by new social movements do not represent a challenge to democracies but a *Crisis of Democracy*. Samuel Huntington (1974, 1981) claims that Western democracies suffer from an "excess of democracy," and unless the demands and political activities of new social movements and similar groups are restrained, the democratic order is vulnerable to collapse.

This volume focuses on this on-going debate over the significance of new social movements for the goals and political process of advanced industrial democracies. The recent development of the modern environmental movement, feminist groups and the peace lobby makes it difficult to gain perspective on the real implications of these movements. In addition, these movements are far from monolithic. One can find evidence of virtually every conceivable political philosophy within these movements, and the behavior of movement activists covers an equal range. Given this diversity, debate on the meaning and implications of new social movements is inevitable.

Our goal is to harness the first wave of systematic, scientific research on these groups to determine the extent of the challenge that new social movements pose to the established political order. We are primarily interested in the political impact of these new social movements, although the findings have broad relevance to other important theoretical and sociological questions. In some ways, in fact, the label "New Social Movements" may be a misnomer, because we are interested in how the boundary of politics has expanded to include social movements that are actively influencing policy on environmental issues, women's rights, and peace and disarmament issues (Offe, 1987a). This book examines how these movements function within the existing structures of interest intermediation, dealing with other interest groups and representatives of the state. In addition, we examine the competing claims regarding the partisan impact of these movements, as a force for partisan realignment or dealignment (Dalton et al., 1984). The behavior of these movements, we believe, provides a key to understanding the evolving political order of advanced industrial democracies.

Approaches to the Study of Social Movements

Social movements have repeatedly acted as a force for social and political change: the agrarian movement and labor movement altered the course of Western development, and the American civil rights movement reshaped basic aspects of social relations in the span of a single decade. Thus the recent emergence of movements dealing with environmental issues, sexual equality, consumer rights, and related issues does not alone constitute a radical departure from the normal course of societal evolution. Still, the emergence and political style of these movements was difficult to explain with existing theories of social protest and collective action.

Early attempts to explain the student protests of the 1960s and the development of new social movements in the 1970s initially turned to the existing literature on collective violence. Perhaps the most prominent example of this genre was Ted Gurr's discussion of *Why Men Rebel* (1970). Gurr presented a model in which frustration induces aggression; that is, feelings of relative deprivation with one's economic or social situation lead to political violence. The civil rights demonstrations of the 1960s were often interpreted in terms of this dissatisfaction model; like de Tocqueville's description of the French revolution, blacks protested when their expectations expanded more rapidly than reality.

This research tradition carried over to the study of student activism and new social movements: students protested because they were radicalized by their lack of social influence or their economic marginality. Certainly, many individuals participated in new social movements to express dissatisfaction with specific policy problems or to criticize general social norms. The West German Green party, for example, drew disproportionate support from the young, unemployed (or underemployed) academics who were frustrated with their limited job opportunities (Bürklin, 1984, 1985). Researchers applied a similar logic to explain the formation of women's groups in reaction to sexual discrimination and the creation of environmental groups in reaction to the excesses of capitalist economies.

Despite the value of the relative deprivation approach in explaining some patterns of political violence, it does not seem to tap the essence of the new social movement phenomenon. Activists in new social movements often hold intense feelings about their cause, but these sentiments fall short of the primordial frustration-aggression emotions that spawned food riots and tax revolts in the eighteenth century and the revolutionary movements of the nineteenth. Moreover, student protestors and environmental activists are not primarily drawn from the ranks of the socially deprived.[1] In stark contrast, the bastions of these new protests were the institutions of traditional rank and privilege: Berkeley, Columbia, Oxbridge, the Sorbonne, Heidelberg, and Berlin. While many women are socially and economically disadvantaged, the stimulus for the women's movement was more often located among affluent feminists. In short, these are all predominantly middle-class movements, whose members benefit from the existing social and political orders.

Empirical evidence also provided uncertain support for the relative deprivation model. Edward Muller's research on the relationship between dissatisfaction and protest found only a weak (and non-linear) empirical link for political activists in the late 1960s (Muller, 1972; Grofman and Muller, 1973). Muller's more extensive analyses of activism in the 1970s failed to produce more convincing support for the relative deprivation model (1979). The evidence marshalled by the *Political Action* study similarly found that feelings of relative deprivation exerted only a marginal impact on the propensity to engage in protests (Barnes, Kaase et al., 1979, chs 13, 14). Subsequent research on the motivations for participation in new social movements has downplayed relative deprivation as an explanatory theory (e.g., Klandermans et al., 1988). Joan Gurney and Kathleen Tierney's critical look at two decades of relative deprivation research thus concludes that "the relative deprivation perspective is itself affected by too many serious conceptual, theoretical, and empirical weaknesses to be useful in accounting for the emergence and development of social movements" (1982, p. 33).

A much different approach to political movements sought to explain citizen action in terms of rational self-interest. This rational choice perspective maintains that revolutions and other forms of "collective" action are not motivated by psychological feelings of deprivation or concern for societal goals – only the hope for private gain motivates individuals to become politically engaged (Olson, 1965; Tullock, 1971). As the leading proponent of this thesis, Mancur Olson elaborated a rational choice model which held that individuals will not participate in large collective actions unless their expected "benefits" exceed the "costs" of their participation. Otherwise, the rational person would abstain and obtain the collective benefits without effort. Political action – ranging from support for a revolutionary party to involvement in an environmental group – is supposedly motivated by this narrow sense of self-interest. Deviations from this model presumably can occur only when coercion (e.g., forced membership) or other external constraints distort this rational calculus of action.

This rational choice perspective is well known among social movement scholars, but it has not attracted much support beyond a group of political economists (e.g., Downing and Brady, 1974; Frohlich et al., 1971; Moe, 1980). Basic flaws in logic and realism hinder the applicability of the theory to new social movements. The rational choice model is most effective in explaining why most people do not participate in groups representing their interests, but it stumbles over the basic question of why a small minority still do participate. The model explains participation in terms of a narrow definition of rationality, focusing on economic or finite rewards. The notion of altruism or collective social gain is not allowable within the model, otherwise the methodological elegance of rational choice calculus is lost.[2]

New social movements apparently violate many of the basic precepts of the rational choice model. The stated goals of most of these movements concern collective goods – protection of environmental quality, improvement in the status of women, and a lessening of international conflict – directly contradicting the rational choice logic of self-interested action. Attempts to rephrase these movements as agents of self-interest generally lack credibility.[3] New social movement protests seldom generate immediate personal gains for the participants (in the terms of rational choice models); few environmentalists or peace marchers can identify the private benefits they reap for their efforts, though the costs are easily recognized. Moreover, empirical research shows that ideological and collective goals outweigh calculations of narrow self-interest in motivating individuals to participate in environmental groups, anti-nuclear protests, and other forms of collective action (Mitchell, 1979; Opp, 1984; Muller and Opp, 1986; Fireman and Gamson, 1979).

Faced with the proliferation of citizen-action groups in the United States that were inconsistent with essential elements of the collective violence and rational choice theories, American sociologists turned to a *resource mobilization* model of political action (Oberschall, 1973; McCarthy and Zald, 1973; Zald and McCarthy, 1987; Jenkins, 1983). Resource mobilization theorists presume that political dissatisfaction and social conflicts are inherent in every society; thus the formation of social movements depends not on the existence of these interests but on the creation of organizations to mobilize this potential. This perspective led researchers to focus their efforts on examining the organizations which mobilize people, money, and other resources in pursuit of a cause, and the factors that affect the creation of such organizational infrastructures.

The resource mobilization theory provides a valuable new conceptual framework for studying social movements. It shifts the focus of our attention from the sources of citizen dissatisfaction to the social movement organizations (SMO) that give meaning and direction to the movement. Through this organizational approach, scholars saw new aspects of social movements. Once a social movement organization was formed, for example, analysts found that the implementation of desired policies had to compete with the organization's desire to maintain itself. In contrast to the amorphous nature of the underlying social movement, many SMOs adapt a hierarchical and highly routinized structure to maximize their efficiency in collecting money, activating members, and mobilizing other resources. Organizational planning means that the tactics of the SMOs are not based on emotional outbursts of frustrated citizens but on conscious calculations of how best to advance the organization's goals – whether through dramatic protests or quiet political lobbying. Moreover, some of an SMO's activities must be directed at organizational maintenance, mobilizing new members and new financial supporters, and not just policy influence. The resource mobilization approach also highlights the importance of individual entrepreneurs in creating and directing these organizations. SMOs often would not exist without the initiative of a single individual or group of people, even though public interest in the cause may be longstanding. In short, the resource mobilization model provides an integrated theory of how organizations are formed, public support is mobilized, organizational behavior developed, and political tactics decided.

The resource mobilization approach was widely adopted in research on social movements in the United States, including new social movements. Ed Walsh (1981, 1989; Walsh and Warland, 1983) very effectively used this theoretical framework to examine the mobilization of local opposition to nuclear power in the wake of the Three Mile Island incident. Jo Freeman (1975), among others, analyzed the American women's movement from a resource mobilization perspective, exploring the linkage between the goals and organizational structure of various women's groups. Resource mobilization was applied to groups ranging from farm workers to anti-bussing groups, to the sanctuary movement, to Mothers Against Drunk Driving.

Although the study of new social movements benefited from the ability of resource mobilization theory to explain numerous aspects of social movement behavior, the apolitical nature of the theory seemingly missed an important element of what was "new" about new social movements. The theory appeared indifferent to the political or ideological content of a movement; it was applied in an almost mechanistic way to organizations of widely differing political and ideological scope, without incorporating these factors within the workings of the model.

European attempts to explain new social movements contrasted with the resource mobilization approach by giving a more prominent role to ideological factors.[4] The leadership corps of many new social movements hold a distinct New Left political viewpoint born during the 1960s student movement. This ideology leads new social movements to challenge many of the consensual goals of Western society, to adopt political tactics that contrast with the traditional neo-corporatist forms of interest group activity, and to employ a new organizational structure as an extension of their ideals for societal reform. Consequently, ideology could not be ignored in discussing new social movements, as it was in resource mobilization theory.

Many of the leading scholars on new social movements maintain that this political identity affects the basis, structure, and tactics of the movement to produce a new form of interest representation in advanced industrial democracies. Thus, new social movements are not just chronologically new; they also represent a qualitatively new aspect of contemporary democratic politics. The claims that new social movements represent a qualitative change in political goals and the pattern of interest representation lie at the heart of this volume.

The Contrast between New and Old Movements

The theorists who maintain that new social movements are unique do not base their argument on a single point, but on a set of factors that in combination tap the essence of such movements. A general pattern of differences in the ideology, origins, structure, style, and goals of these groups is often cited to define what is "new" about NSMs. These factors are what distinguish new social movements from "old" social movements such as the labor or agrarian movement, as well as their own historical predecessors. Researchers still disagree on whether these theorized differences exist and on the explanation of these differences, but this NSM model often underlies research on new social movements.

Much of the new social movement theory has not yet entered the English-language research literature or remains implicit in writings on the topic. Therefore, it is worthwhile to summarize briefly what this literature *claims* is "new" about these movements to provide a frame of reference for the arguments made in subsequent chapters of this volume.

IDEOLOGY

The defining characteristic of new social movements is their advocacy of a new social paradigm which contrasts with the dominant goal structure of

Western industrial societies (Dunlap and van Liere, 1978; Milbrath, 1984; Raschke, 1985; Cotgrove, 1982). For instance, new social movements generally question the emphasis on wealth and material well-being that is prevalent in industrial democracies; instead they advocate greater attention to the cultural and quality of life issues that received less attention in the postwar rush to affluence. The ideology of new social movements also contains distinct libertarian elements. NSMs generally advocate greater opportunities to participate in the decisions affecting one's life, whether through methods of direct democracy or increased reliance on self-help groups and cooperative styles of social organization. These ideological beliefs often lead NSMs into challenging what heretofore were consensual social goals – as when environmental groups oppose the programs of unrestricted economic growth that are favored by both business and labor, when women's organizations work to dismantle patterns of gender relations that have existed for centuries, or when peace groups question the strategic thinking underlying Western defense policy.

The ideology of these movements is the major factor distinguishing them from other traditional European leftist movements and their own historical predecessors. On the one hand, the alternative social goals of these movements often place them in direct competition with the labor movement. In fact, some of the most intense criticisms of the environmental and women's movements initially came from the labor unions and some of Europe's communist parties. In political controversies over environmental and nuclear power issues, NSMs often find themselves opposed by a coalition of both business and labor. The populist and participatory values of new social movements also stand in sharp contrast to the bureaucratized, hierarchical, and neo-corporatist tendencies existing in most established European interest groups (e.g., Schmitter and Lehmbruch, 1979). On the other hand, the reformist elements of their New Left ideology and their anti-establishment orientation separate new social movements from their own historical forerunners, such as the suffrage movement or the nature conservation movement of the early 1900s. NSMs are not simply the type of civic associations concerned with social needs that abound in the United States, because their ideology calls for more fundamental social change.

In the conclusion to this book, after reviewing the evidence of the intervening chapters, we will argue that the ideological orientation of NSMs determines what might be truly new about these movements. Their distinct ideological orientation influences the type of supporters these movements mobilize, their organizational structure, and their choice of political tactics. At the same time, the clash between these new political actors and the prevailing social paradigm creates recurring tensions between maintaining a commitment to fundamental social change and adapting to the pragmatic constraints of contemporary politics. This clash between fundamentalism and pragmatism is a steady subcurrent throughout this volume.

BASE OF SUPPORT

Research on "old" social movements (labor, agrarian, racial) often stresses the class basis of these conflicts. These movements derived from a combination of economic interests and distinct social networks. In other words, they arose to represent the particular interests of a clearly defined social aggregate, and the movement depended on an organizational network that integrated members of the class collective. Social movements were a vehicle for groups that lacked access to political power through other political channels.

While many early leftist accounts of NSMs also emphasized the roots of the movements in capitalist class conflicts (e.g., Melucci, 1981; Habermas, 1981; Offe, 1984), most researchers now question the validity of a traditional class-based explanation for these movements. A distinguishing feature of new social movements is that they lack the narrow special interest appeal to any one social grouping. New social movements are not drawn from the socio-economically disadvantaged or from repressed minorities. The environmental and peace movements garner their support from a socially diffuse group of individuals who share their goals – not from a distinct class, ethnic, or other social stratum. Even the women's movement draws upon a diffuse base of popular support that cuts across gender lines.

New social movements thus signify a shift from group-based political cleavages to value- and issue-based cleavages that identify only communities of like-minded people. The lack of a firm and well-defined social base also means that membership in new social movements tends to be very fluid, with participants joining and then disengaging as the political context and their personal circumstances change.

MOTIVATIONS TO PARTICIPATE

Following from the above point, collective action in old social movements was normally traced to a sense of self-interest consistent with rational choice theory. Farmers mobilized to defend their interests, and workers joined the labor movement to improve their economic position. Thus the goals of these old movements were instrumental, aimed at benefiting the interests of members of a collective, even if society or other social groups must pay the cost.

In contrast, the goals of NSMs often involve collective goods that cannot be restricted to group members and therefore violate Olson's (1965) *Logic of Collective Action*. While some individuals might become active in an environmental group to address a problem in their locale, such instrumental motivations are generally secondary to other goals. Rather than narrow self-interest, it is claimed that participants in NSMs are motivated by ideological goals and pursuit of collective goods (Mitchell, 1979; Muller and Opp,

1986; Rohrschneider, 1988). The expressive and social aspects of participation motivate other individuals to become involved (Parkin, 1968; Barnes, Kaase et al., 1979). This mix of motivations is captured by the common description of opponents of nuclear power as a coalition of "*Lodenmantel* and parkas;" the dark felt hunting coats worn by farmers protesting the location of a power plant in their locale, and the parkas of the university students drawn to the protest by the broad political issue of nuclear energy. The latter group of demonstrators better reflects the core adherents of new social movements, though the movements often attract participants drawn by a variety of motivations (on this point, see Klandermans et al., 1988).

ORGANIZATIONAL STRUCTURE

The organizational pattern of old social movements is often identified with a centralized, hierarchical structure as seen in labor unions, most civil rights groups, and other economic associations. This structure enables these organizations to mobilize their supporters effectively and direct the resources of the organization. Both class-based and resource-mobilization theorists see this as the optimal organizational structure, epitomized in Max Weber's concept of rational bureaucratic authority or Roberto Michels' notion of the "iron law of oligarchy."

The internal structure of many NSMs is often cited as a violation of Michels' iron law. New social movement theorists maintain that these groups prefer a decentralized, open, and democratic structure that is more in tune with the participatory tendencies of their supporters (Brand et al., 1986; Klandermans, 1989a). The fluid organizational structure of new social movements is most visible among locally based citizen action groups or local branches of national organizations (Guggenberger and Kempf, 1984). The small size of these groups and their neighborhood locale make an extensive organizational structure unnecessary and undesirable. Similar organizational tendencies can be observed, however, even in larger NSMs. Joyce Mushaben's account (1989) of the coordinating committee of the West Germany peace movement, for example, stresses the conscious rejection of a centralized organizing structure as an extension of the ideology of the movement (also see Rochon, 1988). Many organizations within the environmental movement – such as NOAH in Denmark, Robinwood in West Germany, and several Friends of the Earth groups – emphasize the participation of the membership, and these anti-oligarchic tendencies seem especially pronounced within Green parties (Kitschelt, 1989; Kitschelt and Hellemans, 1990). Similarly, many activists in the women's movement are openly skeptical of an institutionalized women's lobby that might come to dominate the movement; in the place of such an organizational structure, they prefer local action centers and self-help groups (Freeman, 1975; Ferree, 1987b).

A fluid organizational structure is not only a reflection of the ideology of new social movements; it also arises from the diffuse and fluid social base of these movements. The adherents of new social movements are not integrated into a group-defined social network that can be easily mobilized to support the group's efforts. The type of closed social milieu typical of the working-class movement (or comparable religious networks), including labor union membership and an insulated working-class environment created by the socialist party and its ancillary organizations, is simply lacking in the case of new social movements. A distinct cultural milieu does exist to support the activities of new social movements – counter-culture networks exist in most urban areas – but this network is created by individuals and not controlled or directed by any organization. Indeed, the anti-establishment values of NSM supporters are antithetical to the exclusive, cohesive, clientelistic associations which provide the basis of "old" social movements. Thus the style of elite-directed mobilization that characterized the old social movements is replaced by the fluid structure and persuasive style of the new.

POLITICAL STYLE

Much of the current European research literature on interest groups emphasizes the neo-corporatist tendencies of unions, business associations and similar interests (Lehmbruch and Schmitter, 1982; Grant, 1985). Under the neo-corporatist model, an interest group receives formal or informal sanction from the state, and thus is granted official status as a legitimate participant in the governing process. This participation may involve direct consultation with government ministries as legislation is being drafted, formal representation on government administrative bodies, institutionalized ties with parties that provide the group with representation in parliament, and participation on government advisory commissions. The collaboration between state organs and interest groups is often so close and constant that it is difficult to keep the actors separate.

In contrast, the new social movement approach claims that many NSMs intentionally remain outside the institutionalized framework of government (Alemann and Heinze, 1979; Melucci, 1980; Offe, 1985). Many environmental groups openly reject participation in government commissions and regulatory groups because they feel they may be forced to compromise on their goals. NSMs seemingly prefer to influence policy through political pressure and the weight of public opinion, rather than becoming directly involved in conventional politics.

Just as they supposedly shun conventional participation channels, new social movements seem to relish the use of protest activities (though protests and demonstrations are certainly not new to Western democracies). In historical terms, protest and collective action were often the last, final, desperate acts of individuals, arising from feelings of frustration and deprivation. Collective political action was often a spontaneous event, such as an unorganized crowd staging a food riot or attacking state authorities. While some of the collective actions of "old" social movements were consciously planned, protests were often uncontrolled political outbursts.

The activities of new social movements, however, display a change in the nature of protest as a political form. Protest behavior has become a planned and organized activity in contemporary democracies, and NSMs are at the vanguard of this development (Nelkin and Pollak, 1981; Guggenberger and Kempf, 1984; Tilly, 1985). At some protests, the participants arrive in chartered buses to preplanned staging areas complete with demonstration coordinators and facilities for the media. Protest has simply become another political resource for influencing public opinion and policy-makers.

New social movements also place greater emphasis on the media as a method of mobilizing public opinion. The media allow social movements to extend their reach to the entire public, and the unconventional actions of the movement are often planned for their media impact. Activities such as construction of the women's camp at Greenham Common, building a human bridge across the Rhine river to dramatize water pollution problems, or organizing a mass demonstration are all aimed at attracting media attention and mobilizing popular support for an issue.

The emphasis on autonomy from the political establishment also colors the relationship between NSMs and political parties (Raschke, 1982; Smith, 1984b; Guggenberger, 1980). The established political parties have been hesitant to respond to the political demands of new social movements. This is due, in part, to the uncertain electoral impact of NSMs while the parties remain dependent on their corporatist-industrial constituencies. Alberto Melucci (1980), for instance, holds that new social movements cannot bargain like old social movement organizations, because without a highly structured, cohesive organization that can reliably deliver electoral resources, the NSMs have nothing to offer to the parties.

The apartisan nature of NSMs also occurs because partisan politics is somewhat alien to the style and norms of new social movements. Most political parties are elite-controlled, hierarchic organizations that contradict the organizational style of new social movements. Philip Lowe and Jane Goyder (1983) thus note that most British environmental groups remain relatively aloof from partisan politics, and Russell Dalton (1988b) finds the same traits for environmental groups throughout the European Community. Similarly, the women's movement in several Western democracies openly shuns formal partisan commitments (Lovenduski, 1986).

The unconventional political behavior of NSMs is sometimes traced to the substantial policy conflicts that separate these movements from the business-labor-government coalition of most corporatist systems. Attempts to negotiate with the socio-political establishment often result in a clash of contrasting value paradigms. NSMs also feel that close cooperation with government bodies may lead to cooptation and deradicalization of the movement by dominant corporatist-industrial interests. For instance, a leading West German environmental group resisted relocating their national office to Bonn because they feared that the proximity to policy-makers would contaminate the organization. A strong anti-establishment sentiment runs through new social movements.

The unconventional political style of NSMs also arises from their status as populist, loosely structured interest lobbies. Without a firm social base and formalized members (like unions and churches), NSMs must always be concerned with mobilizing resources. Thus some of the activities of these movements are aimed at organizational maintenance, attracting new members and soliciting funding; and dramatic events are a good way to capture the public's attention. A few years ago, a Dutch Greenpeace ship staged a dramatic escape from Belgian authorities – and then used a television film of the escapade in a telethon that raised several million Guilder for the organization! New social movement organizations are like sharks; they have to keep moving to stay alive.

Few organizations fulfill all the characteristics we have described, but elements of the NSM model are clearly visible across environmental, feminist, and peace organizations. If new social movements continue to develop these traits and expand their influence, this could produce a fundamental change in the stable structure of interest group activities and corporatist policy-making in Western democracies. Certainly, there are substantial reasons for studying these claims about the distinctiveness of NSMs, and assessing the impact of these new movements on the contemporary political order.

Plan of the Book

In recent years new, systematic, comparative, theory-testing research on new social movements has begun. The high visibility of these movements has generated a substantial number of detailed studies of new social movements by American and European researchers. This book draws together some of the best of this research to provide a basis for testing past theorizing about new social movements and for exploring the implications of these movements for Western industrial democracies.

Not all of the claims about the challenges of new social movements can be examined in a single volume. Therefore, we focused our efforts on those aspects of new social movements that we felt posed the most fundamental challenges for the political order of advanced industrial democracies.[5] The structure of this book thus revolves around three themes: the origins of new social movements, their relations with the state and other political actors, and their impact on partisan politics.

The first set of analytic chapters considers alternative explanations of the origins of new social movements. The origins of these movements will, of course, have a broad formative influence on how these movements develop, whom they recruit as supporters, what form they take, and their long-term political prospects. The research literature is perhaps richest when it considers why new social movements have emerged in most Western industrial democracies over the past two decades.

The major source of controversy concerns the question of whether new social movements represent a truly new political force or whether they are only the most recent manifestation of a long-term modernization process in the West (see, e.g., Lipset, 1981; Bürklin, 1984, 1988b; Namenwirth and Weber, 1987; Tarrow, 1983). Karl-Werner Brand's chapter discusses new social movements in the historical context of cyclical theories of social change. He traces the issues espoused by new social movements – environmental quality, women's rights, and peace – back through earlier periods of social mobilization in the 1920s and late 1800s. Although he argues that these periods of social mobilization appear largely independent of long-term, historical economic cycles, Brand also claims that interest in these issues follows a cyclical pattern of social criticism, what he terms a "civilization critique" ("*Zivilisationskritik*"). By drawing attention to this historical pattern, Brand forces us to address the question of how new social movements differ from their historical predecessors.

A number of scholars dispute the cyclical premise of Brand's research and argue that new social movements are historically distinct because they spring from the unique socio-political conditions of advanced industrial societies (e.g., see Offe, 1984, 1985; Bell, 1976). Perhaps the most far-reaching explanation of the new political identity of social movements is based upon Ronald Inglehart's theory of postmaterial value change. In his chapter, Inglehart describes how the social and economic characteristics of advanced industrial democracies are transforming the basic value priorities of Western publics; he then examines the relationship between postmaterial values and popular support for new movements. Analyzing data from a dozen European democracies, he finds that the appeal of these movements is highly dependent on the new political *Weltanschauung* represented by postmaterial values. This, Inglehart argues, is what is "new" about new social movements.

Frank Wilson explores the origins of new social movements from a different perspective. He tests past theorizing that explains cross-national variation in the strength of new social movements as a reaction to the existence of neo-corporatist policy-making structures. In contrast to expectations, Wilson uncovers little empirical evidence that the popular base of new social movements is conditioned by national levels of neo-corporatism. In pursuit of an alternative explanation, Wilson posits a set of structural and political factors that he believes are more important in explaining national differences in the vitality of new social movements.

Max Kaase's chapter asserts that the emergence of new social movements resulted from the convergence of multiple factors. Some factors, such as the political dealignment caused by the post-war modernization wave, fit a cyclical pattern. Value change, the growing politicization of Western publics, and the spread of direct political action techniques represent more fundamental and enduring features of advanced industrial societies. Kaase argues that the eventual implications of new social movements will depend on the relative weight of cyclical and long-term influences in defining the political orientations and cultural networks of these movements. He closes by setting out the agenda for the remaining chapters: what is the relationship between new social movements and the representatives of the socio-political establishment (the dominant interest groups and political parties)?

The next section of the book deals with the political tactics of new social movements and their relations with institutions of the established political order. If new social movements actually represent a new style of citizen politics, as some theorists maintain, this should be most obvious in the behavior of these groups as participants in the political process. These chapters test the claim that new social movements remain outside the institutional framework of government, and attempt to influence policy predominantly through unconventional political means.

The chapters by Thomas Rochon and Bert Klandermans examine one of the most vociferous groups challenging Western governments in the early 1980s: the peace movement. Rochon frames his analyses in terms of the tensions between the anti-establishment ideology and tactical orientations of the movement versus the pragmatic requirements for exerting policy influence. Klandermans's analysis of the Dutch peace movement adopts the theoretical framework of resource mobilization theory to examine the "multi-organizational field" in which the Dutch peace movement functions. Using different methodologies, both authors examine the patterns of alliance and conflict between the peace movement and other political actors. Their results directly address the question of whether peace groups display a marginality from existing social and political institutions.

The women's movements in the United States, Britain, and Sweden are the focus of attention for Joyce Gelb's chapter. She examines how opportunity structures and other systemic factors affect the choice of organizational form and tactics of the women's movements in these three nations. Like Rochon's study of the peace movement, the tension between sociocultural change and political action represents a basic dilemma for women's groups. Gelb recounts how the American, British, and Swedish groups have reacted differently to this dilemma.

Dieter Rucht discusses the difference between new social movements and their historical predecessors in terms of the changing strategies of action in advanced industrial democracies. He also considers how the differing goals of new social movements – power-oriented versus identity-oriented – affect their choice of action repertoires. These theoretical ideas are then illustrated by a discussion of the strategies and tactics of European environmental groups.

The next section of the book is concerned with one of the most visible, and intriguing, aspects of new social movements: their impact on Western party systems. Even though many NSMs are explicitly non-partisan, they raise issues that have politicized and destabilized partisan politics. On the one hand, new political parties have formed in virtually every European democracy to advocate the causes of new social movements. Not only are these new participants in the party system; these parties often represent the unconventional ideology and political style identified with new social movement organizations. On the other hand, many of the older political parties are beginning to respond to this new political challenge. This is a crucial time in the development of new social movements, because the partisan ties now being formed may define the long-term electoral impact of these movements, either creating new bases of partisan division or becoming incorporated into the political structure of Western party systems.

Herbert Kitschelt and Ferdinand Müller-Rommel focus their attention on the new political parties that have arisen from the new social movement milieu. Kitschelt refers to them as "Left-libertarian" parties and Müller-Rommel uses the term "New Politics" parties, but both refer to essentially the same group of parties representing the causes of new social movements.

Kitschelt focuses on the institutional characteristics of these parties, and the implications of these characteristics for established patterns of partisan politics. He maintains that the new social movement origins of these parties have led to a new form of party organization as well as new issue demands. The chapter concludes by considering the implications of this New Left contagion on both the political programs and structural characteristics of the established parties. Müller-Rommel examines the individual-level bases of support for these new parties. His analyses concentrate on the social characteristics of new social movement adherents, and the relationship between support for these movements and voting for New Politics parties.

While Kitschelt and Müller-Rommel emphasize the impact of new parties on Western party systems, Claus Offe and Sidney Tarrow offer a more tempered judgement on the partisan impact of new social movements. Offe discusses the problems the "fundamentalist" ideological core of new social movements pose if these organizations hope to develop an institutionalized, lasting partisan representation. Based on the experience of the West German Green party, Offe holds that new social movements may already be experiencing a "self-transformation" which leads to the institutionalization and moderation of their political views. Sidney Tarrow's chapter considers how the established parties are able to react to the challenge of new social movements, using the Italian party system as his prime evidence. Tarrow pictures the impact of social movements in terms of a protest cycle, with the established parties gradually responding to new social movements by coopting and preempting their issues. Offe and Tarrow both suggest that new social movements may well be beyond their mobilization apex, but that movement-generated impulses are being incorporated into the established system of party politics.

The final chapter of the book attempts to synthesize our findings and offer a prognosis for the future. The contributors to this volume offer sometimes conflicting perspectives, and not all of them can be verified as yet. Many of these disagreements, however, seem to be more a matter of vantage point than of fundamental dissent. Overall, we find compelling reasons to justify the "new" qualifier in new social movements. The ideological bond and political style of these movements represent new political phenomena for Western industrial democracies. In terms of the challenge that NSMs pose to the political order, we find that political systems are now beginning to respond to these movements. It is still too early to determine whether the established parties and systems of interest intermediation will be able to integrate this challenge without altering their own structure, or whether fundamental incompatibilities in ideology and goals will prevent this integration. Though this process is still in flux, we speculate on the ultimate resolution of the challenges new social movements pose to Western industrial democracies.

NOTES

1 Most studies of the relative deprivation thesis concentrate only on economic measures of discontent. Gurr's model also discusses status and power as bases of potential deprivation. These factors may be more effective in predicting support for new social movements that are concerned with explicitly non-economic issues, but these avenues remain unexplored.

2 See, for example, Herbert Simon's (1979) discussion of rational choice models narrowly based on the concept of "Homo Oeconomicus."

3 One possible exception is Hirsch's (1976) analysis of the environmental issue in terms of resource scarcity, but this is not done within a formal modelling framework.

4 See especially Bert Klandermans' (1986) comparison of the different research traditions of American and European social scientists.

5 We also emphasized this approach because Klandermans has recently compiled excellent collections which examine the organizational traits of new social movements (Klandermans, 1989) and the processes of mobilizing movement supporters (Klandermans et al., 1988).

Part II
The Origins of New Movements

2

Cyclical Aspects of New Social Movements: Waves of Cultural Criticism and Mobilization Cycles of New Middle-class Radicalism

KARL-WERNER BRAND

Since the beginning of the 1960s, various and sequential waves of new movements have changed social and political life in Western democracies. Among those which sprang up in the 1960s were public interest and citizen-action groups, community action, neighborhood and self-help groups, civil rights, anti-Vietnam war, and student movements. They were followed in the 1970s and early 1980s by women's, regional, environmental, anti-nuclear power, and peace movements. New left-libertarian and green political parties evolved, and a variety of subcultural life-styles and new "alternative" urban milieus grew rapidly. This development undercut the post-war consensus which had suggested the "end of ideology," based on advancing economic growth and industrial modernization.

In the 1950s and early 1960s empirical findings of a relatively high degree of political apathy were still functionally interpreted as a prerequisite to stability in Western democracies. Then, analysts were confronted by widespread social and political mobilization which appeared not only to overload existing institutions of political participation but also questioned established structures of governance and legitimacy. In the view of conservative observers – or those who had become conservative *vis-à-vis* the new confrontation – a "crisis of governability" was threatening Western democracies.

Only twenty years later, in the mid-1980s, almost all observers agree that the wave of political participation and movement mobilization has ebbed. The utopian visions of the 1960s and 1970s have faded away. The movements' impulses have diffused into cultural and social life. Their themes (environmental protection, disarmament, women's equality, self-help, decentralization) have won a permanent slot on political agendas. Their organizations and the new forms of politics they created are becoming institutionalized. A new normalcy is apparently emerging (for West Germany, see Brand, Büsser, and Rucht, 1986).

How are these phenomena to be interpreted? Current explanations see the emergence of these movements as:

- a reaction to a new kind of problem resulting from the negative side-effects of industrial growth and technological development, which are not considered within the context of existing interest-group and governmental structures (e.g., Offe, 1985);
- an expression of the structural transformation of Western industrial societies which proceeds apace with a critical reorganization of the relationship between society, state, and economy and with the development of a new societal cleavage (e.g., Hirsch and Roth, 1986; Melucci, 1980, 1985; Touraine, 1977, 1981);
- the results of an epochal change in values, a conflict between the "Old Politics", supported by "materialist" priorities, and the "New Politics" based on "postmaterialist" preferences (e.g. Hildebrandt and Dalton, 1978; Inglehart, 1984);
- an expression of revolt by a "new educated class" inhibited in its upward mobility by economic conditions (e.g., Alber, 1985; Bürklin, 1984);
- a new historical variation of periodically appearing waves of anti-modern or romantic-ideological reactions to the functional principles, contradictions, and alienating effects of modern societies (e.g. Berger, Berger and Kellner, 1973; Huntington, 1981).

None of these single interpretations, in my opinion, offers an adequate explanation of the mobilization cycle of "new social movements." Some of the difficulties in satisfactorily explaining this phenomenon are due to national differences in the form and self-understanding of these movements (see Brand, 1985) which also influence the national research perspectives (see Klandermans, 1986). Without taking up these problems in detail, I instead would like to discuss the new movements from a structural and a cyclical angle.

The initial section of this chapter sketches out the structural characteristics of the "new social movements" as actors in a new societal and political conflict arena. This brings the "newness" of these movements into the foreground.

New movements, however, also manifest old characteristics. They reflect not only national continuities in the forms of protest and conflict settlement, but also thematic and class-specific continuities through their various historical waves of mobilization – continuities in the women's, peace, and environmental movements, in humanistic crusades (against slavery and race discrimination, for social reform and human rights), and the multitudinous impulses toward development of alternative (simple and egalitarian) life-styles. All these movements, their actual manifestations as well as their historical predecessors, are specific forms of middle-class radicalism. However, this chapter's primary concern is with a different aspect of continuity, that is, their cyclical lines of continuity.

In this context I follow the thesis that the mobilization waves of new social movements and their predecessors appear in phases of a general cultural crisis, in an atmosphere conducive to a spreading critique of modernization in its various forms. In addition to the past two decades, such periods can be found in the decades around the turn of the century and the 1830s and 1840s. A first empirical reference for this thesis is the fact that the emergence of the new movements in the 1960s, the substantive shifts in their problem conceptualization and patterns of criticism that came about in the 1970s, and the waning of their mobilization power in the 1980s moved apace with an equally clear change in the *Zeitgeist* of these decades. This suggests that it is the prevalent cultural climate of the 1960s and 1970s which supplied the breeding ground for the mobilization cycle of new social movements.[1] The second section clarifies this thesis, as well as the analytical value of "social mood," "cultural climate" or *Zeitgeist*, as a variable in the context of movement research.

The experience of the 1960s and 1970s poses the question whether a similar cultural climate supported the various mobilization waves of the new social movements' predecessors, or whether individual strands of these movements also revived in the context of different social moods. Clearly, such material can be presented here only in a very general form. If historical comparisons support the thesis that the mobilization cycle of new social movements and their predecessors are fed by similar currents of cultural criticism, the question still remains which factors trigger the periodical emergence of such currents. Theories of generational change, of "shifting involvements," or of long economic waves offer suggestive explanation models for such cycles. Given the limits of this chapter, the explanatory power of such theories cannot be discussed here in detail. I will, therefore, confine myself to some concluding comments suggested by the historical findings.[2]

New Social Movements: Sketch of a New Movement Type

Discussing the specifics of "new social movements," most observers agree on the following points (see Klandermans, 1986):

- new social movements are not concerned with questions of distribution, economic power, or political power; rather, they stress questions of the way and the quality of life in modern, industrial societies;

- these movements recruit primarily from the well-educated post-war generation of the new middle class, in changing coalitions with marginalized social groups;
- they do not evolve a new consistent ideological system. Instead, new movements emphasize the "right to uniqueness" within the context of a secular, pluralistic culture and on the basis of postmaterialist values;
- they emphasize principles of autonomous, decentralized organization;
- they have conferred a new acceptability on "unconventional" forms of political participation.

Is this shift in themes, social actors, organizations, and action forms *vis-à-vis* the "old" class and status movements permanent? Are there indicators that new stable cleavages are developing, or at least a new and permanent conflict constellation? Any answers to these questions will be more or less speculative. Nevertheless, existing empirical research on new movements, as well as the visible features of ongoing socio-structural transformation, provide some indications of their social and political perspectives.

Though the thematic conflict field reopened by the new movements is not new in historical terms, it has achieved much greater political importance. This, on the one hand, is related to the leveling of class conflict by welfare state regulations at a relatively high standard of material well-being. On the other hand, it is due to the new global nature of ecological problems, the threat of nuclear self-destruction, and the new dimensions of technological intervention and risks. In addition, questions of the quality of life and of self-realization now receive a higher cultural priority.

Ecological problems, risks of chemical or nuclear catastrophes, and questions of self-determination are not class-specific. But even if the activists and adherents of new social movements are a heterogenous group of "concerned" citizens, the core activists are still the younger, better-educated members of the new middle classes who are either working in the "non-productive", personal service sectors, or are students and groups only marginally integrated in the labor market. It is their sensitivity to questions of self-realization and identity, to the alienating effects of commercialization and bureaucratization of everyday life, and to the ecologically destructive consequences of industrial growth, that constitutes a predisposition for radically opposing the politics of economic growth and technocratic modernization (Cotgrove and Duff, 1980, 1981). The increasing social significance of the personal service sector and the large expansion of higher education since the 1960s have given these groups greater political weight. This indicates the emergence and stabilization of a new post-industrial cleavage.

It is improbable that this conflict line can be completely integrated into existing institutional forms of interest intermediation, even if the established political institutions take up the new issues. Rather, new social movements challenge the premises of established politics. First, they try to change political priorities. Second, they question the prevailing definition of what politics are, penetrating the institutional line between the private and public sector in new arenas of conflict. On the one side, the political relevance of what seemed to be unpolitical becomes a public theme ("The personal is the political"); on the other side, the "privatization" of socially relevant decisions in expertocratic arenas is challenged and brought to the public. This new definition of what is "political" implies, thirdly, the questioning of the forms of institutional politics. New social movements insist not only on the expansion of direct-democratic forms of political participation but also on the widening of opportunities for social self-organization. The cultural pluralism of these movements and their emphasis on autonomy also suggest a more reflexive way of political integration which is more responsive to different cultural norms and varying life-styles. The development of "umbrella" or "rainbow" parties constitutes experimental steps in this direction (see Kitschelt and Müller-Rommel in this volume, Chapters 10 and 11).

The new concept of politics has its socio-cultural basis above all in new alternative milieus, which have developed especially in metropolitan areas since the 1960s. They are communities of like-minded individuals who have undergone similar political socialization, follow a similar life-style, and show similar problem sensibility. Relatively fluid networks of social contacts grow out of frequently changing involvements in diverse local initiatives, campaigns, and political groupings. The extent to which these milieus develop a common political identity depends on the specific national conflict constellation and the overall political climate. But in general, an organizational stabilization of this new political identity seems improbable.

Rather than being bound to new political milieus or parties in a stable way, the core supporters of new social movements show a high degree of susceptibility to issue-specific mobilization. Thus the development of critical issue publics becomes a prerequisite for effective mobilization processes within the new areas of conflict. Whereas the rise and decline of such issue-specific publics and campaigns largely depend on the attention cycles of the mass media, the new alternative milieus and the changing network of movement organizations, citizen initiatives, alternative groups and subcultures generally maintain the material and social infrastructure for further mobilization.

The structural indeterminacy of the new cleavage, the relatively fluid, minimally institutionalized, and culturally pluralistic character of the new movements, and the missing sensual obviousness of many problems and risks initiating the protest – all these factors make the new movements relatively open to changing social and political contexts, and to the shift of problem sensitivities and the overall cultural climate.

My thesis is that this openness to and dependence upon changing social moods has substantially influenced the development and the public resonance of the new movements, their rise in the 1960s, their thematic shifts in the 1970s, and their decline in the 1980s. This will be seen in more detail in the next section.

From the 1950s to the 1960s: Changing Social Moods and the Mobilization Cycle of New Social Movements

Looking back over three decades, it seems evident that the changing social climate of these times had a persistent influence on the development of new social movements, on changes in their thematic emphases, standards of criticism, utopian models, and mobilization opportunities. At the same time, then, it is surprising that the role of social moods has until now hardly been studied systematically within the context of social movement research.

What I call the *Zeitgeist*, "social mood," or "cultural climate" of a given period means the specific configuration of world-views, thoughts and emotions, fears and hopes, beliefs and utopias, feelings of crisis or security, of pessimism or optimism, which prevail in this period. This *Zeitgeist* creates a specific sensitivity for problems; it narrows or broadens the horizon of what seems socially and politically feasible; it directs patterns of political behavior and life-styles; it channels psycho-social energies outward into the public or inward into the private sphere. Thus it provides or deprives social movements of essential public response.

Some recent publications by American authors point to the cyclical appearance of a cultural climate which promotes an overall social and political mobilization. Hirschman (1982) conceptualizes these cycles as a periodic alternation between times of "private interest" and "public action." Schlesinger (1986) reconstructs these continuing shifts in 30-year cycles of American history. Huntington (1981) proposes a 60-year cycle of "creedal passion periods." Namenwirth and Weber (1987) postulate a long-term cycle of value change of about 150 years and a short-term cycle of about 50 years on the basis of a quantitative content analysis of American party platforms and British "Speeches From the Throne." It still remains questionable, however, whether there exists a general cultural climate which is equally supportive or inhibitive to all kinds of social movement. The mobilization waves of labor movement and middle-class radicalism by no means always converge. Thus the assumption that for different kinds of social movements there are also differing favorable or unfavorable conditions within the cultural context seems more tenable. The following discussion focuses on a special variation of such social moods – the spread of cultural criticism or modernization critique – which provide new social movements as well as their predecessors with an especially favorable sounding-board.

Before turning to empirical evidence for this thesis, however, I want to introduce the concept of cultural criticism as it is used here more systematically. What I call "cultural criticism" or "modernization critique" means a heterogeneous pattern of critique of fundamental aspects of modern life, such as commercialization, industrialization, political centralization, bureaucratization and democratization, cultural rationalization and pluralization. I refer to this concept in a threefold sense.

First, critique of modernization can feed on pre-modern, agrarian, religion-based world-views. This kind of anti-modernism develops during the stages of transition from traditional to modern societies. It can take the form of criticism "from below" referring to popular traditions, and the form of criticism "from the top" referring to old patterns of legitimacy and social order.

Second, critique of modernization can appear in a form that is best described with the German term *Zivilisationskritik*. This form of cultural criticism is no longer embedded in an agrarian, pre-modern way of life. Rather, it springs from an already, at least to some extent, modernized society which experiences a rapid change in social structures and what Nietzsche termed a "re-evaluation of all values." Anti-modern attitudes of this kind have a pessimistic touch. They get their special characteristics by a general feeling of a loss of sense and orientation and by fears of status deprivation, moral decay, and social decline. Compensatorily, a sentimental attachment to pastoral traditions, to rural life with its simple virtues, spreads; an idealized nature becomes the source of moral and physical recovery. Moralistic and puritanic traits also become stronger.

Third, critique of modernization can take the form of an artistic and intellectual criticism of alienation, which appears in two variations: one that is primarily aesthetical, and another that is primarily moral-idealistic. The latter is aroused by the discrepancy between universal, humanistic values and the harsh reality of economic exploitation, political oppression and social misery; or in a more general sense, by the contrast between moral principles and the functional imperatives of capitalist, industrial, and bureaucratic development. The aesthetic-countercultural variant takes classical form in the European Romanticism of the early nineteenth century. Initially a revolt against the Enlightenment, against obsolete rules and conventions, utilitarianism, and the mechanistic belief in progress, romanticism is the expression of a new subjective sensibility. In addition, it is deeply impressed by the cataclysmic events of the French Revolution and the dissolution of the old order. Thus it displays a tension between cultural avantgardism (an aesthetic cult of individuality) and a feeling of homelessness, a yearning for "real community," for a reconciliation of the world through love, "poetization of life," or religious feeling. Seen structurally, the appearance of this kind of cultural criticism is connected with the rise of a new class of autonomous artists and "intellectuals."

These patterns of modernization critique usually appear together, in various mixtures, in times of widespread cultural criticism. In Western countries, however, the pre-modern variant of anti-modernism has progressively lost ground from the middle of the nineteenth century onward.

Coming back to the cultural context of the mobilization cycle of new social movements, a typical sequence of basic social moods can be determined from the 1950s to the 1980s in virtually all Western democracies: from the conservative 1950s with their emphases on private and material values, to the technocratic reform enthusiasm, the optimistic cultural-revolutionary thrust and moral radicalism of the 1960s, changing to the sobering 1970s which saw a growing crisis-consciousness and the spread of pessimistic anti-modern moods, finally giving way to the neo-conservative, "postmodern" *Zeitgeist* of the 1980s. It is the social mood of cultural criticism of the 1960s and 1970s – a mix of moral-idealistic and aesthetic-countercultural critique of modernization, on the one hand, and a more pessimistic pattern of *Zivilisationskritik*, on the other hand – which gives the new movements, as part of a comprehensive social mobilization process, their stimulus. Both decades, however, create differing sensibilities to problems of capitalist democracies and the industrial way of life which open up differing perspectives for social action.

In each country, of course, this general sequence of prevailing social moods is modified by the specifics of national politics, economic development, and political culture. In particular, there are very different historical burdens and starting points for post-war development in countries such as the Federal Republic and the United States. These national variations are also dependent on the differing ability of the new movements to gain public acceptance for their definition of problems and reality. In this chapter, however, only the general trends are at issue. Focusing on the question of how these changing cultural opportunity structures influence the development of new social movements, these general shifts in the cultural climate will now be discussed in more detail.

The 1950s and early 1960s were times of generally stable economic growth which parallel a previously unknown increase in material standards of living. Standardized, durable mass consumer goods (televisions, refrigerators, washing machines, automobiles) became available to almost everyone. Supermarkets and self-service stores expanded; advertising became omnipresent and a prime growth sector. Structural economic changes increased the proportion of white-collar jobs. Science and technology became socially dominant forces, also in the public's consciousness. Old class contrasts appeared to erode. The rise of the "affluent society" heralded the "end of ideology." Functional thinking, belief in technical progress, privatized-material orientations, and a moral conventionality moulded the world of the 1950s. Anti-communism and the cold war reinforced this domestic political complacency. That applies to the America of Eisenhower no less than to the West Germany of Adenauer or Britain's "Butskellism," supported by a new social welfare state consensus.

The 1960s saw a dramatic shift of personal attention and energies to the public sphere. The consensus about the prevailing *petit-bourgeoisie*, privatistic values broke down. The complacency of the 1950s gave way to a critical view of and moral outrage about the shadowy side of the "affluent society." The still-existing and considerable sector of poverty, continuing race and minority-group discrimination, and decay of metropolitan areas appeared as scandals against the background of a continually improving standard of living and a basic, unbroken belief in modern progress. Measured against the universally propagated ideals of Western democracies – freedom, self-determination, equal opportunities – established structures of power and inequality appeared unbearable. So did idealistically embellised military interventions in foreign countries. This criticism which began to divide younger and older generations was intensified still further by the growing importance of consumption and leisure time, which stimulated the rise of a new hedonism and life-style oriented toward self-fulfilment, sexual freedom, and spontaneity. Measured against modernism's promises of happiness, the successor generation found life boring, empty, and alienating in a society focused on acquisition and functionality. The break in values taking place in this generation since the beginning of the 1960s – gradually at first and then increasingly confrontational – was indeed a cultural revolution penetrating all spheres of daily life.

The social movements of these years took the idealistic and utopian promises of the "affluent society" at their word, measured them against (inferior) reality, and emphatically demanded the promise as reality. But the still unbroken, optimistic belief in progress also nourished trust in technocratic solutions to existing problems "from the top" or the fundamental changeability of existing structures "from below". This tension between, on the one hand, various facets of emerging critique concerning the dark sides and deficits of advanced capitalism, and, on the other hand, the firm belief in the fundamental possibility of realizing its humanistic ideals, motivated the political reform phase of the 1960s (and in Europe also that of the early 1970s). This tension also underlay the exuberance of the cultural revolutionary uprising and the impetus for various emancipatory, consumption-critical, and romantic anti-capitalist movements of this period. Moreover, political reform programs and the pressure of social movements mutually strengthened one another.

In the 1970s the mood darkened. General optimism about progress and belief in the rapid changeability of political and social structures disappeared. Failure of utopian-anarchist expectations for revolutionary change furthered dogmatic tendencies and the development of many cadre-political sects within the New Left, and induced some to terrorist strategies. A more broadly effective reaction, however, involved the diversion of interest in social macro-structures to the closer proximity of everyday life. This subjective turn took on many different forms. It led to increasing interest in self-experiences and therapies of all sorts. Individual psychic and physical well-being moved to the center of attention, and a new health cult developed. A growing need for transcendental experiences and spirituality stimulated broad interest in Eastern religions and meditational techniques.

Apart from this subjective turn, there were also clearer articulations of interests related to encroachments on the quality of life, such as continuing environmental pollution and threats to health through noise, toxic substances, or nuclear radiation. Citizen initiatives and local self-help groups sprang up. Industrial conflicts became militant. Regional movements, rebelling against "internal colonialism," demanded cultural and political autonomy. Thus the "turn inward" also had a politically offensive impact. It created new conflicts around daily needs, questions of individual identity, and subjective concern about the burdens of economic, political, and cultural modernization. The low responsiveness of established political institutions to these new problems caused a rapid loss of legitimacy for political parties and government (in the American case accelerated by the Vietnam war and Watergate). That loss was paralleled by a higher evaluation of autonomous forms of interest representation.

These tendencies received a dramatic touch through a series of crisis experiences which fundamentally shattered the belief in technical progress and the trust in technological reform efforts. First, and most fundamentally, the limits of industrial growth burst into public consciousness in the early 1970s (at the latest with the 1973 oil crisis), mirrored in discussions about population explosion, exhaustion of natural resources, and threatening ecological catastrophes. Second, there was the additional experience of world-wide economic recession, paralleled by increasing unemployment and inflation which could not be effectively countered through the panacea of the 1960s, the touted Keynesian instrument. Third, the growing mood of crisis and loss of perspective was strengthened by disappointment with the low effectiveness and non-intentional counter-productive effects of the implemented policy reforms. All of that initiated a pessimistically tinged, anti-modern mood which has determined general sentiments since the mid-1970s. By this time, a general critique of technology, bureaucracy, and rationalism found broad public resonance. The urban middle class was caught up in a new longing for a simple, healthy, and natural way of life. There was also greater nostalgia for pre-modern ways of life and fascination with mystical, holistic, non-scientific ways of experiencing.

The mood-change toward anti-modern *Zivilisationskritik* produced ambivalent political results. On the one side, it substantially contributed to a conservative trend and a new moralism and traditionalism. On the other side, it strengthened ecological crisis-consciousness and interest in alternative life-styles. The ecological and anti-nuclear power movement focused the new uneasiness into mass protests. The anti-modern-tinged ecological protest, however, did not link up everywhere with the existing counter-cultural and emancipatory currents of protest. In only a few countries a new political polarization emerged as comprehensive as that in the Federal Republic of Germany (see Brand, 1985). In Germany, the anti-modernist mood of the 1970s was still aggravated by a deepening economic crisis and East–West tensions in the early 1980s; it solidified itself into a general catastrophism, giving impetus to a struggle against the deployment of middle-range missiles and to the new party of the Greens.

A further change in the *Zeitgeist* is obvious in the 1980s – a "postmodern" mood spreads. In it the fundamental critique of modernization loses much of its impetus. Concerns of the new social movements are taken up by parties and partially institutionalized. Alternative milieus lose their clear-cut oppositional identity. A new realism and pragmatism spreads. The historicism of postmodern architecture couples with the pluralism of life-styles: "anything goes." Yuppiedom, the unconcealed hunt after money and status symbols, stands unconnected next to poverty and mass unemployment. Neo-conservatism holds sway over the intellectual political climate. The Left, already deprived of its attraction by the prevailing anti-modernism of the 1970s, loses still more ground. Trade unions' negotiating power is persistently weakened by permanently high unemployment, forced economic and technological structural changes, and by massive governmental strategies of confrontation. Patriotism, orchestrated with new conservative self-confidence, unites with a nostalgic renaissance of the 1950s.

Although this cultural and political climate deprives the new movements of broad public resonance, while also causing internal disintegration, it cannot be expected that the various fields of autonomous movement activities, the manifold networks of citizen action and self-help groups, of movement organizations, and counter-publicities will be completely absorbed by institutional channels of interest intermediation. Despite all changes, the problems addressed by the movements of the 1960s and 1970s remain virulent and meet increased public sensitivity. A new, more solid social consensus is not yet in sight. Thus, new social movements will change character, losing much of their mobilization strength, but they will not disappear.

This, surely, does not hold for every kind of movement, in particular not for those addressing questions of social injustice and bread-and-butter issues. I am refering here only to movements primarily concerned with the quality of life and the new quality of risks in modern, industrial societies, with cultural emancipation, and questions of life-style. My argument is that the upsurge of these new social movements in the 1960s and 1970s, their radicalism, utopian strength, and broad societal resonance were borne by a revival of broad currents of modernization critique which fundamentally shattered the technocratic consensus of the post-war decades. This emphasis on cyclically changing cultural variables does not question the enormous significance of political opportunity structures for the development of individual movements and protest cycles (Tarrow, 1983). But these political opportunity structures are embedded in, and influenced by, the prevalent cultural climate.

If this thesis is tenable, it should also be demonstrable with historical materials. Therefore, the next section takes up the question of whether mobilization phases of the predecessors of new social movements appear at the same time as the periodic appearance of similar broad currents of cultural criticism, of whether they also appear at other times, disproving the thesis.

Recurring Waves of Modernization Critique and the Precursors of New Social Movements

Figure 2.1 shows – for descriptive purposes – the periods marked by the upsurge of widespread cultural criticism (modernization critique), as well

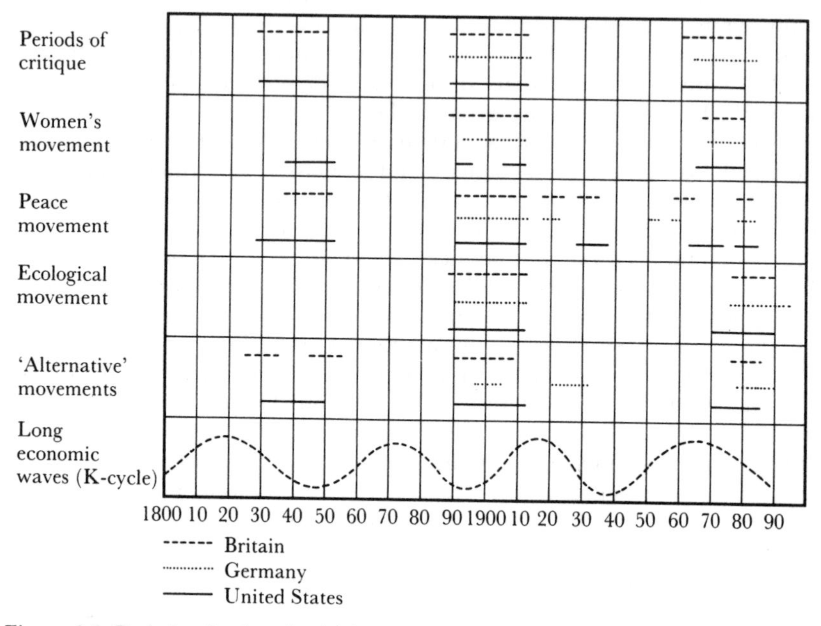

Figure 2.1 Periods of cultural criticism, phases of high movement mobilization, and long-term economic waves in Britain, Germany, and the United States, 1800-1990
Source: Compiled by the author. Long term economic waves are from Gordon et al., 1982.

as the mobilizing phases of some of the precursors of the new social movements in the nineteenth and twentieth centuries for three countries (Germany, Great Britain and the United States). The figure includes the women's movement, the peace movement, and the environmental movement, as well as the broad spectrum of "alternative movements." Although radical-democratic movements of young middle-class intellectuals or humanitarian crusades (such as the anti-slavery crusade) in the nineteenth century could also be regarded as predecessors of the new social movements, the selected movements appear to be typical of the specific mixture of emancipative, moral, and romantic-idealistic motives that inspired the movements of the 1960s and 1970s. Finally, the figure shows, for possibilities of comparison, the development of long-term economic waves that have become known as Kondratieff cycles (Gordon et al., 1982). Assuming that the industrial development of Western, capitalist societies follows long up-swing and down-swing phases which last some 40 to 60 years is still controversial (see Freeman, 1983). It seems to be, however, sufficiently supported for use as a heuristic model.[3] The question is whether or not these long economic waves coincide with the periodical appearance of widespread cultural criticism and new middle-class radicalism, thus suggesting some kind of causal linkage.

The mobilization phases of the selected movements cannot always be dated precisely to the exact year. Only occasionally is the beginning or the end of a phase of increased mobilization indicated by clear-cut events. The development of the movement organizations, for example, does not always coincide with the development of the movement's following and public resonance. Most historical studies agree, however, in the dating of their ups and downs.[4] Certainly, the exact timing of the spread of anti-modern or modernization-critical moods is more difficult. The degree to which such currents influence the way of thinking and feeling of a given period has to be reconstructed from contemporary documents, from *feuilleton* and public speeches, literary and artistic trends, the development of new scientific (and anti-scientific) schools and last, but not least, from the anxieties and utopias underlying the various social reactions to the problems of their times. For the purpose of this discussion, however, I date these phases by means of a secondary analysis of relevant cultural and socio-historical studies.

MOBILIZATION WAVES OF THE PRECURSORS OF NEW SOCIAL MOVEMENTS

Whereas the *women's movement* in England and Germany did not take organizational shape before the middle of the nineteenth century, eliciting only gradually broader responses, the women's movement in the United States experienced a first peak in its formative years in the 1840s, initially in close association with the anti-slavery crusade. In all three countries, however, feminism developed into a true mass movement only during the two decades surrounding the turn of the century (Banks, 1981; Evans, 1977; Flexner, 1975; Rendall, 1984; Rowbotham, 1973; Schenk, 1981).

Organized pacifism begins with the end of the Napoleonic wars, when peace societies were founded almost simultaneously in England (1816) and the United States (1815). In these two countries the peace movement showed a first broad mobilizing phase in the 1830s and 1840s. But pacifism, too, became a mass movement only around the turn of the century, and in Germany as well. The years after World World I brought a widespread revival of anti-military feeling and pacifist movements in Europe. Whereas in Germany they flourished only up to the mid-1920s, there was a new upsurge of pacifism in Britain at the end of the 1920s, which lasted till the confrontation with Fascism began to absorb the intellectuals' political commitment. In the United States, a new wave of pacifism spread in the 1930s, combining with the struggle against economic exploitation, and coming to an end before World War II. The Cold War atmosphere of the post-war era proved uncongenial to the expansion of pacifist ideas, in both Britain and the United States. Only in West Germany did the moral burdens of the Nazi regime and the still fresh remembrance of the catastrophic war destructions cause an upsurge of protest against rearmament and the deployment of nuclear weapons in the 1950s. The British CND, however, starting at the end of the 1950s, heralded the subsequent mobilization cycle of the new social movements (Brock, 1968, 1970, 1972; Riesenberger, 1985).

Environmentalism as an organized movement can be traced back to the early Victorian campaign against cruelty to animals and for bird protection, to the spread of natural history societies from the 1830s onward, or to the beginnings of a romantically inspired protection of natural monuments. Toward the end of the century, the increasing anti-urbanism and the nostalgic idealization of wilderness and rural life gave these efforts the impetus of a true mass movement, speeding up legislation of nature and monument preservation, animal and plant protection, and the establishment of national parks. Outdoor activities, such as cycling, hiking, and camping came into fashion. The popularity of the outdoor movement increased still further in the inter-war period; landscape and city planning, as well as safeguarding of natural resources, obtained a higher political priority. It was not before the 1960s, however, that environmental issues triggered another sweeping public mobilization (Hays, 1958; Linse, 1986; Lowe, 1983; Lowe and Goyder, 1983; O'Riordan, 1971; Sheail, 1976).

Whereas these movements showed a gradual (even though discontinuous) growth in the nineteenth and twentieth centuries, *alternative movements* manifest much less organizational continuity. The 1830s and 1840s were a time of social criticism, mass agitations and liberal reform campaigns in both Anglo-Saxon countries. In this context, alternative movements experienced a first, modest bloom as well. Under the influence of Owenism and Fourierism, an unprecedented wave of commune-building spread. The temperance movement, too, had its first peak during the 1840s. Again, it was not until the turn of the century that the spectrum of alternative movements fanned out into a multitude comparable only with their development in the 1960s and 1970s: the "thousand blossoms" of utopian-anarchist and agrarian-socialist movements, the back-to-the-land and back-to-nature movements, libertarian community experiments, sectarian "life-reform" movements (temperance, vegetarianism, nudism, naturopathy, anthropology, rational dress, school reform and garden city movement, etc.), counter-cultural youth movements and Bohemian settlements, the nostalgic revival of folklore, and rustic arts and crafts. Whereas this fascination with alternative ways of life found a quick end in World War I, in Germany youth movements, commune-building, and millenarian simple-life movements revived anew in the 1920s because of the fundamental crisis of political and economic life (Hardy, 1979; Krabbe, 1974; Linse, 1983; Marsh, 1982; Nash, 1967; Sieferle, 1984, Zablocki, 1980).

What general patterns evolve in the development of all of these movements?

- A first feature is the largely parallel nature of the mobilization phases of the individual movements. Only the peace movement shows independent, additional mobilization phases.

- A second is the largely parallel nature of these mobilization cycles in all three countries (in so far as the individual movements exist already).
- A third feature is the appearance of these mobilization cycles in specific historic periods and specific intervals: a beginning (in the Anglo-Saxon countries) during the period of 1830–50, a second all-embracing wave around the turn of the century (in Germany still continuing in the twenties), and a third still more sweeping wave in the 1960s and 1970s. The intervals between these periods of increased mobilization come to approximately 60–70 years.

These periods coincide with times of an emergent, pervasive mood of cultural criticism. This will be illustrated more in detail in the following section.

Describing Periods of Cultural Criticism (Modernization Critique)

In the nineteenth and twentieth centuries, moods of cultural criticism swept in three waves across Europe and the United States: in the period of romanticism at the beginning of the nineteenth century which achieved its broadest effects in both Anglo-Saxon countries in the 1830s and 1840s; in the two decades around the turn of the century (in Germany a marked mood of *Zivilisationskritik* still prevailed in the 1920s); and in the 1960s and 1970s. The different historical and national conditions give these critiques of modernization a different thematic direction and a different political weight in each period and country.

The 1830s and 1840s are considered a turbulent period in both Anglo-Saxon countries. The contrasts between the old agrarian and the new industrial era, and between the aristocracy and the rising middle-class elite, overlapped with the tensions arising from the increasing social problems of early industrialization and the experience of cultural crisis disseminated by the accelerated spread of capitalism. Although the latter called forth anti-modern reactions which are fed by pre-modern, agrarian traditions as well as by a general romantic mood, these reactions were embedded in a dominant optimistic belief in moral and social progress which was carried by liberal and democratic groups forcing their way to power. Thus, the high sensibility to the problem of the time gave birth to a comprehensive social, political and moral reform zeal, still fired by the gradually ebbing spirit of the preceding evangelical revival. This combined with the prevailing romantic element in the thoughts and feelings of that time, to give the various forms of social criticism, utopianism, revolt, and escapism their particular characteristics. It strongly influenced the philosophy of Transcendentalism; and it ignited as well a new bellicose national self-consciousness, for instance in Ireland and the Southern United States (Rose, 1981; Clark, 1955).

The romantic sentiment in Germany reached its zenith around 1800 and found its strongest political expression in the Anti-Napoleonic liberation war of 1813–15. The political restoration initiated by the Vienna Congress in 1815 caused a swift change of the romantic impulse for reconciliation into an idealization of traditional, hierarchical bonds and the sentimentality of the *petit-bourgeois* idyll of the "Biedermeier." The radical-democratic movements of intellectuals and *petit-bourgeois* classes originating in the French revolutions of 1830 and 1848 hardly showed any romantic specifics.

Towards the turn of the century a more pervading wave of anti-modernism spread in all three countries. In Europe, the third quarter of the nineteenth century was, like the period following the Civil War in the United States, a time of turbulent economic expansion which accelerated the transition from agrarian to industrial mass society. *Laissez-faire* liberalism, social Darwinism, fascination by positivist science and technology and a solid belief in material progress dominated the general way of thinking and feeling.

During the 1880s, however, a dissatisfaction with modern culture and industrial civilization spread on both sides of the Atlantic. The prevailing mechanistic conception of the world and the worship of material progress appeared increasingly hollow and empty in the eyes of a new middle-class generation. They rejected the artificiality and prudery of the Victorian or Wilhelminian conventions they sensed as an unreal façade. They also had a clearer view of the ugly side of industrialization, of the mass misery in the slum quarters of the cities, of the social, moral and psychological costs of progress. A realistic and naturalistic social criticism became dominant in the literature of the 1880s and 1890s. The feeling of over-civilization and spiritual homelessness paved the way for a wave of cultural criticism and anti-modernism. A new fascination with the fantastic, with occultism and spiritualism, grew. "Modern doubt" and a "morbid self-consciousness" expressed the feeling of *fin-de-siècle*, as did the melancholic aestheticism of Decadence. Growing anti-urbanism accompanied a revival of pastoral sentiments and a nostalgia for country life, folklore and handicrafts (Hynes, 1968; Krabbe, 1974; Lears, 1981; Marsh, 1982). In the United States, the return to the old republican values of the agrarian democracy promised the solution of the social, political and moral problems of over-civilization (Hofstadter, 1962). At the same time, vitalistic tendencies expanded, leading to an upsurge of enthusiasm for outdoor recreation and sports, a revitalized interest in wilderness, and the spread of a new "muscular spirit" (Higham, 1970). The erosion of conventional morality, however, also called forth massive religious counter-reactions. At the beginning of the nineteenth century, "purity crusades" in England gained a following that grew by leaps and bounds (Hynes, 1968).

All in all, the years before the turn of the century had a markedly pessimistic character, tinged by an atmosphere of crisis, self-doubt, introspection, nostalgia, and melancholy. In the early 1900s, the balance shifted to a regenerative activism in both Europe and America: in the United States to the reform zeal of Progressivism, in Edwardian England to an optimistic, liberal mood of breaking all traditional bonds. In Wilhelminian Germany the feelings of cultural awakening and of moving on the brink of a precipice remained closely interwoven with each other.

The social criticism and the idealistic reform impulse of the pre-war period gave way to a technical functionalism – the "new functionality" – and a withdrawal to private interests in the 1920s. In Germany, however, the political and economic turmoils of the 1920s caused a cultural atmosphere of uncertainty and disorientation which formed a fertile ground for the widespread criticism of modern civilization and a proliferation of anti-modern reactions, *völkisch* nationalism, and salvationism (Linse, 1983).

Conclusions

The historic materials presented here appear to support the thesis that the mobilization cycles of new social movements and their precursors find an exceptionally fertile ground in times of spreading cultural criticism. Such times heighten public sensitivity to the problems of industrialization, urbanization, commercialization, and bureaucratization. They temporarily upset the hegemony of the materialistic conception of progress, thus giving way to a broad spectrum of anti-modern reactions and reform movements. These phases also provide a short-lived opportunity for the creation of radicalized visions of a non-alienated, communal, and egalitarian way of living. Peace movements similar profit from such moods, though they also emerge independently for a variety of reasons, such as the reaction to increased international tensions, or as an attempt to ban the experienced horrors of war.

The movements discussed here do not benefit equally from every kind of cultural criticism. Optimistic variants of a romantic-idealistic mood favor emancipatory, egalitarian, and cultural-revolutionary movements. Pessimistic moods of *Zivilisationskritik* favor escapist and nostalgic back-to-nature movements, environmental protection, self-help, and health and therapeutic movements. However, most themes of this middle-class radicalism experience only a shift of accent by the change from optimistic to pessimistic moods of modernization criticism and vice versa. Within the women's movement, for example, the accent may shift from the stress on universal demands as for equality of rights and self-determination to the stress on gender differences and a specific "female superiority."

Our findings regarding long economic waves suggest that optimistic versions of cultural criticism appear rather in times of economic prosperity (as between the turn of the century and World War I or in the 1960s); pessimistic versions appear in phases of economic down-swing and depression (as in the 1830s and 1840s, in the last two decades of the nineteenth century, in part in the 1920s, and again in the 1970s). But, obviously, the periodic appearance of anti-modernism and the new middle-class radicalism it feeds does not run parallel to the Kondratieff cycle (see figure 2.1). So, one can conclude that neither economic up-swings nor down-swings cause the specific periodicity of recurring waves of cultural criticism. Whether such a periodicity exists at all remains an open question which cannot be answered on the basis of only three historical cases. If such moods spread, however, the economic situation strongly influences the kind and direction of criticism. But this is not all that can be said. There exists a more hidden link between socio-economic processes and these cultural and movement cycles which reveals itself if we look closer at the structural and political characteristics of the three periods in question.

The first wave of cultural criticism emerged in a transitional period, in which the old agrarian structure of society became more and more obsolete, whereas the outlines of the new bourgeois society were still fluid, bringing about new social misery as well as magnificent future prospects. In contrast, the other two waves appeared in fully developed capitalist societies. They were preceded by a long period of economic growth which triggered a rapid and comprehensive process of industrialization of production and everyday life. In both periods the material living conditions improved enormously for vast parts of the population. For the sons and daughters of the new middle classes – especially the liberal and educated – growing up in the boom phase, however, the material promises of industrial progress lost much of their attraction. The consensus on material progress broke down. Instead, the dark sides and the deficits of modernization, as well as the blocked possibilities of further emancipation, came more clearly into view.

But this is only one aspect, the postmaterialist side of the coin. In all three historical cases the emergence of new middle-class radicalism, nourished by the upsurge of cultural criticism, goes hand in hand with a more dramatic breakdown of the established basic consensus of social integration and political regulation caused by unintentional side-effects of economic and social modernization. In the first case, the upheaval of the 1830s and 1840s, the struggle between the supporters of the old, traditional order and the new, bourgeois society gave way to a new liberal order. In the second case, the rapid processes of industrialization and urbanization produced not only unprecedented social mobility and social strains concentrated in rapidly growing urban slums, it also gave birth to an increasingly militant and politically self-conscious working class eroding the hegemonic liberal consensus of the blossoming "Age of Capital" (Hobsbawm). Over a long period of bellicose imperialism, social struggles, economic depression, and the fascist challenge, a new class consensus of Keynesian politics and pluralist democracy established itself in the 1930s and 1940s. The latest wave of cultural criticism goes hand in hand with the breakdown of this post-war consensus in view of social and ecological "limits to growth" and industrialization, of a new technological revolution fundamentally transforming the social structure, and a cultural revolution equally radical in transforming the traditional socio-cultural patterns of orientation and behavior.

What conclusions can be drawn from this? If the latter two cases (the first showing a more diffused pattern) may be generalized upon, cultural criticism and new middle-class radicalism spread in the wake of a period of rapid industrial growth and social transformation, favored by the establishment of a new model of political integration and development promising progress and the solution of longstanding problems and social conflicts. Losing much of its attraction for a successor generation and showing by this time its negative side-effects and new structural problems more publicly, this model, however, is being questioned more and more. The level and the intensity of conflict increase, disseminating feelings of crisis and alienation, but, at the same time, opening up new horizons of alternative models of development and social life.

These macro-sociological conditions of periodically emerging waves of cultural criticism and new middle-class radicalism should not be misunderstood as a breakdown model (Tilly, 1978) of social movement analysis. Rather, it tries to combine a structural and phenomenological approach of discontinuous social change (as is advanced for example by Bornschier (1988) and Imhof and Romano (1988), or by neo-Marxists such as Hirsch and Roth (1986)) with the symbolic-interactionist tradition of collective behavior research and the resource mobilization approach, although processes of movement mobilization and organization have not been at issue here. Nor does this explanation model necessarily contradict Hirschman's theory of "shifting involvements" or its combination with a concept of generational cycles, as in Schlesinger's model of a 30-year cycle of American history. Neither concept, however, explains the specific phenomenon being discussed here. Huntington's "creedal passion periods" correspond with the periods of cultural criticism we have identified. This points to underlying common processes in Western societies not sufficiently taken into account by Huntington's emphasis on the peculiarities of American political culture.

The revival of a new wave of modernization critique in the 1960s and 1970s makes it clear that its appearance does not represent a transitional phenomenon that will disappear with the establishment of a full-fledged industrial society. Rather, it accompanies the dynamic of capitalist and industrial development which does not come about gradually, but in discontinuous processes of crisis, disorganization and restructuring. It is not before the 1960s and 1970s, however, that the specific kind of new middle-class radicalism which is fed by the periodic upsurge of modernization critique becomes a major political force in the restructuring processes of a new model of social integration and political development.

NOTES

1 The concept of a mobilization cycle used here relates closely to Tarrow's concept of "cycles of protest" (1983), but with some differences. First, I concern myself in this chapter only with a specific section of the overall social movement sector, that is, with protest cycles of new social movements. This part of the movement sector certainly can overlap with other parts, e.g., industrial conflicts and racial tensions. Second, this concept of a mobilization cycle refers to the entire period of the 1960s and 1970s (and, in European countries, the early 1980s) in which

– with national differences, including one or two cycles of protest – these movements flourished better than in preceding decades and most likely in the years following. Third, Tarrow contends that cycles of protest occur infrequently. In contrast, I am concerned with mobilization cycles of a specific type of middle-class radicalism which, as I attempt to show, indicate greater regularity.

2 The theses offered for discussion here are part of an ongoing research project. In a historical (nineteenth- and twentieth-century) three-country comparison of Germany, Britain, and the United States, this project pursues the question of what is new about new social movements.

3 The figure shows an approximate dating following the evidence in the literature on long-wave theory (see Freeman, 1983). The timing of up-swing and down-swing phases by different scholars largely coincides.

4 One substantial difference in the timing of mobilization waves is provided by O'Riordan (1971) and Lowe and Goyder (1983), who identify three instead of two peaks in the history of environmental movements. In addition to the upsurge of environmentalism around the turn of the century and in the 1960s–1970s, they observe another peak in the 1930s. In the United States, however, this was not a mass-based movement, but a new political emphasis on rational planning of resource use. In the case of Britain, this refers primarily to the spread of groups organizing outdoor activities.

3

Values, Ideology, and Cognitive Mobilization in New Social Movements

RONALD INGLEHART

Why do people participate in new social movements? Their behavior reflects a variety of factors operating on several levels. When people act in these movements, it reflects an interaction between (1) objective problems, (2) organizational networks, (3) relevant motivating values, and (4) certain essential skills. The absence of any one of these factors can inhibit the movement from emerging.

On one level, new social movements reflect the existence of objective problems such as the degradation of the environment, the exploitation of women, the coldness and impersonality of industrial society, or the danger of war; people rarely engage in political action unless there is some problem to solve. But on another level, it is difficult for isolated individuals to engage in effective political action: political participation is facilitated by the existence of social networks or political organizations that coordinate the actions of many individuals. Furthermore, people do not act unless they want to attain some goal: the existence of problems and organizations would have no effect unless some value system or ideology motivated people to act.

The boundary between an ideology and a value system is not always sharply defined; both are belief systems that may lead to a coherent orientation toward a whole range of specific issues. But the term "ideology," on one hand, is generally understood to refer to an action plan propagated by a specific political party or movement: it is adopted consciously as the result of explicit indoctrination. A "value system," on the other hand, is absorbed through one's socialization as a whole, particularly that of one's early years. An ideology might be adopted or rejected from one day to the next through rational persuasion. Values are less cognitive, more effective, and tend to be relatively enduring. They may motivate one to adopt an ideology. Thus, a given ideology, *per se*, would not necessarily be associated with enduring intergenerational differences, except in so far as they reflected differences in the underlying values.

Even when we take account of problems, organizations, values and ideology, we still have not considered all the major influences on participation in new social movements. Effective political action requires the presence of certain skills among the relevant individuals; even severe problems or a superb organization would be unlikely to mobilize a population consisting of illiterate and apolitical people. The term "cognitive mobilization" refers to the development of political skills that are needed to cope with the politics of a large-scale society; relatively high or low levels of these skills have been shown to be an enduring characteristic of given individuals and of given political cultures (Inglehart, 1977; 1990, ch. 10).

Finally, there is a significant interaction between values and cognitive mobilization. Values, we have argued, have a relatively strong affective component; they may be present even when their cognitive implications have not been worked out. "Preferences need no inferences," as Zajonc (1980) puts it. Considerable thought and effort may be required to develop the logical connection between given values and the appropriate political stance. Consequently, the impact of values on political behavior tends to be greatest among those with relatively high levels of education, political information, political interest, and political skills: in short, among those with high levels of cognitive mobilization. For the remainder of the public, one's values and attitudes (and behavior) may show little constraint. The potential influence of values remains latent until a situation arises that makes their implications salient. Accordingly, we are likely to find more constraint between one's basic values and active participation in a movement, than between values and relatively passive expressions of attitudes. For active participation implies that the given topic has become salient; one has thought about it and worked out the cognitive implications of one's central values. In survey research, one can elicit responses concerning almost any topic; but these responses often reflect superficial reactions produced on the spur of the moment or "nonattitudes," as Converse (1970) puts it. The fact that given survey responses may be shallow, produced almost at random, does not mean that relevant values do not exist, however; under appropriate conditions, when the individual is given the time and incentive to work out the implications of one's basic orientations; or, when the given topic becomes a salient part of one's life, the observed connection between values and behavior may be quite substantial (Inglehart, 1985b).

As this chapter will demonstrate, the materialist/postmaterialist dimension has played a crucial role in the recent rise of new social movements. Clearly, the emergence of new values has not been the sole factor involved. Objective problems, organizations, and ideologies have all been at work. And the emergence of the new social movements owes much to the gradually rising level of political skills among mass publics, as education has become more widespread and political information more pervasive. But the emergence of new value priorities has also been an important factor. The rise of the ecology movement, for example, is not simply due to the fact that the environment is in worse shape than it used to be. Partly, this development has taken place because the public has become more sensitive to the quality of the environment than it was a generation ago. Similarly, it does not seem that women are more disadvantaged today than they were a few decades ago; but it does seem clear that women in advanced industrial society place greater emphasis on self-fulfilment and on the opportunity to have a career outside the home.

Schmidt (1984), Müller-Rommel (1984), and Bürklin (1984) all found that support for the West German Green party, to a very disproportionate extent, came from postmaterialists. We believe that this represents merely one case of a much broader phenomenon. Postmaterialist values underlie many of the new social movements.

Postmaterialists emphasize fundamentally different value priorities from those that have dominated industrial society for many decades. The established political parties that control electoral politics in most Western societies emerged in an era when social class conflict dominated the political agenda; and existing alignments still largely reflect this orientation. But the old alignments do not adequately represent such new issues as the women's movement, the environmentalist movement, or the opposition to nuclear power. Because they seek goals that the existing political parties are not well adapted to pursue, postmaterialists are likely to turn to new social movements. In the takeoff phase of industrial revolution, economic growth was the central problem. The postmaterialists who have become increasingly numerous in recent decades place less emphasis on economic growth and more emphasis on the non-economic quality of life. Their support for environmentalism reflects this concern – with an explicit reference to the quality of the physical environment; and (less overtly, but at least equally important) with a concern for the quality of the social environment: they seek less hierarchical, more intimate and informal relations between people. It is not that the postmaterialists reject the fruits of prosperity – but simply that their value priorities are less strongly dominated by the imperatives that were central to early industrial society.

Similarly, the rise of the West European peace movement in the 1980s reflected many factors, and can be analyzed on a number of levels. It can be traced to specific political decisions made by specific political leaders, in a specific strategic context. The presence of postmaterialists would not automatically have generated the movement in the absence of these other factors. But it does seem clear that the emergence of postmaterialism was one of the key conditions that facilitated the development of the peace movement and that enabled it to mobilize larger numbers of supporters than any of its various forerunners, from the early days of the Cold War through the Vietnam era.

Why were postmaterialists so much likelier to be active in the peace movement than those with other values? Is it because they are more afraid of war? The answer to the latter question is no. Postmaterialists are concerned with war, but they are no likelier to feel that World War III is imminent than the rest of the public. Instead, the linkage between postmaterialism and the peace movement seems to reflect two main elements, one of which is a relative sense of security. For postmaterialist values develop from a sense of economic and physical security, and the latter part of the syndrome includes a sense of national security as well as domestic security. Postmaterialists are likelier than materialists to take national security for granted. Accordingly, they are more apt to feel that the American presence in Europe is unnecessary and that additional arms are superfluous.

The other side of the coin is that postmaterialism has emerged in a setting in which war seems absurd. Since the end of World War II, it has seemed that the only war likely to take place in Western Europe would be a total war involving both of the superpowers. In the thermonuclear age, the costs of such a war would almost certainly outweigh the gains by a vast margin. Indeed, if the Nuclear Winter hypothesis is correct, it would wipe out human life in the entire Northern hemisphere, and possibly the Southern one as well. By any cost/benefit analysis, this is not a paying proposition.

This has not always been the case. Throughout most of history, it has been at least conceivable that the material gains of a given war might exceed the material costs. In an economy of scarcity, it was even possible that under extreme conditions, a given tribe or nation's only hope for survival might lie in a successful war to seize a neighbor's land or food or water supply. In advanced industrial society, the cost/benefit ratio has swung far in the opposite direction. On one hand, the costs of war have become very high; and on the other hand, the benefits are relatively low: with a high level of technology, there are easier and safer ways to get rich than by plundering one's neighbors (as the recent history of Germany and Japan illustrates). From a postmaterialist perspective, war seems absurd.

The rise of the new social movements is not a result of values alone: to some extent, the emergence of these movements also reflects explicit ideological indoctrination. But to pose the question as one of values or ideology is a false choice. It is both. Moreover, the rise of new values constitutes a key element in any explanation of why a new ideological perspective has arisen. For the ideology of the new social movements is not simply the traditional ideology of the Left. Except in the very general sense that the Left (then as now) constitutes the side of the political spectrum that is seeking social change, the traditional and contemporary meanings of Left are very different: the Old Left viewed both economic growth and technological progress as fundamentally good and progressive; the New Left is suspicious of both. The Old Left had a working-class social base; the New Left has a predominantly middle-class base. To a great extent, the spread of new values and the rise of new issues has already reshaped the meaning of Left and Right. To mass publics, the core meaning of "Left" is no longer simply state ownership of the means of production and related issues focusing on social class conflict. Increasingly, it refers to a cluster of issues concerning the quality of the physical and social environment, the role of women, of nuclear power and nuclear weapons. The meaning of Left is changing – imperceptibly but continuously.

In this analysis, we will use the respondent's self-placement on a Left/Right scale as an indicator of one's ideological affiliation.[1] Responses to this item tap a general sense of identification with the parties and the issue positions of the Left (Inglehart and Klingemann, 1976). We will use this item to measure the degree to which given individuals support the new social movements because of previous ideological influences: they are environmentalists because they have always been on the Left. Clearly, this procedure somewhat exaggerates the influence of ideology on support for the new social movements, since the meaning of the Left has already changed, under the impact of new social movements. Today, people are on the Left because they are environmentalists rather than the other way around. We will nevertheless use this somewhat exaggerated estimate of ideological influence, in order to obtain a conservative estimate of the impact of materialist/postmaterialist value priorities on support for the new social movements. As we will see, this impact is substantial, even when we control for the effects of one's Left–Right ideological orientation.

The Rise of Postmaterialist Values

Our central interest here is the impact of materialist/postmaterialist value priorities: the tendency to give top priority to economic and physical security ("materialist" values), versus the tendency to give top priority to self-expression and the quality of life ("postmaterialist" values).

The hypothesis of an intergenerational shift from materialist toward postmaterialist values is based on two key concepts: (1) people value most highly those things that are relatively scarce; but (2) to a large extent, one's basic values reflect the conditions that prevailed during one's pre-adult years (Inglehart, 1971, 1977, 1990).

The scarcity hypothesis is similar to the principle of diminishing marginal utility in economic theory. The recent economic history of advanced industrial societies has significant implications in the light of this hypothesis. For these societies are a remarkable exception to the prevailing historical pattern: the bulk of their population does not live under conditions of hunger and economic insecurity. This fact seems to have led to a gradual shift in which needs for belonging, esteem and intellectual and aesthetic satisfaction became more prominent. We would expect prolonged periods of high prosperity to encourage the spread of postmaterialist values; economic decline would have the opposite effect.

But it is not quite that simple. There is no one-to-one relationship between economic level and the prevalence of postmaterialist values, for these values reflect one's subjective sense of security, not one's economic level *per se*. While rich individuals and nationalities tend to feel more secure than poor ones, these feelings are also influenced by the cultural setting and social welfare institutions in which one is raised. Thus, the scarcity hypothesis alone does not generate adequate predictions about the process of value change. It must be interpreted in connection with the socialization hypothesis.

One of the most pervasive concepts in social science is the notion of a basic human personality structure that tends to crystallize by the time an individual reaches adulthood, with relatively little change thereafter. This, of course, doesn't imply that no change occurs during adult years. In some individual cases, dramatic behavior shifts occur, and the process of human development never comes to a complete stop (Brim, 1966; Mortimer and Simmons, 1978; Levinson, 1979; Brim and Kagan, 1980; Riley and Bond, 1983). Nevertheless, human development seems to be more rapid during pre-adult years than afterward, and the bulk of the evidence points to the conclusion that the likelihood of basic personality change declines after one reaches adulthood (Glenn, 1974, 1980; Block, 1981; Costa and McCrae, 1980; Sears, 1983; Jennings and Niemi, 1981; Jennings and Markus, 1984).

Taken together, these hypotheses imply that, as a result of the historically unprecedented prosperity and the absence of war in Western countries that have prevailed since 1945, the post-war generation in these countries would place less emphasis on economic and physical security than older groups, who had experienced the hunger and devastation of World War II, the Great Depression, and perhaps even World War I. Conversely, the younger birth cohorts would give a higher priority to non-material needs such as a sense of community and the quality of the environment.

These hypotheses were first tested in the 1970 European Community surveys, carried out in Great Britain, France, West Germany, Italy, The Netherlands and Belgium. Representative national samples of these publics were asked which goals were most important if they had to choose between various alternatives. These alternatives included some that emphasized economic and physical security and others that emphasized the non-material quality of life. Those whose top two choices related to economic and physical security were classified as "materialists;" those whose top two choices emphasized self-expression or other non-material values were classified as "postmaterialists;" and those who chose one of each were classified as "mixed" types.

As hypothesized, we found tremendous differences between the values of old and young. Figure 3.1 depicts this pattern in the pooled sample of six West European publics interviewed in our original survey. Among those who were older than 65 in 1970, materialists outnumbered postmaterialists by more than twelve to one. Among the post-war generation (those born after 1945), postmaterialists were about as numerous as materialists. Moreover, postmaterialists were especially numerous among the student population. Here they clearly constituted the dominant influence – all the more so because they tend to be more articulate and politically active than material-

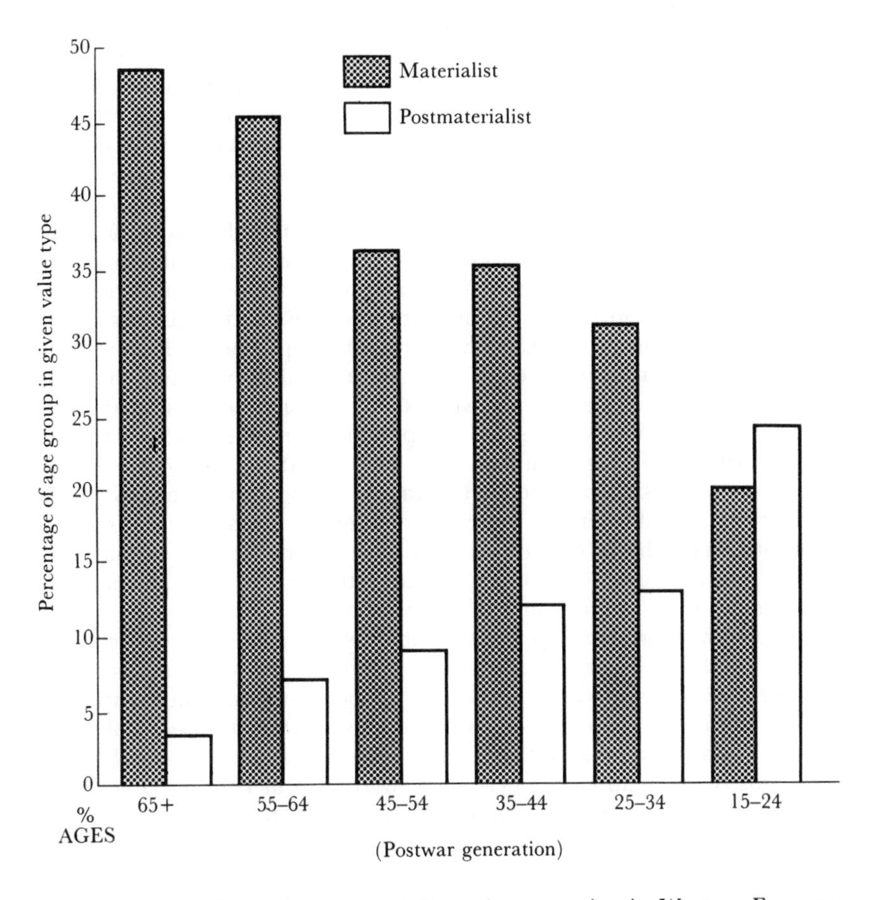

Figure 3.1 Materialist and postmaterialist value types in six Western European nations by age group, 1970
Source: Combined results of the British, French, West German, Italian, Belgian, and Dutch studies in the 1970 European Communities Survey.

ists. Among the rest of the public, materialists outweighed postmaterialists heavily; but postmaterialism prevailed in the student milieu. It seems that a major factor in the rise of student protest in the late 1960s was the fact that postmaterialists first emerged in substantial numbers at that point, as the post-war generation began to enter the adult population.

Subsequent surveys carried out in the United States and more than twenty other countries confirm the pattern found in 1970. There are dramatic differences between the goals emphasized by old and young. But one immediately wonders: do these differences reflect a process of intergenerational value change, as we hypothesized? Or do they simply reflect a life-cycle effect? At any given point in time, the young are less materialist than the old, but they become more materialist as they age; if this is so, then the 20-year-olds of today will be just as materialist as the 50-year-olds are now, when 30 years have gone by. The former interpretation implies that a major historical change is taking place; as older cohorts die off and are replaced by younger ones, the goals prevailing in the society as a whole will shift toward postmaterialism. The latter interpretation does not imply that society will change at all.

Do our findings reflect enduring intergenerational differences, or simply a life-cycle effect? Recent research has shown that these values are stably anchored in the orientations of Western publics, drawing on two kinds of complementary evidence: (1) panel data from the Political Action study has been analyzed using the LISREL structural equation technique to help control for measurement error. The results indicate that a latent materialist/postmaterialist orientation persisted stably among the American, Dutch, and German publics throughout the period from 1974 to 1981 (Inglehart, 1985b, 1987b; cf. Jagodzinski, 1984; Mohler, 1986); (2) cohort analysis based on successive cross-sectional surveys carried out from 1970 to 1984 demonstrates that relatively materialist orientations were a stable characteristic of older cohorts, while relatively postmaterialist orientations were a stable characteristic of younger cohorts throughout the 16-year period covered by these surveys in six Western nations.

Data from more recent surveys enable us to update this cohort analysis through 1987.[2] Figure 3.2 shows the results. Each cohort's position at a given time is calculated by subtracting the percentage of materialists in that cohort from the percentage of postmaterialists. Thus, the zero point on the vertical axis reflects a situation in which the two groups are equally numerous (this is approximately where the cohort born in 1946–1955 was located in 1970). An index of −45 would result if, say, 50 percent of a given cohort were materialists and only 5 percent were postmaterialists (with the rest being mixed types); the oldest cohort was located slightly below this point in 1970. Figure 3.2 summarizes the results from more than 160,000 interviews. It demonstrates rather conclusively that the age-group differences observed in 1970 reflect long-term intergenerational differences rather than life-cycle effects.

Each cohort retains its relative position with striking consistency throughout the 17-year period. The 1946–1955 cohort is less materialistic and more postmaterialistic than any of the older cohorts at every point in time. The only cohort that is even less materialistic is another postwar cohort, born between 1956 and 1964 – a group that was too young to be interviewed in 1970 but that becomes an increasingly important part of our sample from 1976 on. As figure 3.2 demonstrates, each of the older cohorts proves to be more materialist than all of the younger ones at every time point, with only a few minor anomalies. The inter-cohort value differences are extremely stable. Moreover, there is no indication whatever that each cohort becomes more materialist as it ages as would be the case if these differences

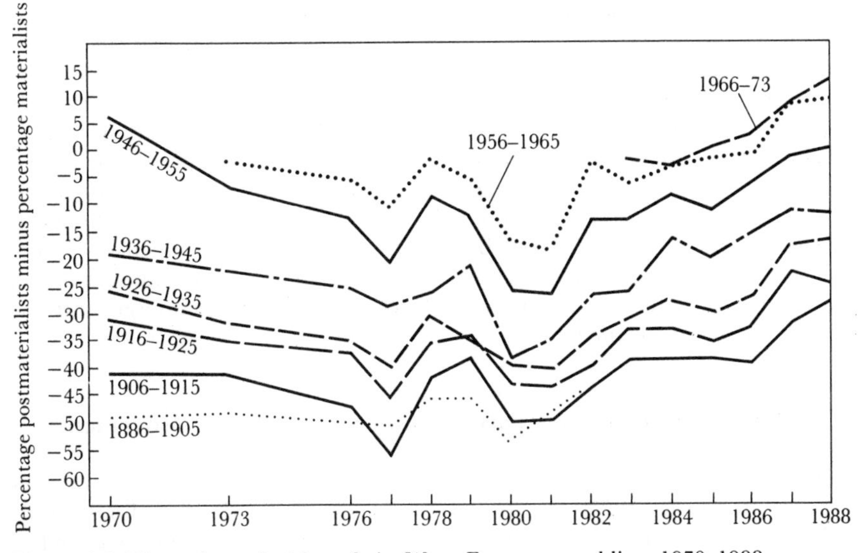

Figure 3.2 The value priorities of six West European publics, 1970–1988
Source: European Community Studies and *Eurobarometer* surveys.

reflected life-cycle effects, as Böltken and Jagodzinski (1985) have argued was the case. At the end of the 16-year period, virtually all of the cohorts were fully as postmaterialist as they were in 1970. Indeed, there is a slight upward tendency: most of the cohorts are less materialist in 1987 than they were in 1970. There are also some significant short-term fluctuations: each cohort shows a brief downward swing in 1977 and again in 1980–1. These fluctuations reflect period effects that are largely due to the impact of inflation. Böltken and Jagodzinski assumed that the net shift downward from 1970 to 1980 reflected life-cycle effects; they had no explanation for the upward swing in 1978 and 1979, but treated it as an idiosyncratic period effect. We argued that there was no evidence of life-cycle effects, attributing the downswings of the mid-1970s and early 1980s to period effects primarily linked with inflation. In Böltken and Jagodzinski's analysis (which stopped with the 1980 data), a life-cycle interpretation was plausible. But the fuller time series makes their interpretation untenable (Inglehart, 1985a, 1985b). By 1985, inflation had subsided to the 1970 level. With period effects held constant, there is no sign at all of the gradual conversion to materialism that would be present if a life-cycle interpretation were applicable. The pattern reflects intergenerational value change.

We find stable intergenerational value differences. This implies that, other things being equal, we will witness a long-term trend toward postmaterialist values, as one generation replaces another. A good deal of intergenerational population replacement has already taken place from 1970 to 1987. During these years, most of the 1886–1905 birth cohort died off. They were replaced by a much more numerous cohort born after 1955. In 1970, the post-war generation comprised only 20 percent of our sampling universe (those citizens over 15 years of age); in 1987, it constituted about half the sample.

There has been a corresponding shift in the distribution of value types. In 1970, materialists outnumbered postmaterialists in these six nations by a ratio of 3.4 to 1. In 1987, this ratio had fallen to less than 2 to 1. Contrary to predictions that postmaterialism would disappear as a result of the economic crisis, the underlying process of intergenerational change continued to function throughout the period from 1970 to 1987, though its effect was sometimes masked by negative period effects. When short-term forces returned to normal, the results were manifest. A substantial net shift toward postmaterialism had taken place, most of it due to intergenerational population replacement.

Postmaterialism and Participation in New Social Movements

Participation in these movements is strongly related to whether one has postmaterialist or materialist values, in each of the 12 nations of the European community, as table 3.1 demonstrates. This table is based on responses to questions assessing support and potential membership in new social movements that were included in *Eurobarometers* 17, 22 and 25.[3] Table 3.1 shows the percentage saying they are members of, or might join, "the ecology movement," "movements concerned with stopping the construction or use of nuclear power points," and "anti-war or anti-nuclear weapons movements."

The proportion saying that they would join these movements varies greatly from one nation to another. It is consistently lowest in Belgium, which is characterized by the most apolitical culture in Western Europe (Inglehart, 1990). Rates of participation in the ecology movement are highest in Greece, and it is significant that metropolitan Athens probably has the most severe pollution problems of any major city in the European community. Participation in the anti-war movement is highest in West Germany, the country that was scheduled to receive the largest contingent of Pershing and Cruise missiles of any NATO member – a fact that probably contributed to the emergence of a particularly strong and widespread anti-war movement in that country. Thus, examining the data in cross-national perspective suggests that situational factors are very important, and that we must take account of long-term factors (such as the political culture of a given society), intermediate ones (such as the severity of pollution in a given country in a given period), and short-term ones (such as the NATO decision concerning the installation of intermediate range missiles, and its subsequent implementation). Such factors almost certainly contribute to the wide range of cross-national variation in membership and potential membership in new social movements.

But it is equally clear that, in addition to such situational factors, the presence of materialist/postmaterialist values constitutes another major com-

Table 3.1 Membership or potential membership in new social movements by materialist/postmaterialist values in 12 nations, 1986

Member or potential member of ecology movement

Values	*Belgium*	*Denmark*	*France*	*Portugal*	*Britain*	*Italy*	*Ireland*	*Lux.*	*W. Germany*	*Neth.*	*Spain*	*Greece*
Materialist	4	8	8	8	10	15	27	26	13	28	31	40
Mixed	9	9	11	17	21	31	32	35	34	35	42	52
Postmaterialist	10	26	23	21	45	38	43	57	64	52	72	78
Total	8	11	12	12	22	26	31	35	37	38	42	50

Member or potential member of anti-nuclear power movement

Values	*Belgium*	*France*	*Portugal*	*Italy*	*Lux.*	*Denmark*	*Neth.*	*Britain*	*Spain*	*W. Germany*	*Ireland*	*Greece*
Materialist	1	5	7	9	9	14	10	9	21	9	30	41
Mixed	9	6	16	19	17	13	12	17	30	30	40	53
Postmaterialist	3	13	24	21	24	29	31	37	55	59	51	77
Total	5	6	12	15	15	16	17	18	30	34	38	50

Member or potential member of anti-war movement

Values	*Belgium*	*France*	*Portugal*	*Denmark*	*Britain*	*Neth.*	*Lux.*	*Italy*	*Spain*	*Ireland*	*Greece*	*W. Germany*
Materialist	3	5	9	14	8	10	19	15	24	30	43	32
Mixed	9	10	19	15	17	17	22	30	36	40	56	58
Postmaterialist	5	28	32	30	41	40	36	40	62	53	77	82
Total	6	11	15	17	18	22	23	25	35	38	53	59

Source: Eurobarometer 25 (fieldwork in April 1986)

ponent underlying support for the new social movements. In each of the twelve nations, postmaterialists are far more likely to be members or potential members of these movements than are materialists. In any given country, postmaterialists are at least twice as likely to be members as are materialists, and the ratio is often three to one, or even four or five to one. This relationship persists over time, moreover, as table 3.2 demonstrates in regard to the anti-war movement. Though levels of approval (top panel) and levels of reported and actual membership (lower panel) may rise or fall in response to given events in given countries, those with postmaterialist values consistently remain far more likely to give their approval – and their active participation – than are those with materialist values. Consistently, year after year, across these five nations as a whole, those with postmaterialist values are at least five times as likely to be members of the anti-war movement as are those with materialist values. Though the pure postmaterialist type constitutes only one-eighth of the public, they consistently furnish an absolute majority of the movement's activists.

The Impact of Values, Ideology and Cognitive Mobilization

What is the relative impact of ideology, cognitive mobilization and materialist/postmaterialist values in encouraging support for these movements? The answer depends on whether one is focusing on attitudes or behavior. In regard to general approval of the ecology movement, Left/Right self-placement is a considerably stronger predictor of one's attitude than materialist/postmaterialist values. Moreover, the respondent's level of cognitive mobilization is totally unrelated to this attitude.[4] But in regard to behavioral intentions and actual behavior, cognitive mobilization is an important predictor and materialist/postmaterialist values are the most important predictor of all.

As table 3.3 demonstrates, postmaterialists are significantly more likely to approve of the ecology movement than are materialists, but the difference is relatively modest (gamma = .077). Self-placement on a Left/Right ideological scale shows a much stronger linkage with approval of the ecology movement (gamma = .209) – though to some extent this may reflect the tendency for ecology-minded people to place themselves on the Left. But the linkage between ideology and behavior is only slightly stronger than the linkage between ideology and attitudes. Those who place themselves on the Left are almost twice as likely to be members of the environmentalist movement as those who place themselves on the Right (gamma = .261). The linkage between values and behavior, on the other hand, is far stronger than the linkage between values and attitudes. Those with postmaterialist values are more than five times as likely to be members of the environmental movement as those with materialist values (gamma = .427).

In moving from the realm of attitudes toward environmentalism to the realm of pro-environmentalist behavior, we are moving from a relatively

Table 3.2 Support for peace movements, by value type, 1982–1986

	France			*The Netherlands*			*West Germany*			*Italy*			*United Kingdom*			*Five-nation average*[a]		
	1982	*1984*	*1986*	*1982*	*1984*	*1986*	*1982*	*1984*	*1986*	*1982*	*1984*	*1986*	*1982*	*1984*	*1986*	*1982*	*1984*	*1986*
Strongly approve																		
Materialist	31	28	26	35	27	21	20	10	10	72	39	52	25	14	26	43	26	28
Mixed	38	32	31	36	33	27	26	23	21	72	58	62	28	23	28	47	33	35
Postmaterialist	53	44	53	61	59	52	66	57	52	79	70	67	44	40	50	63	52	55
Member of peace group																		
Materialist	0	<1	0	1	0	0	1	0	1	1	<1	1	1	1	1	1	<1	<1
Mixed	<1	<1	0	1	1	1	4	2	1	3	1	1	1	2	1	1	1	1
Postmaterialist	3	1	1	6	7	3	10	7	4	6	10	7	6	6	11	5	6	5

[a]Figures for the five-nation average are weighted according to the population of each country. Data are available at all three time points only for these five countries.

Source: Eurobarometer 17 (fieldwork in April 1982), *Eurobarometer* 22 (November 1984) and *Eurobarometer* 25 (April 1986).

Table 3.3 Attitudes and behavior toward the ecology movement by Left/Right ideology, cognitive mobilization, and value priorities, 1986 (percentages)

		Group membership		
	Strong approval	*Might join*	*Already member*	*Vote for Green party*[a]
Left/Right self-placement				
Left and Center-Left (1–4)	43	36	2	17
Center (5–6)	33	23	1	4
Right, Center-Right (7–10)	29	17	1	2
(Gamma)	(.209)	(.261)		
(N)	(8,979)	(7,980)		(976)
Cognitive mobilization				
Low, medium-low (1–2)	37	19	1	4
Medium (3)	32	28	1	8
High, medium-high (4–5)	41	37	3	13
(Gamma)	(.015)	(.305)		
(N)	(10,025)	(8,974)		(1,019)
Value priorities				
Materialist	37	15	1	1
Mixed	33	25	1	4
Postmaterialist	46	45	3	23
(Gamma)	(.077)	(.427)		
(N)	(9,530)	(8,505)		(929)

[a]The vote for the Green party is based only on the West German survey.
Source: Combined data from 12 nations surveyed in *Eurobarometer* 25 (fieldwork in April 1986 (N = 11,831).

"soft" indicator that contains a large component of spur of the moment response to a relatively "hard" indicator that refers to specific activities one either has, or has not, done. Only a small minority (about 1 percent) of the public claim to be members of the ecology movement. But this small active group consists of people who have given the ecology movement a great deal of time and thought. The relationship between materialist/postmaterialist values and active participation in the movement is very strong. The same is true of electoral behavior concerning the largest and most salient of the ecology parties, the West German Greens: the linkage between votes and values is extremely strong (postmaterialists are 23 times as apt to vote for them as are materialists).

Moving from attitudes to behavior does not have a comparable effect on the relationship with Left/Right ideology. This variable is not much more strongly linked with behavior than it is with attitudes. This reflects the fact that the Left/Right dimension (unlike the materialist/postmaterialist dimension) has been a major fixture of West European politics for many decades. Practically everyone can place oneself on this dimension – but some people do so out of traditional loyalties, and others as a reflection of their orientations toward current issues. Traditionally, the Left has been favorable to change-oriented movements, and those who identify with the Left tend to give at least lip service to the new social movements. Yet more active involvement in these movements is not linked with a markedly higher level of constraint between behavior and Left/Right ideology. For materialists remain a major component of the Left – and for them, a closer examination of the relationship between one's beliefs and the goals of the new social movements does not have the same unequivocal result that it does among postmaterialists. From the viewpoint of the materialist Left, it is not unambiguously clear that the environmentalist and anti-war movements will enhance one's top priorities – economic and physical security.

The impact of cognitive mobilization, on the other hand, does show a dramatic increase as we move from attitudes to behavior and behavioral intentions. In their attitudes, those with high levels of political skill are not necessarily more favorable to the ecology movement than those with low levels. But high levels of political skill are virtually a prerequisite for active participation in such a movement, and cognitive mobilization shows a strong linkage with our behavioral indicators.

These findings apply to the anti-nuclear power movement and the anti-war movement as well as to the ecology movement. Table 3.4 compares the linkages between our three independent variables on one hand and attitudes and behavior and behavioral intentions towards each of these three types of movement on the other hand. For each of these three kinds of movement, across the 12 nations of the European community, one's Left/Right ideological location is the strongest predictor of attitudinal approval. But when it comes to behavioral intentions and actual behavior, cognitive mobilization is an equally strong predictor, and materialist/postmaterialist values a much stronger predictor than is Left/Right ideology.

For the rise of these new social movements has reshaped the meaning of the old, familar Left/Right ideological spectrum, as much as it has been shaped by it. One major component of the meaning of Left and Right today is simply its connotations with specific political parties that have traditionally been considered to be on the Left or on the Right. Originally, these Left/Right locations reflected, above all, the fact that certain parties represented the workers (on one hand) or the privileged classes (on the other hand). Conventionally, the communists occupied the Left-most position on the spectrum, followed by the socialists and then the liberals and various other more conservative groups. But this conventional component of the Left/Right dimension has an ambiguous relationship to the new social movements. As table 3.5 demonstrates, the communist electorate is not markedly more supportive of the ecology and the anti-nuclear movements than is the electorate of the extreme right. Instead, one finds a curvilinear

Table 3.4 Attitudes and behavior toward three new social movements by Left/Right ideology, cognitive mobilization, and value priorities, 1986

	Approval of movement	*Membership or potential membership in movement*
Correlation with Left/Right ideology		
Ecology movement	.209	.261
Anti-nuclear power movement	.233	.339
Peace movement	.295	.299
Average	.246	.300
Correlation with cognitive mobilization		
Ecology	.015	.305
Anti-nuclear power movement	.066	.279
Peace movement	.026	.295
Average	.036	.293
Correlation with value priorities		
Ecology movement	.077	.427
Anti-nuclear power movement	.048	.433
Peace movement	.116	.465
Average	.080	.442

Table entries are gamma correlation coefficients.
Source: Data from combined 12 nations surveyed in *Eurobarometer* 25 (fieldwork in April 1986).

relationship, with the socialists being more favorable to these movements than the communists. By far the most dramatic rise in the support for social change comes when we reach the ecology parties – which most observers initially placed somewhere near the middle of the Left/Right spectrum. Today, one might argue that they really should be placed to the Left of the communists – which would then create a relatively clear, monotonic relationship between the Left/Right dimension and support for the new social movements. There is no reason why we could not adopt this convention – but we should be sure to bear in mind that if we did so, it would represent a redefinition of the Left/Right spectrum in the light of new facts, rather than the reshaping of the new movements by the traditional Left.

The degree to which the Left has been reshaped by such new social movements, rather than the other way around, is put in clearer perspective when we compare the deep-rooted communist parties of France and Italy with the communist parties of the new party systems that have emerged since democratic institutions were restored in Greece, Spain and Portugal in the 1970s. Thus far, the new movements have had only a relatively modest impact on the long-established and heavily bureaucratized commu-

Table 3.5 Membership and potential membership in new social movements by voting intention in ten nations

	Membership or potential membership	
	(%)	*(No. of cases)*
Ecology movement		
Communist	23	399
Socialist	28	1,924
Ecologist	69	254
Liberal/Giscardians	19	499
Christian Democrats/Conservatives/Gaullist	17	1,854
Extreme Right (MSI, NF, etc.)	21	93
Anti-nuclear power movement		
Communist	15	404
Socialist	20	1,941
Ecologist	53	242
Liberal/Giscardians	6	497
Christian Democrats/Conservatives/Gaullist	7	1,850
Extreme Right (MSI, NF, etc.)	11	90
Peace movement		
Communist	30	400
Socialist	33	1,977
Ecologist	60	244
Liberal/Giscardians	16	505
Christian Democrats/Conservatives/Gaullist	17	1,847
Extreme Right (MSI, NF, etc.)	10	94

Table entries are the percentage who are members or might join each social movement.
Source: Based on combined 10-nation data from *Eurobarometer* 21 (fieldwork in April 1984), weighted according to population in each nation.

nist parties of France and Italy; within these countries, the electorates of various other parties are more supportive of the ecology movement than are the communists. But in the newer party systems of Greece, Spain, and Portugal, the Left/Right partisan space coincides more closely with support for the new social movements, with the communist party being the most pro-ecologist party in the system. These newer communist parties emerged in an era when environmentalism and the anti-nuclear movements were already major issues, and they have been relatively responsive to them.

Membership in New Social Movements: A Multivariate Analysis

The new social movements reflect a variety of factors, as we have argued. Not all of them can be readily measured by survey research, but let us attempt to estimate the relative importance of those that can. Some of these variables are correlated with others; age, for example, is linked with both materialist/postmaterialist values and with Left/Right self-placement. Is it possible that both of the latter variables show strong linkages with membership in the new social movements, simply because younger people are most likely to join them? Similarly, both cognitive mobilization and postmaterialist values are linked with relatively high levels of income and education; are they both simply surrogate indicators of social class? In order to answer such questions, and gain a clearer understanding of the relative impact of given variables when we control for the effects of others, we performed a series of Multiple Classification Analyses (MCA). Membership and potential membership in given social movements were the dependent variables in these analyses (dichotomizing between those who said they were members or might join the given movement, and those who said they would not join). As predictor variables, we used materialist/postmaterialist values, Left/Right self-placement, cognitive mobilization, age, income, religiosity, and closeness to a political party. A number of additional demographic variables were included in preliminary analyses, but those which had neither theoretical significance nor empirical importance were eliminated from our final version of these analyses, reported in tables 3.6 and 3.7.

Table 3.6 shows the percentages who say they are members of the ecology movement, or might join it, within the respective categories of each of our seven predictor variables, for the European Community as a whole, along with the summary statistics from a Multiple Classification Analysis based on these variables. Table 3.7 provides comparable information concerning membership and potential membership in the peace movement. The results from our analysis of the anti-nuclear power movement are so similar to these results that we will not present them here.

Materialist/postmaterialist value priorities prove to be our strongest predictor of participation in both the ecology movement and the anti-war movement across these twelve nations. This holds true of both the zero-order relationship (indicated by the eta coefficients in tables 3.6 and 3.7) and the partial relationship, controlling for the effects of all the other predictors included in these analyses (as indicated by the beta coefficients in these tables). The impact of values is not a spurious one; it cannot be attributed to the fact that postmaterialists are younger, better educated, more inclined to the Left and so forth. One's values remain the strongest predictor of activism and potential activism in both of these new social movements, even when we control for the effects of these other variables.

Left/Right self-placement and cognitive mobilization also have substantial impact, controlling for the effects of other variables. Their impact is not just an indirect manifestation of social class conflict. Indeed, these results demonstrate rather vividly just how little social class conflict has to do with the new social movements. For income proves to be a feeble predictor of activism in these movements. Moreover, it shows the wrong polarity: the upper income groups are more likely to belong to these presumably Left-oriented movements than are the lower income groups. If we view this phenomenon as reflecting these groups' relative levels of cognitive mobilization, it is perfectly in keeping with our expectations. But it is not at all what one would have predicted if one viewed these movements as manifestations of the class struggle.

One's age does have a significant impact on activism in these movements: it is the third strongest predictor of membership and potential membership in the ecology movement, and the second strongest predictor of membership in the anti-war movement. This holds true when we control for the impact of materialist/postmaterialist values (which are, of course, related). Age

Table 3.6 Predictors of membership/potential membership in the European ecology movement, 1986

Predictor	*Percentage*	*Eta*	*Beta*
Value priorities			
Materialist	15		
Mixed	26		
Postmaterialist	48	.228	.153
Left/Right self-placement			
1–2 (Left)	40		
3–4	37		
5–6 (Center)	24		
7–8	18		
9–10 (Right)	17	.184	.129
Age of respondent			
15–24 years	37		
25–39	31		
40–54	24		
55 years or older	15	.182	.122
Cognitive mobilization			
Low	15		
Medium-low	23		
Medium	29		
Medium-high	39		
High	43	.184	.092

(cont.)

Table 3.6 (cont.)

Predictor	*Percentage*	*Eta*	*Beta*
Self-defined religiosity			
1–2 (Low)	33		
3–4	33		
5–6	24		
7–8	24		
9–10 (High)	20	.109	.043
Closeness to a political party			
Very close	28		
Fairly close	31		
Merely sympathiser	26		
Not close to any	23	.063	.042
Family income			
Lowest quartile	21		
Second quartile	25		
Third quartile	27		
Highest quartile	32	.083	.028
		R =	.331

Percentages in the table are of those saying they are members or might become members of the ecology movement. The eta coefficients measure the simple bivariate correlation between each variable and membership in ecology groups; the MCA beta coefficients represent the independent non-linear correlation of each variable while controlling for the other predictors.
Source: Based on 12-nation data from *Eurobarometer* 25 (fieldwork in April 1986).

plays a significant role even apart from generational differences in value priorities and Left/Right self-placement. It seems likely that life-cycle effects are involved, though we cannot be sure without longitudinal data. Religiosity has a fairly pronounced zero order relationship with membership in the new social movements: only 20 percent of those who consider themselves "very religious" are members or potential members of the ecology movement, for example – as compared with 33 percent of those who consider themselves "not at all religious." But this linkage fades into marginal significance when we control for the effects of the other variables (including the fact that the more religious respondents are less likely to be postmaterialists, less likely to place themselves on the Left, and tend to be older and lower on cognitive mobilization).

Closeness to a political party plays even less of a role in determining membership or potential membership in the new social movements. Those who say they are very close to some political party are somewhat more apt to be active in these movements than those who are not close to any party, but the impact of this variable fades to an almost insignificant level when we control for the effects of the other variables. This negative finding has considerable theoretical significance. It demonstrates the fact that these new social movements are genuinely new: the established political parties, which for decades have been the most important agency mobilizing political participation, play only a marginal role in these new movements. The new social movements represent a different type of political participation, one that is less elite-directed than has generally been true of participation in the past, and one that is shaped to a far greater degree by the individual's values, ideology, and political skills. The new social movements are new, not only in their goals, but also in their political style and in the factors that mobilize their activists.

Table 3.7 Predictors of membership/potential membership in European peace movement, 1986

Predictor	*Percentage*	*Eta*	*Beta*
Value priorities			
Materialist	15		
Mixed	27		
Postmaterialist	51	.251	.175
Left/Right self-placement			
1–2 (Left)	43		
3–4	39		
5–6 (Center)	26		
7–8	18		
9–10 (Right)	16	.200	.136
Age of respondent			
15–24 years	40		
25–39	33		
40–54	22		
55 years or older	17	.195	.139
Cognitive mobilization			
Low	15		
Medium-low	25		
Medium	31		
Medium-high	37		
High	47	.183	.087

(cont.)

Table 3.7 (cont.)

Predictor	*Percentage*	*Eta*	*Beta*
Self-defined religiosity			
1–2 (Low)	36		
3–4	34		
5–6	26		
7–8	25		
9–10 (High)	20	.125	.043
Closeness to a political party			
Very close	32		
Fairly close	32		
Merely sympathizer	28		
Not close to any	24	.066	.043
Family income			
Lowest quartile	24		
Second quartile	29		
Third quartile	27		
Highest quartile	30	.050	.036
			R = .337

Percentages in the table are of those saying they are members or might become members of the peace movement. The eta coefficients measure the simple bivariate correlation between each variable and membership in peace groups; the MCA beta coefficients represent the independent non-linear correlation of each variable while controlling for the other predictors.
Source: Based on 12-nation data from *Eurobarometer* 25 (fieldwork in April 1986).

Conclusion

In the preceding chapter, Karl-Werner Brand presents a cyclical interpretation of the rise and fall of new social movements. His contribution is valuable because it emphasizes the fact that support for these movements *does* rise and fall, rather than taking off in an ever-ascending trajectory, as some of the movements' more naïve supporters may initially have imagined. Instead, support for any given movement reflects a complex interaction between long-term and short-term forces, between the movement's leaders and its mass base, and between characteristics of the movement itself and the immediate situation in which it finds itself.

The present chapter has pointed out a number of such factors, and then analyzed the relative importance of various individual-level factors ranging from materialist/postmaterialist values to social class, religion, political party loyalties and cognitive mobilization. The presence of materialist or postmaterialist values proves to be the most important single influence on whether a given individual will support new social movements.

Clearly, materialist/postmaterialist values are *not* the only factor involved, however; for, as Brand points out, the proportion of postmaterialists in the populations of Western countries has grown only gradually during the past 20 years, whereas support for any given movement may rise or fall quite suddenly. People's behavior reflects their immediate historical context, as well as their underlying values. But it is equally clear that the gradual evolution of a sizeable pool of postmaterialists has been a crucial factor in the emergence of the new social movements – indeed, it is largely responsible for the fact that they are perceived as *new*. For the forerunners of the ecology movement (for example) in the 1840s and again around the turn of the century were based on relative handfuls of activists. The contemporary movements can mobilize hundreds of thousands of activists in a single demonstration; over three million West Germans participated in demonstrations against new NATO missiles during a single week in 1983. Moreover, the historical forerunners of these movements in the nineteenth and early twentieth centuries reached phases of high mobilization during periods of economic *downturn* (as figure 2.1 in Brand's chapter illustrates). The contemporary movements flowered during a period of high *prosperity*; they reflect postmaterialist motivations, rather than the traditional protest linked with deprivation.

To view the new social movements as simply cyclical phenomena leaves both of these crucial developments in obscurity, providing no explanation for either their unprecedented size or their distinctive motivating forces. For the concept of a cycle suggests that in the long run one returns to the point where one started. It ignores the fact that there are long-term trends, as well as cyclical fluctuations.

It is important to remember that short-term fluctuations occur – and to probe into the reasons for these fluctuations. Specific historic situations change; a movement may overshoot its goals, engendering a reaction: social processes rarely move in a straight line. But it is equally important not to lose sight of the fact that long-term changes also occur. The social movements of the 1970s and 1980s are not the same as those of the 1840s or of 1900. The emergence of postmaterialist values among a substantial segment of Western publics, together with the process of cognitive mobilization, give new scope and new impetus to movements that earlier history foreshadowed only faintly.

NOTES

1 Left/Right self-placement was measured by the standard *Eurobarometer* question: "In political matters, people talk of the 'Left' and the 'Right.' How would you place your views on this scale?" (The respondent is handed a scale consisting of ten boxes with the word "Left" at the left end and the word "Right" at the right end.)

2 The following analyses are based upon data derived from the *Eurobarometer* surveys conducted by the Commission of the European Community (EC). These public opinion surveys have been conducted semi-annually since 1974 in all the EC member states. The original survey data are available from the Inter-university Consortium for Political and Social Research at the University of Michigan in Ann Arbor.

3 Approval of new social movements was measured by the question: "There are a number of groups and movements seeking the support of the public. For each of the following movements, can you tell me whether you approve (strongly or somewhat) or you disapprove (somewhat or strongly)?" Table 3.2 displays the percentage expressing strong approval. A second question assessed membership in these groups: "For each of these groups could you tell me whether you are a member, or might join, or would certainly not join the group?" Table 3.1 displays the percentage saying they are now members or might join.

4 The cognitive mobilization index is created by combining educational level and the frequency of political discussion with friends (Inglehart, 1977, p. 339). The lowest level on this index is composed of individuals who left school before the age of 16 and who "never" discuss politics; the highest ranking on the index consists of those whose schooling continued until they were 20 or older and who "frequently" discuss politics.

4

Neo-corporatism and the Rise of New Social Movements

FRANK L. WILSON

The sudden and unexpected appearance of new social movements in the late 1960s and 1970s challenged the prevailing interpretations of group politics in industrialized democracies. Advocating such diverse causes as ethnic minority rights, women's equality, environmental protection, and university reform, these new movements appeared in nearly all Western democratic countries. But their political influence has varied widely. In some countries, new social movements came into being, accomplished a few immediate goals, and then disappeared. In other countries, the new social movements achieved their initial objectives and have gone on to establish and pursue other goals. In some countries, they have had little lasting impact on the political scene; in others, their activities portend profound restructuring of the political system.

Among the few explanations for the variable political impact of new social movements is the structure of relations between traditional interest groups and the state. The failure of existing socio-occupational and advocacy groups to adequately represent new issues and even the shifting interests of their usual clienteles are often seen as promoting the rise of the new movements.

For years, the prevailing paradigm for the study of group politics in Western democratic countries was pluralism. In recent years, however, many observers have detected a change in interest group/government interaction toward a pattern of neo-corporatism. This neo-corporatist pattern is seen by many as fostering the emergence of new social movements. Before discussing the alleged association of neo-corporatism with new social movements, it is useful to review briefly the basic features of these two alternative patterns of interest group/government relations with special attention to their consequences for group formation and the choice of political tactics for outsider or challenging groups.

Pluralism, Neo-Corporatism, and New Social Movements

The appearance of new groups in the 1960s and 1970s was neither surprising nor disturbing to pluralists. It was seen as part of the normal generative process which brought new groups into the political process (Pizzorno, 1981, pp. 278–82). Pluralist theory stressed the spontaneous appearance of new groups in response to new issues – such as environmental protection or women's rights – or to challenge established groups that have lost touch with their rank and file, as did the tax revolt groups or the small shop owners. To be sure, the sudden flourishing of group life was surprising in those European countries, such as France or Italy, where individualism or other political attitudes had hindered the development of voluntary associations (Rose, 1954; Gallagher, 1957; Almond and Verba, 1963). In these countries and elsewhere, the reaction to the rise in associative life was generally positive, with most observers applauding the increased public involvement in civic affairs.

On the other hand, the tactics used by the new movements did cause consternation among pluralists. The social movements of the 1960s and 1970s seemed interested in dramatic media events designed to embarrass the government rather than to actually change the shape of immediate policy. The pluralists' vision of a relatively open political process led them to view the new groups' recourse to unconventional tactics as aberrant. The disruption of the ordinarily tranquil political scenes in Britain, the Netherlands, Denmark, or West Germany by groups that seemed to prefer advocating their causes in the streets to pressuring government or representatives prompted fears that the disorderly new movements would bring political instability.

Concerns grew as several of these movements adopted an anti-system posture. Even those groups that did not have anti-systemic goals spurned offers to become involved in the established processes. These new movements were clearly not conforming to the normal pattern of emerging groups anticipated in the pluralist model. Thus, if their origins can be explained by pluralist theory, the tactics and outlook of the social movements are difficult to account for under it.

This deviation from the pluralist model is deliberate. In their rhetoric, leaders of new movements acknowledged the pluralist cycle of group formation and absorption into the political system. But they indicated also their determination to avoid that fate for their groups. The new social movements of the 1970s did not intend to be coopted into a political process that they felt would inevitably dilute if not destroy their ideals. With all their might, they would resist Lowi's (1971) "iron law of decadence" whereby all established groups eventually lose their militancy and become blatantly conservative.

An alternative model for the emergence of new social movements and their unconventional tactics is neo-corporatist theory. The neo-corporatist model stresses the exclusive relationships between a handful of privileged groups and the state.[1] Instead of the multiplicity of relevant interest groups predicted by pluralism, the neo-corporatist model posits the presence of a single group for each interest sector: labor, employers, farmers, veterans, and so on. That one group is viewed as the only legitimate vehicle for that sector. There is thus little competition from rival groups within the various sectors.

These representational monopolies are jealously guarded through the connivance of both the established groups and the state. The entry of new groups – either associations that would champion new causes or schisms from existing official groups – into the political arena is limited by formal and extralegal restraints. The established groups seek to perpetuate their monopolies by coopting or vigorously opposing new groups claiming to represent the same interests. The government may limit the organization of new groups by requiring the licensing of interest associations and restricting entry of groups deemed by the government to be "unrepresentative." It may refuse to consult with all but the officially recognized groups or accord privileged access only to them. The government often subsidizes the privileged groups and denies such funding to unofficial organizations. The official groups are given control over valuable services – such as insurance, financial advice and loans, or even social security benefits – that help them maintain their hold on their clientele. As a result, membership in these approved groups is virtually mandatory for the affected clienteles.

Interest group/government interaction is well structured and institutionalized under corporatism. In contrast to the largely informal and *ad hoc* pattern of interaction under pluralism, the groups have "a formalized and substantial share in formulating and administering government policy" (Martin, 1983, p. 99). Government devolves policy-making authority within certain issue areas to statutory bodies. In some cases, the corporatist bodies are quasi-legislative organs based on functional representation rather than territorial districts. More frequently, they are councils, committees, boards, or work groups that bring together government and group representatives to make authoritative policy decisions. They are not simply consultative bodies; they actually make the decisions either directly or indirectly. They then become directly involved in administering the policies since such groups "are more able than government to deliver policies effectively" due to their technical knowledge, manpower, or legitimacy in the eyes of those affected (Richardson and Jordan, 1979, pp. 149–50).

The effectiveness of the groups' participation in these corporatist bodies renders other forms of action unnecessary. Groups do not have to work with party leaders or engage in legislative lobbying or even make appeals to public opinion. Other forms of pressure – such as demonstrations or political strikes – are not only unneeded, but are avoided out of fear that they might damage the harmony and spirit of accommodation that is the essence of the corporatist process.

In effect, corporatism is seen as a restraint on uncontrolled participation.

By restricting involvement to official groups committed to accommodation, corporatism channels participation into acceptable forms. In addition, neo-corporatism encourages group leaders to discipline their members to accept the "general interest" negotiated by the elites instead of pressing for their individual interests (Crouch, 1983). It thus becomes "the ideal solution to the central problem of modern capitalism: the maintenance of order . . ." (Crouch, 1978, p. 215). In practice, these restrictions affect working-class organizations more than employers' groups and, in the view of many critics of corporatism, they are specifically designed to do so (Offe, 1981). Corporatism is seen as an effective way of promoting governability in an era of government overload and expanding levels of participation (Schmitter, 1981). Even a critic of neo-corporatist relations concludes grudgingly that "the relative ruliness and effectiveness of the outcome is impressive" (Schmitter, 1981, pp. 317–18). The corporatist framework channels political participation and facilitates negotiating collaborative policy agreements that avoid straining the state's financial resources and political coalitions even in times of economic difficulty.

Neo-corporatism and New Social Movements: Theoretical Links

Only a few who write on corporatism have directly addressed the link between neo-corporatism and social movements (Beer, 1982; Helm, 1980; Pizzorno, 1981; Schmitter, 1981). However, implicit in many descriptions of the strengths and weaknesses of this form of interest group/government interaction are elements of an explanation for the rise of new social movements and the unconventional tactics they use. In a like manner, studies of new social movements often allude to the pattern of interest group politics in explaining their rise.

First, neo-corporatism contributes to the development of stodgy group leaderships that are unresponsive to the needs and concerns of their grassroots memberships. The elitist character of corporatism inevitably produces tensions both within existing associations and among latent outside groups that feel that their concerns are neglected. As group elites interact, they come to see each other's views more clearly and to moderate the demands from the grassroots of their organizations (Panitch, 1979, pp. 139–45). Their view from the top may lead them to ignore the concerns or demands of their members. They are protected from the rank and file by their monopoly, mandatory membership in their associations, and the secrecy of corporatist negotiations.

The participation of the group leaders in formulating a consensual accord of all participating interests builds the obligation for the leaders to then defend that consensus, even when it departs from the demands of their members. The hierarchical and centralized structures that are necessary for effective implementation of decisions reached by the leadership stifle participation.[2] Consequently, group leaders inevitably face challenges from rivals within their organization who claim to be more attuned to the interests of the rank and file (Sabel, 1981). Indeed, one author suggests that neo-corporatism promotes less conflict resolution and more conflict displacement, with the level of conflict within and between groups shifted to the micro level (Wassenberg, 1982). In the extreme, there are revolts at the base, unauthorized action or wildcat strikes, schisms, and the formation of new groups to challenge the monopoly of the traditional group.

Second, rigidities in neo-corporatism facilitate the rise of new social movements. Corporatism thrives in societies where "government policy decisions are more or less confined to the area of a prevailing consensus" (Harrison, 1980, p. 182). It is less able to adapt to handle new issues where the consensus is not yet defined or to address intractable existing problems. Corporatism avoids the new (and old) issues that are outside the area of consensus. Those concerned with these issues find little response to their concerns from the established groups, whether labor unions, farm groups, employers' associations, or the political parties that support corporatist collaboration. The only option is the formation of a new group to press for the neglected issue. This is especially true in corporatist states with leftist governments. The rigidity and unresponsiveness of the political system to new policy demands opens up new opportunities for new movements (Kitschelt, 1986). The static nature of the neo-corporatist bargain leads to the constant presence of "new groups-in-formation" (Harrison, 1980, p. 211), even if there are institutional barriers to the creation of new associations.

Corporatist policy-making tends to produce compromises among the participating groups at the expense of higher taxes, bigger government, and inefficiency. With these privileged participants engaged in a "benefits scramble" (Beer, 1982), the interests of consumers, taxpayers, or the environment are unrepresented. The consequence is the rise of new movements to force policy-makers to take these interests into account. Thus Wilensky (1976) finds the tax revolt of the 1970s linked with corporatism.

Third, the closed corporatist policy-making processes lead to distinctive tactics for excluded groups. The usual channels for group influence – the corporatist bodies – are not open to excluded groups or new movements. These groups must operate in a setting of "residual pluralism" with limited opportunity for success (Schmitter, 1983, pp. 917–19). Frustrated by their inability to gain a hearing for their concerns by the policy-makers, outside groups resort to unconventional methods such as demonstrations, boycotts, political strikes, sit-ins, and even violence to draw the attention of the public or the policy-makers to their concerns. They may hope that their actions will be effective enough to sabotage policies they were unable to influence. They usually recognize the futility of their efforts but hope that attention drawn to the issue by their spectacular conduct and the abuse they receive from the defenders of the status quo will attract public sympathy. The emphasis of the movements is not so much on practical results in policy shifts as on media events that will shape opinion over the long run.

This disdain for the usual political tactics carries over into a general disregard for the entire political order (Beer, 1982). The hostility the movements meet in presenting their concerns encourages the development of an anti-system orientation. They want not only to change the content of the decisions but also to replace the decision-making process.

There are several paradoxical consequences of corporatism on social movements. Despite the emphasis on restricting the entry of new groups, corporatist dynamics presumably foster the emergence of new groups both from grassroots revolts in existing associations and from outsider movements seeking to attract attention to issues excluded from the agenda of corporatist processes. In addition, while corporatism is seen as promoting stability and order by channelling participation into acceptable forms, it ends up fostering unconventional tactics on the part of the outsider groups. Finally, corporatism comes from the desire for control and system stability, but in practice it nonetheless forces an anti-system orientation on non-sanctioned groups which see disrupting the system as the only way of achieving influence. Thus, while corporatism is seen by some as promoting stability and governability, others see it bringing into play forces tending toward the disruption of the system.

This summary suggests that we might expect to find that differences in the extent to which new social movements develop will vary among countries according to the type of interest group/government interaction. The more corporatist the interest group representation, the more likely it will be that new social movements will have an important impact on politics. In more pluralist settings, new social movements will be less important.

Group Representation and the Incidence of New Social Movements: An Empirical Test

The move from these theoretical statements to the real world is not an easy one. One problem is the absence of reliable measures of either the extent of corporatism or the incidence of new social movements. In addition, there is the problem of assessing the health of the corporatist system. While corporatist theorists suggest that new social movements might be expected to emerge as pressures from unfulfilled or new demands build, they contend that when the corporatist system is functioning well, its control mechanisms will prevent the emergence of dissident movements. What follows in this section is a very tentative attempt to test the effect of the type of group/government interaction on the development of new social movements.

THE EXTENT OF CORPORATISM AND INCIDENCE OF NEW SOCIAL MOVEMENTS

The first problem is establishing the degree to which various European countries have moved toward the corporatist model of interest group/govern-

Table 4.1 Classification of countries by degree of corporatism, according to four different analyses

Wilensky	*Schmitter*	*Lehmbruch*	*Schmidt*
Strong corporatism			
1 Belgium	1 Austria	1 Netherlands	1 Austria
2 Netherlands	2 Norway	2 Norway	2 Norway
3 Norway	3 Denmark	3 Sweden	3 Sweden
3 Sweden	3 Sweden	4 Austria	
5 France	3 Finland		
6 Finland			
7 Austria			
Medium corporatism			
8 Italy	6 Netherlands	5 Switzerland	4 Belgium
9 Denmark	7 Belgium	6 Finland	5 Denmark
10 Germany	8 Germany	7 Denmark	6 Finland
11 Switzerland	9 Switzerland	8 Germany	7 Germany
		9 Belgium	8 Netherlands
		10 Ireland	
Weak to no corporatism			
12 Britain	10 Canada	11 France	9 Britain
12 Canada	10 Ireland	12 Italy	10 Canada
12 Ireland	10 USA	13 Britain	11 France
12 USA	13 France	14 Canada	12 Ireland
	14 Britain	15 USA	13 Italy
			14 USA

Source: Modified from Cawson (1986; p. 99). Based on Wilensky (1976, p. 50); Schmitter (1981, p. 294); Lehmbruch (1985, p. 13); and Schmidt (1982, p. 245).

ment relations. There is disagreement among corporatist theorists over rankings of countries according to the extent of corporatism. Table 4.1 compares several rankings with some dramatic differences. France, for example, is regarded as among the most corporatist states in Wilensky's ranking but among the least corporatist in the view of the others. Even the same observer gives different rankings in various writings.[3] Often the reasons for these rankings are unspecified; when they are specified they are not readily quantified or similarly defined by all. For the purposes of this paper, I will use Schmitter's classification expanded to include Italy among the weak corporatist states. Not only is Schmitter's listing comprehensive, it has been used widely by others in assessing the causes and consequences of neo-corporatism. In addition, my own assessment of the extent of corporatism is close to his.

MEMBERSHIP IN SOCIAL MOVEMENTS

The next problem is determining the incidence of new social movements. The difficulties of identifying what constitutes a social movement and of gauging its impact have prevented the development of an acceptable ranking of countries according to the incidence of new social movements. By their nature, social movements are transient. To have a significant impact on democratic politics, social movements must do more than simply achieve a few specific objectives: they must make some lasting changes in attitudes, behavior, and/or policies.

One indicator of the impact of social movements can be found in public attitudes toward such groups. A Eurobarometer poll in 1986 measured European public attitudes toward ecology movements, anti-nuclear power movements, and peace movements (see Inglehart in this volume, chapter 3). Table 4.2 presents the percentage of respondents indicating strong approval of three types of social movements. There is little indication that support is stronger for these movements in countries where corporatist relationships may have stifled means of access to the political process for those advocating issues beyond the corporatist consensus. In fact, Italy, a "consensus" choice as among the least corporatist, is the country with the highest approval for these movements. Germany, with greater corporatism, has the lowest approval for all three causes.

Table 4.3 shows the respondents' inclination to join social movements. The actual members are very few in all countries; the highest reported membership was among Dutch respondents who said they were members of ecological movements – only 2.7 percent. The table combines those who say they are already members and those saying they might join a social movement.[4] Again, there appears to be no relationship between the extent of corporatism and interest in joining social movements. Two of the most corporatist states have the lowest levels of interest in joining social movements; the Irish, in a weak corporatist state, show the second highest interest in affiliating with social movements.

POLITICAL TACTICS

There is often the expectation that people in corporatist systems may be more willing to engage in unconventional means of political action because the more conventional forms are either unavailable or unproductive. Table 4.4 looks at willingness to engage in unconventional political actions in twelve countries according to their approximate level of corporatism. These polls suggest little if any association between level of corporatism and willingness to engage in unconventional political action. Certain "pluralist" countries – the USA, Ireland, and France – display the greatest citizen propensity to use unconventional political actions. But the people in the

Table 4.2 Sympathy for new social movements by level of corporatism, 1986

	Percentage indicating strong approval			
	Ecology movements	*Anti-nuclear movements*	*Peace movements*	*(N)*
Strong or medium corporatism				
Denmark	22.7	37.9	34.8	1,043
Netherlands	51.4	25.1	30.6	1,001
Belgium	31.7	29.4	33.7	1,007
West Germany	12.6	13.0	23.4	987
Average	29.6	26.4	30.6	
Weak to no corporatism				
Ireland	21.5	46.2	47.5	1,002
France	28.4	17.5	28.8	1,003
Britain	20.2	26.8	27.7	1,379
Italy	43.5	35.3	53.9	1,102
Average	28.4	31.5	39.5	

Text of question: "There are a number of groups and movements seeking support of the public. For each of the following movements can you tell me whether you approve (strongly or somewhat) or disapprove (strongly or somewhat)?" Percentages are based on all respondents.
Source: Eurobarometer 25 (fieldwork in April 1986).

least corporatist states (Britain and Italy) were about as ready to use such tactics as were citizens of the most corporatist states.

There are many factors influencing the individual's perception of the repertoire of political action other than the structure of interest group politics. Traditional patterns of political behavior in a country have a strong influence on the kinds of action deemed appropriate in the present. For example, in Britain, the expectation that groups will be heard and considered provided that they present their views in a moderate and responsive way has served to curb recourse to more disruptive forms of political action (Ridley, 1984). The French historical propensity to protest (Tilly, 1986) is more likely to explain acceptance of these tactics than is the form of interest group representation.

Expectations about efficacy affect people's choice of forms of political action (Wolfsfeld, 1986). In addition, Barnes, Kaase et al. (1979) contend that the overall pattern of political values has a strong impact on the selection of acceptable forms of political action. Those who have a postmaterialist orientation are more likely to be receptive to using unconventional means of political action than those with materialist values, irrespective of the nature of the political structures or of the interest group/government

Table 4.3 Membership in new social movements by level of corporatism, 1986

	Percentage already member or might join		
	Ecology movements	*Anti-nuclear movements*	*Peace movements*
Strong or medium corporatism			
Denmark	6.8	10.9	12.2
Netherlands	31.1	14.2	18.7
Belgium	5.4	3.6	4.5
West Germany	25.7	22.0	39.6
Average	17.3	12.7	18.8
Weak to no corporatism			
Ireland	19.3	26.5	26.5
France	10.4	5.6	9.3
Britain	18.2	16.1	16.6
Italy	18.2	10.5	17.9
Average	16.5	14.7	17.6

Text of question: "There are a number of groups and movements seeking support of the public. For each of the following movements can you tell me whether you are a member or might probably join or would certainly not join?"
Percentages based on all respondents as in table 4.2.
Source: Eurobarometer 25 (fieldwork in April 1986).

relationship. There is also a gap between willingness to engage in unconventional political acts and actual use of such tactics. The Barnes, Kaase et al. (1979, pp. 548–9) study found that while the Dutch and Austrians declared themselves to be more willing to use unconventional political actions, they reported that they had done so no more frequently, and sometimes less frequently than did the British and Americans. Thus, the relationship between corporatist structures and political tactics associated with new social movements appears to be insignificant.

In a recent article, Kitschelt (1988a) argued that corporatism is part of a syndrome responsible for the rise of what he calls "left-libertarian parties." How accurately this indicates the importance of new social movements is uncertain since "left-libertarian parties" include some communist parties, ecologists, New Left parties, and even the centrist element of the Swedish bourgeois bloc. Apart from this definitional problem, it is not clear that votes for such parties are reliable indicators of support for the values and goals of new social movements. The voting strength of "left-libertarian parties" reflects electoral rules, party systems, issues, personalities, and even the Left/Right division spurned by new social movements more than it gauges the overall political impact of new social movements. Thus, Müller-

Table 4.4 Acceptance of unconventional political action by level of corporatism

	Percentage expressing acceptance of:				
	Lawful demonstrations	*Boycotts*	*Unofficial strikes*	*Damaging property*	*Occupying buildings*
Strong or medium corporatism					
Austria	44	29	15	3	10
Denmark	49	31	29	3	10
Finland	57	41	29	4	16
Netherlands	42	33	14	3	21
Belgium	42	23	13	4	13
Germany	44	34	13	2	11
Switzerland	65	44	18	4	32
Average	49	34	19	3	16
Weak to no corporatism					
Ireland	50	37	26	3	9
France	52	43	32	4	28
USA	68	62	28	4	24
Britain	41	35	24	3	12
Italy	43	30	10	1	14
Average	51	41	24	3	17

Text of question: "I am going to read out some different forms of political action that people can take, and I'd like you to tell me, for each one, whether you have actually done any of these things, whether you would do it, might do it, or would never under any circumstances, do any of them."

Figures represent those answering that they "have done", "would do" or "might do" these actions as a percentage of all respondents.

Sources: Data for European Community countries are from Hastings and Hastings (1984, p. 546). Figures for the remaining countries are from the first wave of the 'Political Action' (ICPSR No. 7777) study (Barnes, Kaase et al., 1979, pp. 548–9; ZA-codebook, pp. 131–8).

Rommel (1985b, pp. 62–3) demonstrates that the voting support for the German Greens is more related to Left/Right orientations than to the value system linked to new social movements. Even if Kitschelt's voting totals for Left-libertarian parties were accepted as a measure of the impact of social movements, the effect of the system of interest intermediation still is only a part of the syndrome of economic development, social policy, corporatism, and Left party governments that he sees as explaining these parties' vote.

Public attitudes toward social movements, propensity to join such causes, and the inclination to use unconventional political tactics are more direct and accurate ways of assessing the impact of social movements. And, as seen above, the evidence from these measures raises questions about the strength of any alleged association between the overall pattern of interest group politics and the incidence of new social movements. The expectations of the critics of corporatism that this pattern of government/group interaction would foster the development of new social movements do not seem to be realized so far.

Alternative Explanations of the Rise of New Social Movements

Since the different patterns of interest group/government relations do not account for the varying impact of new social movements, it is useful to look for alternative explanations. In doing so, I will draw on the experience in France where new social movements have had a relatively minor political impact.

VITALITY OF EXISTING REPRESENTATIVE BODIES

One key factor in explaining the impact of new social movements appears to be the strength and adaptability of traditional representative bodies. This includes the vitality of the major political parties and the principal occupational interest groups. Where these usual representative bodies are successful in capturing new issues, the ability of new movements to emerge is reduced. On the other hand, where parties and interest groups age and fail to adapt to new social circumstances or ignore new issues or political cleavages, rival movements are likely to develop to challenge or even replace the established parties and groups.

The French case illustrates this quite well. The 1960s and 1970s were years of party revival and reform in France. While parties elsewhere were presumably experiencing decline, the French parties were revitalizing (Wilson, 1979, 1988). In a very real sense, the French parties emerging from this era were new parties that had captured the attention and many of the aspirations of the French voters. Both Right- and Left-wing parties established factory-based party units. The parties dominated civic–action and neighborhood groups, setting up their own or absorbing earlier, non-partisan groups. Similarly, the trade union movement was experiencing considerable ferment, with the CFDT active in championing new causes such as feminism, worker self-management, neighborhood and community action, and environmental awareness. With these traditional representational groups still promising to address the new issues, there was less need for new movements than in countries where parties and occupational groups were less receptive to the causes of the 1970s.

In Britain, the Labour party's radicalization caused internal strife during the 1970s and 1980s but it also attracted the interest of potential sympathizers of alternative movements. They joined the fray in the party and as Labour moved Left it succeeded in absorbing many of the issues that elsewhere spawned new movements. This was especially evident in the Labour party's dominance of the British peace movement, in contrast to the movement's non-partisan and even anti-partisan character elsewhere in Europe.

Another related factor is the nature of the governing coalition. Most of the contemporary social movements have a natural affinity for the Left. Where leftist parties were in opposition during the 1970s, the gestation period for most contemporary social movements, there was a tendency for the supporters and even the militants of new causes to place their hopes on the ultimate electoral victory of the Left. Such was the case in France where the Left had been out of government for twenty years. In seeking a broad coalition, the French Socialists wooed the supporters of these causes by including pledges to respond to their concerns if elected. Interviews with leaders of women's movements, consumer advocates, ecologists, and like-minded groups in 1979 revealed their expectation that things would improve if the Left won power.[5] Even if they doubted the Left's commitment to their full agendas, they believed that the Left would move toward some of their goals. They could still hope for answers to their demands coming through existing parties and normal political channels.[6]

In contrast, in those countries where the Left was in power, the advocates of new issues tended to feel that the leftist government fell far short of responding to their concerns. The West German case illustrates this inability of governing leftist parties to retain the loyalty of those committed to new civic action and environmental issues. The compromises of the SPD–FDP government, supported also by the trade union movement, were viewed as evidence of the lack of sincere attachment to these new concerns on the part of these traditional left-wing organs. This encouraged the formation of separate movements and led these new movements to discount the capacity of the existing democratic system to meet their goals.

THE NEED FOR A MOBILIZING ISSUE

The more successful social movements have been organized around an issue that is either already in the policy decision-making process or that the movement succeeds in forcing to a decision. The attention attracted to the issue gives stature to the group and assists it in recruiting followers who see the prospect of immediate policy results. In the United States, the Equal Rights Amendment served that purpose for the feminist movement; in West Germany, the construction of nuclear energy plants and pollution damage to the Germans' beloved forests served to mobilize public support for the environmental movement. In France, however, there was no such mobilizing issue. The anti-nuclear energy issue failed to gain public backing (Nelkin and Pollak, 1981); abortion reform came rapidly enough to prevent that issue from serving to rally women around a feminist movement (Lovenduski, 1986); a broad consensus on maintaining France's *force de frappe* undermined the peace movement. The absence of such a central cause retarded the development of all social movements.

The importance of even a single mobilizing issue is amplified by the tendency for new social movements to share their resources. The resources generated by one movement may be used by other related movements (Freeman, 1983). A successful movement tends to spawn others through this resource sharing and by showing the possibility of affecting public policy. Many of these movements, although each is oriented toward specific and different issues, share the same general socio-political outlook. As an illustration, the German Greens have come to be seen as the semi-official voice of all new social movements and not just of the ecologists.

THE OPENNESS OF THE POLITICAL SYSTEM

Another set of factors influencing the impact of new social movements are those related to the open or closed nature of the political system. One type of openness relates to the likelihood that the efforts of the movement will affect policy outcomes. A second type of openness has to do with the ability of new movements to capture public attention. A third related issue is the extent of public alarm about the tactics used by the movement. Among the important determinants of movement success is the political opportunity structure in their polity (Kitschelt, 1986; Opp, 1986; Rucht, 1985). This refers to the possibilities for successful political action on behalf of the cause. Where political systems are closed, with few opportunities for protest to affect the policy process, social movements have less success in attracting followers and activists. Where political systems are open to the influence of protesters and outsiders, movements are more likely to flourish.

France under the Fifth Republic has acquired the reputation of being a strong state capable of resisting demands from outside groups (Hayward, 1976; Kuisel, 1981; McArthur and Scott, 1969). Its strong political executive, disciplined parliamentary majority, centralized administration, and Rousseauian notion of the state's duty to the general will have permitted the regime to ignore even noisy groups pressing for self-serving demands when the dominant political elite has so chosen. Few other democratic governments are able to act as decisively and to be as free from the need to consult powerful interests as the French government. An important consequence of this strong state is the ability of the French government to ignore social movements. This was well illustrated in the case of the anti-nuclear energy movement of the 1970s. In contrast to the success of the movement across the Rhine in West Germany, the French movement failed to slow the French government's ambitious plans for nuclear energy (Kitschelt, 1986; Nelkin and Pollak, 1981). The French government was able to pursue its objectives over the protests of social movements. This ability to resist pressure discourages potential supporters for social movements of all types.

In West Germany, a weaker state with multiple levels and points of access, a less centralized and less aloof administration was forced to indefinitely postpone much of the government's nuclear energy program. Success attracted attention to the anti-nuclear movement and gave it credibility as an effective political force. The achievement encouraged its activists and aided in recruiting new supporters. Advocates of many different causes could hope for similar success if they mobilized for political action.

Second, social movements depend upon capturing the attention of the public for their success. They are best able to do so where the political arena is not already filled with actors involved in interesting activities. In such a setting of *ennui* with ordinary politics, the press and public will eagerly turn their attention to the novel and the unusual. Where the political scene is dominated by a close struggle between sharply dividied rivals or where the established parties are undergoing significant or controversial changes, the public's attention will not be as easily diverted to challenging groups.

This is again illustrated by contrasting the impact of social movements on parties and interest groups in France and Germany. During the 1970s and into the 1980s, French public attention was focused on the struggle for power between the Left and Right in a series of increasingly close electoral battles. The stakes in this contest were perceived to be important, with each side offering sharply different views of how society ought to be structured. The rebuilding of the Socialist party, maneuvering between Communists and Socialists in the union of the Left, and rivalries between Gaullists and Giscardians, kept a hold on public attention. Those inclined toward political engagement wanted to become involved in these important national political debates rather than "waste" their efforts in marginal and impotent social movements.

In West Germany, however, the party scene was much more tranquil. The more dynamic period of party change and reform had come earlier, in the 1950s and early 1960s. Party leaders and party politics in the 1970s were uninspiring. With little excitement from the established parties, the press and the public gave more attention to newly emerging political groups and movements. In addition, the governing Social–Liberal coalition looked in practice much like the CDU alternative. The prevalence of consensus politics produced an atmosphere of disillusionment in which social movements thrived (Chandler and Siaroff, 1986).

Finally, by political necessity and ideological preference, the new movements tended to have recourse to direct, extra-parliamentary politics. In settings where such tactics were unusual, such as Germany, Britain, the Netherlands, or Scandinavia, this attracted even greater public attention to the movements. The forms of political action provoked introspective analyses of what might be happening to society which brought attention and sometimes sympathy to the social movements. They also led to often panicky searches for ways to involve citizens and the new movements in order to thereby defuse the threats to public order seen in the growth of direct political action.

In countries where there were already well-developed traditions of direct political action, such as France or Italy, the consequences were less conducive to the growth of the new movements. The use of direct action did not have the shock effect on public opinion that it did in more placid political settings. For example, in France the rise of new social movements did not mean any perceptible increase in unconventional participation. Such action has been part of the standard political repertoire for centuries (Tilly, 1986). Farmers spread manure on autoroute curves; workers occupy their plants; truckers clog narrow downtown streets; shop-owners shutter their stores to visiting public officials. This is the ordinary content of politics, so that the media-oriented strategies of new social movements are neither novel nor threatening.

The social movement had developed as a common form of popular contention in France by the middle of the nineteenth century. The emergence of a new set of such movements in the 1970s thus prompted little comment in the press and certainly did not provoke the kind of concerns about their consequence for the fabric of society that they elicited in other countries. Nor did their tactics seem unusual or alarming. The French government has long experience in handling such disorderly bodies without endangering overall stability and without having to pay much heed to the interests thus manifested.

Conclusion

The phenomenon of new social movements has affected politics in most Western democratic countries. But the impact of these movements has varied sharply from one country to the next, with some countries' political processes much more clearly disrupted than others'. The expectation that the type of interaction between interest groups and the government would affect the incidence and impact of new social movements does not seem to be supported by available data. The argument by both corporatist theorists and their critics that corporatism would encourage the rise of new social movements is not borne out in Western Europe.

Instead, based on the evidence drawn from the French experience, the success of new social movements seems more related to other political factors. Among the most important are the vitality of established political representative bodies such as parties and interest groups, the ability of movements to capitalize on an important issue with potential for mobilizing a broad range of supporters, and the vulnerability of the political decision-making process to pressure from outside or dissident groups.

The significance of this argument is that even though corporatist politics now seem to be receding in many countries, we may not have seen the end of the challenge of new social movements.[7] This does not necessarily mean that the current challenge of the new social movements will continue to have a disruptive effect on Western democracies. Instead, it means that the challenge will be met not by altering interest group/government interaction but rather by other means. As traditional parties or interest groups struggle to respond to the challenge, their efforts at renewal may enable them to regain the public's attention; as governments and the public learn to accept and cope with unconventional political tactics, their use will attract less attention to the movements; as political parties adjust to the competition from these groups by addressing the neglected issues that the movements advocate, they may limit the movements' ability to capitalize on popular new issues. Indeed, this process of adjustment seems to have proceeded quite far in some countries that only a few years ago seemed threatened by the rise of new social movements.

NOTES

1 There are as many definitions for neo-corporatism as there are authors who use the term. However, the most common use reflects Schmitter's (1979) distinction between neo-corporatism and pluralism. For a critical discussion of the definitions of neo-corporatism, see Wilson (1983).

2 See, however, Lange's (1983) claim that there is little difference in the responsiveness of groups in corporatist and pluralist settings.

3 Note, for example, two other rankings by Lehmbruch (1984, p. 66; Lehmbruch and Schmitter, 1982, pp. 16–23).

4 By combining these two categories, there are some unusual results. For example, the number of Germans who say they are or might be members of a peace movement (1.4% and 38.2%) far exceed the number saying that they strongly approve of these movements (23.4%).

5 In interviews in 1982 and 1984, most leaders of these groups expressed their disillusionment with the reality of Socialist government.

6 Britain offers another example. Although Labour was in power, the sharp shift of the party to the Left after its defeat in 1979 permitted the Labour party to preempt the appeal of new social movements it did not control.

7 See the tacit admission of such a reversal of the formerly proclaimed trend toward corporatism in the latest writings of corporatist theorists in their notions of "sectoral" or "meso-corporatism." See Cawson (1986) and Lehmbruch (1985).

5

Social Movements and Political Innovation

MAX KAASE

This volume's proclamation of an era of new social movements needs to be placed in perspective. In September of 1986, 23 percent of the West German voting-age population considered themselves supporters of the anti-nuclear movement, 28 percent were adherents of the peace movement, and 12 percent belonged to the women's liberation movement (Zentralarchiv, 1987, pp. 104–5). There is, however, considerable overlap: 8 percent subjectively belonged to all three movements, 12 percent subscribed to two out of the three, and 14 percent to only one out of the three. Thus, 34 percent of the public support at least one of the movements, but two-thirds of the West Germans did *not* consider themselves supporters of any of the three movements. We shall return to these and similar data for other European countries later in the chapter.

As authors of other chapters in the book point out, the above-mentioned movements, particularly in the European scholarly and public debate, have become known as "new" social movements. Even recently, prominent students of social movements have not used the "new" affix at all (e.g., Zald and McCarthy, 1987). Thus, one question to be addressed is whether this affix has a specific theoretical meaning. Heinz Eulau's general observation here suggests at least some caution:

> Nothing is so beguiling in social science as the word "new" . . . I have survived or am surviving the "new functionalism", the "new institutionalism", the "new political economy", the "new marxism", or, more recently, the "new corporatism". What the word "new" really tells us, however, is that new phenomena attacked by the new approach are not so new, after all, and that the new approach is a revival of some "old" ways of seeing and dealing with things. . . . There is a good deal of revivalism in social science as there is in religion, and this is hardly something to cheer about, even though the new apostles are unwilling to acknowledge it, either out of ignorance or sheer perversity. (Eulau, 1989, pp. 19–20)

As a starting point for the following reflections, it seems worthwhile to begin with a definition of social movements in general which is sufficient to serve as a reference when dealing with the social movement phenomenon. Raschke in his book on social movements defines them "as mobilizing collective actors which pursue the goal of producing, preventing or taking back somewhat more basic social change. They pursue these goals with a certain amount of continuity based on high symbolic integration, little role specificity and by means of variable forms of organization and action" (1985, p. 77).

If one tries to extract the main elements of Raschke's definition, then the following concepts emerge:

1 Type of actor	– collective
2 Goal(s)	– orientations towards more or less basic change (direction: open; can even entail no change)
3 Organization	– certain amount of continuity
	– high degree of symbolic integration (we-feeling)
	– little role specificity
4 Type of action	– direct action (prevailing)

Raschke's conceptualization is somewhat eclectic, and theoretically less stringent than, for example, the resource mobilization approach (for the most recent statement of this approach see Zald and McCarthy, 1987; also see the chapter by Klandermans in this volume, chapter 7). However, given the over-emphasis on the organizational aspects of social movements in resource mobilization theory (for a friendly critique see, e.g. Gamson 1987, pp. 6–7), this approach seems less suited to deal with particular aspects of the proclaimed "newness" of the movements mentioned above. Therefore, in the following reflections, Raschke's definition is used as an organizing principle for dealing briefly with the concept of *new* social movements. This decision reflects epistemological pragmatism in the light of the unsatisfactory status of social movement research in general.

It was previously cautioned that the affix of "new" social movements might have little theoretical meaning. Roth and Rucht (1987a, p. 15) make a virtue out of a vice when they argue: "As far as the concept (of new social movements) is open in the sense that it permits a successive and promising clarification of what the meanings of 'new', 'social' and 'movement' are, things do not look bad. What is born out of sheer embarrassment must not permanently remain like that." For systematic reasons not to be discussed further in this chapter, it may well be that the "new" affix is just a German or European cultural specificity. Klandermans and Tarrow (1988, p. 17) are therefore correct in arguing for more *comparative* social movement research and in expressing at least a note of hesitation: "Only by comparing old and new movements can we discover empirically whether the 'newness' in the latter is inherent, developmental, or situational."

The lack of systematic cross-national evidence on the existence, organization, actions, and belief systems elements of social movements makes it difficult to generalize across countries on the social movement theme. Raschke's effort to integrate the "new" movements into his conceptual scheme is therefore very much based on the German experience, although he regards them as systematic corollaries of the general development of Western democracies into postindustrial societies. The new social movements inhabit specific locations on the core elements of his definition of social movement in the following way:

1 There exists a broad variety of independent movements which, however, are integrated through a joint cultural theme, a specific paradigm of modern living. In this sense, the existing partial movements are bound together to one overarching movement (social movement industry would be the closest equivalent in the resource mobilization approach).

2 The actors operating within the separate movements share a common value horizon – postmaterialism.

3 The frequent interlocks between the social movements' "networks of networks," as Neidhardt (1985) calls them, is an indicator of the affinity between persons and programs.

4 These movements lack a coherent ideology, and are characterized by thematic multiplicity and rapid thematic change. Nevertheless, their mobilizing themes are not a random choice from the possible issue agenda. Rather, they are mostly related to the modern living paradigm, or in political terms the "new politics" (Hildebrandt and Dalton, 1978). Furthermore, new social movements display a low degree of organizational structure, resort frequently to direct action tactics and are internationally oriented (Raschke, 1985, pp. 74–5, 411–36).

In order to broaden our perspectives in thinking about contemporary social movements, the decision must remain open for empirical assessment of the ideological location and specific goal orientations of these movements in cross-national terms. This is particularly relevant because there is reason to expect that movements which differ on those dimensions will relate differently to the political process in advanced industrial democracies.

Present scholarly thinking on social movements emphasizes the role of organizational continuity, as contrasted with earlier perspectives which thought of social movements in life-cycle terms of growth and decay (e.g. Heberle, 1967; Rammstedt, 1978). It is this continuity notion which makes the social movement theme particularly relevant for political science. The question here is whether contemporary social movements are representative of a permanent change in intermediary structures of interest articulation and aggregation in Western democracies in the sense of amending the hitherto dominant collective actors of interest groups and political parties (see Donati, 1984; Melucci, 1984a; Nedelmann, 1984; Gundelach, 1984). A logical consequence of such an approach would be to systematically analyze the interactions between social movements on the one hand and established interest groups, political parties and political authorities on the other hand, as Klandermans does in his chapter in this book.

Let us continue to dwell on this problem for a little while. There is now some theoretical work (Donati, 1984; Neidhardt, 1985; Roth, 1987) as well as some scattered empirical findings (Melucci, 1984b) which point to the important role of personal networks of communication for the continuity of social movements and particularly for their mobilization into action (Melucci, 1984a, pp. 819, 829). The point to be made in terms of theories of social differentiation is that these networks are becoming a permanent feature of modern society and therefore have to be looked at as much as causes as well as consequences of social movements. A further qualification to be made *vis-à-vis* the social movements of earlier days is that contemporary social movements are more culturally oriented than power-oriented and gain their cohesiveness from cultural orientations (this is a major point in Raschke's book where he distinguishes between power-related and culture-related social movements; Raschke, 1985, pp. 109–16).

The Milan study (Melucci, 1984b) finds that social movement elites mainly derive from the cultural sector of big cities, that the most resourceful networks produce mainly information and culture (the "alternative" media), and that "this new form of 'cultural' representation has taken the place of political representation of previous movements" (Donati, 1984, p. 854). Thus, it cannot come as a surprise that the existing networks are *not* a result of joint feelings of political deprivation. Instead, "movement recruitment, rather than being a function of the ideology or the deprivation of the individual, is structured by sociospatial factors and meaningful social inter-relations" (Donati, 1984, pp. 843–4).

Over extended periods of time these networks may operate in political latency and, through engagement in cultural affairs, function as training grounds for later political action. Nevertheless, movement elites must be interested, to be politically successful once engaged, to find issues which help to convert bystanders into active participants. Therefore, the question of the ideological and belief system underpinnings of contemporary social movements remains in the fore.

The strong emphasis on cultural matters of contemporary social movements agrees well with claims that these movements dwell around post-materialist issues discussed by Ronald Inglehart (chapter 3 in this volume), like participation, self-actualization, even hedonism. The socio-structural and intellectual resourcefulness of the movement elites, and to a lesser degree, the participants, feed the notion that contemporary social movements are socio-political innovators pressing for the kinds of changes necessary to overcome the threats of nuclear war, pollution and the exhaustion of natural resources (e.g. Brand et al., 1986). There may also be reasons to believe that the social movements of our days are an unintended consequence of the modernization process with the rationality paradigm as its core element. In this sense, they may reflect an anti-modernity and anti-rationality stance and in the same vein be regarded as the rearguard actions of disappearing modernism, to be followed by a new epoch postmodernism (Vester, 1985, p. 11; Huyssen and Scherpe, 1986).

The Socio-structural and Ideological Bases of Contemporary Social Movements

As fascinating as such speculations may be, much can be said in favor of a down-to-earth effort to locate contemporary social movements in the present system of social cleavages and political representation. Social movements, it is argued, are collective actors of some continuity, but also with considerable fluidity, the latter being particularly true with regard to the *membership* in social movements. One way to analyze social movements as collective actors is to study their organizational structure, elite composition and socio-political goals, the latter through content analyses of written documents and movement media. Unfortunately, such information is not available except for individual movements in individual countries. A more limited, but slightly better-documented approach to the empirical analysis of contemporary social movements is the study of movement membership, understood here in the widest possible sense of the term.

We shall not be too concerned with the problem of what membership in a given movement entails. In resource mobilization theory terminology, the citizenry at large is divided into opponents, bystander publics, adherents and constituents, the latter being the ones who provide resources for a social movement organization (McCarthy and Zald, 1977, p. 1221). Obviously, individuals can take quite different positions on that membership continuum over time (Kuechler, 1984). Given the commitment of contemporary social movements to avoid Michels' iron law of oligarchy at all cost – what Raschke calls small role specificity – it is very difficult to apply in empirical studies a membership concept of the kind usually found in analyses of interest groups or political parties (e.g., dues-paying member). Whatever empirical evidence is available with regard to social movement membership must be critically evaluated against this problem.

As mentioned in the introductory section of this chapter, three questions were asked in the 1986/7 German national election study conducted as a three-wave-panel in September of 1986 and January (pre-election) and February (post-election) of 1987 regarding adherence to the anti-nuclear movement, the peace movement and the women's liberation movement.

Table 5.1 Change in individual adherence to new social movements over time, 1986–1987 (percentages)

	September 1986 (t_1)		
January 1987 (t_2)	*Adherent*	*Non-adherent*	*Total*
Adherent	19.9	9.3	29.2
Non-adherent	14.3	56.5	70.8
Total	34.2	65.8	100.0
(N=1,519)			
	January 1987 (t_2)		
February 1987 (t_3)	*Adherent*	*Non-adherent*	*Total*
Adherent	17.9	8.5	26.4
Non-adherent	10.8	62.8	73.6
Total	28.7	71.3	100.0
(N=1,291)			

Table entries are percentages based on the total sample at each timepoint.
Source: 1987 West German Election Study conducted by the Forschungsgruppe Wahlen, Mannheim (Zentralarchiv, 1987).

Table 5.1 summarizes the information on individual adherence to *at least one of the three movements* over time.

The first conclusion drawn from these data concerns the changes in marginal distributions over the three timepoints. From September 1986 to January 1987, there is a five percentage point *increase* in the number of non-adherents. This trend continues to the post-election survey (it should be noted that only 0.5 percent of the overall 7.8 percent increase in non-adherents can be accounted for by panel attrition). A substantive interpretation easily at hand relates these changes to the election effects and its crystallizing of support toward the political parties (for a detailed analysis and corroboration see Pappi, 1988, pp. 16–26).

Looking at individual changes over time in movement adherence, one has to keep in mind the very generous definition of stability used to compute this table: anyone indicating subjective membership in at least one of the three movements at two consecutive timepoints was classified as a stable adherent. This decision is justified because the high positive correlation between membership in the three movements indicates a latent common dimension in the sense of a single movement sector (more than half of the adherents "belonged" to at least two movements). Nevertheless, this obviously maximizes the chances to be assigned to the adherent category.

The second conclusion from the table involves the short-term instability in new social movement adherents. Both time comparisons reveal that about one-fifth of the sample are in the "change" category. Thus, one is reminded of Kuechler's (1984) analysis on the same topic, which found a substantial individual turnover in peace movement adherence. Future analyses will have to establish to what extent these data reflect measurement unreliability or true change due to campaign effects. Considering, however, that movements seem to "lose" a substantial proportion of their members at any given point in time (for which they largely compensate through membership gains), this suggests that social movement membership is rather volatile. In addition, the *structural* similarity in membership composition of the separate movements (age, education, issues preferences, party preferences, value concerns, ideological self-definition) indicates that there is a reservoir of like-minded persons who can be recruited into social movement action through movement networks (Pappi, 1988). This finding emphasizes the need to look at the mobilization process regarding social movement participation.

One of the most interesting aspects of social movements is the way they interlock with established actors in the political process. Once again, with the exception of Klandermans' and Rochon's analyses in this book (chapters 7 and 6), there is preciously little information on such links. An MA thesis at the University of Mannheim (Pfenning, 1987) on the party organization of the Green party in the state of Rhineland-Palatinate showed a very interesting system of linkages between that party and local chapters of social movements of the kind discussed here. Elite members of such movements regularly participate in local party meetings as non-party members in order to exchange information and to coordinate joint actions like demonstrations of the party and the movements. Of course, we presently lack the empirical evidence to generalize these findings, although a more general validity can reasonably be expected.

The affinity between the new social movements and the Green party is one between collective actors. Our survey data permit us to see to what extent this affinity also exists on the level of individual citizens (table 5.2). These data support the notion that adherence to new social movements is basically an orientation that is open to every political group, although there are obvious strongholds of positive social movement identification on the Left side of the party spectrum. This finding is, of course, related to the fact that contemporary social movements form around an issue cleavage, which by now is fairly well embedded in the party spectrum of the West German party system. Looking at the column percentages reveals that the Greens are most closely linked of all the parties to the new social movement industry. This is no wonder considering the fact that the Greens were born from that specific industry. The SPD, which is second regarding the strength of the link to the new social movements, is split in the sense that it is overwhelmingly the "new," postmaterialist SPD left which displays this affinity to the new social movements.

Table 5.2 Adherence to new social movements and party-preference, 1987 (percentages)

	Voting Intention in 1987 (Second Ballot)				
Adherent of:	*CDU/CSU*	*SPD*	*FDP*	*Greens*	*Other parties*
Anti-nuclear power movement					
Yes	6.2 (11.4)	31.7 (59.8)	12.9 (2.2)	75.7 (25.6)	23.5 (1.0)
No	93.8 (52.8)	68.3 (39.1)	87.1 (4.6)	24.3 (2.5)	76.5 (1.0)
Total	100	100	100	100	100
(N)	(747)	(760)	(70)	(136)	(17)
Peace movement					
Yes	13.1 (19.8)	36.0 (55.0)	21.4 (3.0)	76.5 (21.0)	37.5 (1.2)
No	86.9 (52.7)	64.0 (39.4)	78.6 (4.5)	23.5 (2.6)	62.5 (0.8)
Total	100	100	100	100	100
(N)	(747)	(758)	(70)	(136)	(16)
Women's liberation movement					
Yes	5.5 (18.9)	14.7 (51.2)	10.1 (3.2)	41.7 (25.3)	18.8 (1.4)
No	94.5 (46.8)	85.3 (43.0)	89.9 (4.1)	58.3 (5.1)	81.2 (0.9)
Total	100	100	100	100	100
(N)	(743)	(756)	(69)	(132)	(16)

Basic table entries are column percentages, excluding cases with missing data on the social movement and voting intention questions; values in parentheses are row percentages.
Source: 1987 West German Election Study conducted by the Forschungsgruppe Wahlen, Mannheim (Zentralarchiv, 1987).

We have already mentioned that, quite in accordance with Raschke's expectation (Raschke, 1985, p. 74), there is a distinctive propensity to consider oneself an adherent of more than one social movement. Our data also lend support to another generalization put forward by Raschke (1985, pp. 414–10): in terms of socio-structural affiliation, adherents to any of the three social movements studied here are, in comparison with the average population, much younger and much better educated. Particularly interesting is the finding that in the most movement-prone group of citizens, those below 35 years of age with high education, females are substantially more engaged (only 25 percent non-adherents) than males (44 percent non-adherents). In part, this is a reflection of the fact that the women's movement is included among the set of three movements in the survey. In addition, one may also wonder whether this finding corroborates an analysis by Kaase and Marsh (1979, p. 184) that "many women (may) report low participation rates in conventional politics not merely because of their traditional inactivity conditioned by lower educational levels, but from a sex-based lack of identification with conventional politics. . . . Young women generally declare a readiness to be mobilized in protest activity while shunning the grey-suited male-dominated world of politics."

Of course, it has to be kept in mind that the 1987 German Election Study has just identified *one* type of orientation towards social movements (the adherents). Further studies will have to differentiate the public with regard to its orientation towards new social movements in order to examine more closely the membership concepts offered by the resource mobilization theorists (see, e.g., Klandermans and Tarrow, 1988). Furthermore, we will have to see to what extent the data on West Germany can be generalized across the broader set of Western democracies. At least some additional information is available.

The *Eurobarometer* surveys (a semiannual series of representative national samples of the European Community member countries) included data on four social movements for the Federal Republic, France, Great Britain, Italy and the Netherlands in 1982, 1984 and 1986. Two questions were asked in regard to the nature protection movement (e.g., World Wildlife Fund), the ecological movement, the anti-nuclear movement and the peace movement: (1) how strongly does the respondent approve or disapprove of each movement, and (2) does one consider oneself a member, a supporter or a non-supporter?

Following Watts' (1987, p. 49) secondary analysis of these data, these two questions can be combined into an additive index which provides a rough approximation of an individual's closeness to the various movements. To simplify the data structure the index proposed by Watts is condensed into three categories: positive orientation ("activists" plus "supporters" plus "sympathizers"), indifference, and negative orientation ("critics" plus "opponents"). Several conclusions emerge (data not shown, see Inglehart's chapter in this volume, chapter 3):

First, while there is a certain amount of aggregate change over time in the marginal distributions of those categories, in none of the five countries for any of the four movements is a noteworthy general trend visible between 1982 and 1986; the levels of support and opposition remain fairly stable. Here it has to be kept in mind that the Chernobyl nuclear accident happened *after* the majority of fieldwork for the 1986 study.

Second, the *level of support* for the various movements differs between countries, but particularly between movements. The nature protection movement is supported everywhere and at all times by at least two-thirds of the citizenry. Lack of support here is mostly indifference, hardly ever opposition. To borrow a term from electoral sociology, one could talk of a *valence movement.*

Regarding the three remaining movements, we are facing quite a different situation. The anti-nuclear movement is, again unanimously, least supported of all movements studied here. But this relatively low level of support does not imply that it is not supported at all. Rather, it is regarded as highly ambivalent, even controversial within those five countries at all times, thereby resembling what in electoral sociology is called a position issue. However, with the exception of Italy where there is a higher level of support than in most of the other countries, support for both the ecology movement and the peace movement appears very much polarized. In terms of party preferences of social movement supporters in 1986, all three movements seem embedded in a Left/Right party framework, or since all five countries had conservatively dominated governments, in a government–opposition cleavage (see Watts, 1987, pp. 62–3).

In ideological terms measured by self-placement on a Left/Right scale, closeness to the three new social movements is, with few exceptions, very strongly related to the Left side of the political spectrum. There is little reason to believe that this finding is an artifact of the conservative dominance in governing the five countries. Rather, this evidence points to the phenomenon that the socio-democratic Left parties of the West are slowly changing their profile from the Old Left (industrial workers) to the New Left (people with postmaterialist value preferences; see Inglehart, 1987a). Thus, it does not come as a surprise that closeness to the new social movements nearly uniformly displays a positive correlation with Inglehart's postmaterialism index (see chapter 3 in this volume).

The affinity to the political Left of the new social movements clearly is not a German singularity. There seems to exist an issue cleavage, the so-called New Politics (Hildebrandt and Dalton, 1978), which was discovered and politicized by the emerging social movements. This cleavage contributed to the growth of the new social movements because it was left unattended too long by the established political parties. By now, however, inclusion of this issue cleavage into the Western party systems has proceeded at a fast rate, either by adoption through established parties or through the appearance of new, ecological parties (like the Greens in West Germany). This is one of the reasons why voter analyses of the ecological parties are also relevant for the study of contemporary social movements in Western Europe (see Müller-Rommel, 1982; 1985a; 1985b; and chapter 11 in this book).

The data presented here deserve interest because they document the broad orientational basis in Western mass publics for the new social movements. The German concern with this phenomenon, the data show, is justified and understandable because its socio-structural and ideological contours emerge particularly distinctly in that country. The *Eurobarometer* data further reveal that the youthfulness, higher education levels, postmaterialism, Leftness, and proximity to non-institutionalized political actions of the adherents to the new social movements is a fairly common feature, too. What remains to be seen is whether the new social movements will continue to thrive in the future, given the fact that the issue cleavage on which they have built their support is becoming an integral part of established politics in the Western democracies.

This consideration necessarily leads back to our initial concern that the *organizational* dimension of the new social movements is not well researched in comparative or longitudinal terms. Only when this situation is improved it will be possible to assess reliably whether the social movement elites, the movement entrepreneurs building on sociospatial networks and networks of networks, will be successful in keeping social movements as a more permanent feature of modern cultural, social, and political life in the West (based on analyses of the American situation Zald (1987) answers this question very much in the affirmative).

The Changing Political Process and Contemporary Social Movements

This section is concerned with conditions that may have contributed to the emergence of contemporary social movements. As intermediary organizations with some minimal continuity, social movements obviously compete with other intermediary organizations, especially with political interest groups and with political parties, for scarce resources not the least of which is the material and nonmaterial engagement of individual citizens. This competition, however, is not necessarily a zero-sum game. Rather, the stock of both material and of nonmaterial resources has increased over time. The economies of Western developed countries are richer and technologically much more advanced; people are more educated, more mobile, and have more leisure time. How much of those resources are they going to spend in politics in the future?

The concept of a limited political role for the individual citizen was empirically and normatively in vogue in the 1950s and early 1960s. Probably the most lucid and data-rich statement of the matter is to be found in the *Civic Culture* study (Almond and Verba, 1963). Both authors consider that type of democratic order as ideal which enables citizens to participate competently, but which also provides for an institutional structure that permits citizens to leave politics unattended for a certain amount of time without having to fear that loyalty leads to deprivation. This is what Gamson (1968) called a "blank check," a check issued by those governed to the governors because of the citizens' belief that political authorities will be responsive to political needs in anticipation of citizen sanctions, be they voice, exit, or both.

Historically speaking, the equal, free, and secret ballot was the main breakthrough to obtain an institutionally fixed political role for *all* citizens. The present-day controversy about too limited a political role for the citizenry easily ignores the fact that this role was only achieved in the 1920s, and in some democratic countries was not fully implemented until the 1950s and 1960s. It is well known and need not be further discussed that the emergence of the modern political party, and here particularly the institutionalization of the workers' movement into working-class parties, was the decisive condition for the successful mobilization and integration of hitherto politically excluded citizens (see, e.g., Marshall 1950; 1964).

One of the most fascinating and stimulating observations by political sociologists (Lipset and Rokkan, 1967a) was the relatively high degree of stability found in Western party systems from the 1920s to the 1960s. It seemed that the early social cleavages and their translation into party systems had developed a life of their own and did not wither away through time, in spite of vast technological and social changes. Ever since the Lipset–Rokkan piece was published, a scholarly debate has inquired whether this stable relationship between major social cleavages and political parties is beginning to erode or not (see Dalton et al., 1984; Crewe and Denver, 1985).

Why should such a dealignment take place? Various conditions and factors have to be considered here. The first set of factors relates to the technological and economic development in Western democracies. The invention of new modes of industrial production resulted in an enormous increase in individual and societal affluence. It further resulted in changes in occupational structure which lessened the importance of existing organizations of interest aggregation, in particular of the trade unions.

Secondly, the implementation of modern mass communication systems has substantially altered not only the political process in general (for details see Kaase, 1986a), but also the way in which people within societies and societies among themselves are connected. Mass media increasingly define what social and political reality is. Thus, it is only logical that individual and collective political actors develop strategies of communication, agenda-setting, and persuasion in which the mass media play an instrumental role.

Thirdly, increased affluence has paved the way for a redefinition of advanced education as a public good which is – differences in aptitudes not withstanding – basically available at little or no cost for everybody. The broadening educational basis of modern societies has had the side effect that high formal education is no longer a sufficient condition for high-quality jobs and for high-salary jobs. We have, in short, reached the social limits of growth (Hirsch, 1976). But we have, in addition, enlarged the cognitive basis of these societies. One result is the encompassing cognitive mobilization (Dalton et al., 1984, pp. 14–15; Dalton, 1988a, pp. 18–24). As one example, in the Federal Republic from 1960 to 1985 political interest among the electorate has increased from about 30 percent to about 65 percent.

Empirical findings regarding the above-mentioned dealignment, that is the weakening of ties between cleavages and parties, are still equivocal (see Dalton et al., 1984). There seems to be agreement that net volatility – the aggregate changes in party strength at two consecutive elections of the same kind – has increased over the past thirty years, and that the number of individual voters who switch parties in consecutive elections has grown (Crewe and Denver, 1985). Also, for the United States and for Great Britain a substantial decrease in party identification has been shown (Saarlvik and Crewe, 1983; Abramson, 1983).

One corollary of this development is the increase in issue voting and the emergence of issue publics. Obviously, elections always involved issues, but issues corresponded more closely to the positions of parties in the ideological space. What has changed is that there are more and more complex issues, ideology and partisanship have lost some of their unifying force (in the sense of constraining belief system elements), and citizens are now belonging to more and to more varied personal networks. Thus, issue publics are no longer homogeneous in partisan terms, and the importance and life-span of any given issue determines how relevant it may be for triggering thoughts of bypassing the parties and established interest groups in order to get an issue on the political agenda and to get quick(!) results. Here we find probably one of the most influential conditions for the emergence of contemporary social movements.

One final related observation pertains to the change in participatory orientations in mass publics. Barnes, Kaase et al. (1979), Kaase (1984a), Jennings and van Deth (1990) and many others have shown that the return of ideology into politics in the late 1960s has included a quest to go beyond what institutionalized politics has long offered as the main mode of political involvement for the individual citizen: voting and related activities. What Kaase (1984b) has called the participatory revolution has opened up new channels of political influence. Participation in social movements is one such channel.

Of course, the future direction of the contemporary social movements depends a lot on the significance one is willing to attribute to the various factors discussed above. In a functionalist perspective which unfortunately comes dangerously close to being circular, it can be argued (e.g. Raschke, 1985, pp. 430–1) that social movements emerge because the established system of interest articulation and aggregation is not working properly. If it works (again), the argument should continue, then social movements will go, at least recede.

In contrast, if one emphasizes the broad structural changes which modern democracies have experienced over the last three decades, then one will see social movements (and/or their equivalents) as phenomena to stay. In the next section, we will try to assess more thoroughly what the present and future political role of social movements is going to be.

The Present and Future of Contemporary Social Movements

Students of social movements seem to agree that movements strive for some kind of basic societal and/or political change. A closer look at this core element of many social movement definitions quickly reveals its theoretical ambiguity. No criteria are offered to specify what a *basic* societal or political change is.

If one nevertheless tries to work with this definitional element, it becomes clear that *basic political* change can reasonably only mean the change in regime form; for instance, from an authoritarian/totalitarian to a democratic political order or vice versa. Substantial empirical research on this matter proves that in present-day Western democracies the change from a democratic to another type of political order is not supported by groups of any relevant magnitude. Quite the contrary is true: there is a widespread acceptance of and identification with the democratic order (Lipset and Schneider, 1983; Kaase, 1986b; Fuchs, 1989; Westle, 1989). It is not surprising, then, that contemporary social movements have no conception of an alternative political order they want to achieve. In political terms, contemporary social movements are, at best, partial reform movements in that they want to change only elements of the democratic institutional structure (Heberle, 1967, p. 8).

The situation is much more difficult if one tries to find out what a *basic societal* change is. If one looks at the crudest of economic typologies – market economy versus planned economy – then much of what is said about the political order of democracy will also hold for the acceptance of the market economy, provided it is embedded in a system of welfare state regulations (Alber, 1986). What is left, then, is to look at changes at a lower level, for example, at changes in life-styles which are deemed necessary by some if the human race is to survive. What are such basic changes that the new social movements claim to be absolutely necessary?

The major new political issue around which almost all others (nuclear plants; nuclear weapons) can be grouped, is the environment, an issue which is only slowly being integrated into the postmaterialist syndrome. Considering the fact that this issue area became relevant only in the early 1970s, we can observe a surprising amount of awareness and change in citizen attitudes and behaviors (the German case is documented in Kaase, 1986c). If our observations are correct that this issue area is either becoming integrated into the agenda of established political parties or has led to new political parties (or both), then a major force feeding contemporary social movements is fading away. A similar argument exists for the women's liberation movement; no political party interested in a majority of the popular vote can afford to ignore the demand of women for more equality.

In sum, those who interpret the emergence of the new social movements as a result of deficiencies in the established political process must expect a lessening political role for those movements in the future.

Quite the contrary is true if one, as an opposing paradigm, sees social movements arising not so much from collective deprivations, but rather from processes of societal differentiation which require a corresponding differentiation on the side of political institutions and of collective political actors, most noteworthy in expanding beyond political parties. The network approach to the study of social movements claims that social differentiation has led to value communities with high interpersonal interaction density. These value communities can exist without overt societal and political actions for extended periods of time, as kinship and friendship networks do. These networks are stabilized, the argument continues, through entrepreneurs whose qualities must entail the ability to search the market for themes suitable for mobilization into action. This political logic could, for instance, explain why concrete issues brought to the fore by new social movements can vary considerably over time (see Raschke, 1985, p. 420). They vary because movement entrepreneurs try out issues exactly for their ability to mobilize, and this ability, due to events, fads, and changes in emotional status of the individuals and groups involved, does not remain constant over time. Needless to say, these issues do not vary randomly, because social movements and personal networks unite persons who share a certain value horizon. Shared value horizons also make for cooperation between separate movement organizations. This is why the peace theme was ideally suited to activate the whole New Politics movement industry.

Social movements as eventual manifestations of latent sociospatial networks with culturally derived identities are obviously competing for resources and decisions with interest groups and political parties, as we observed before. It is their flexibility in membership definition, issue stances and place, time, and means of action, plus the emotional warmth that derives from the network element of the movement, and the opportunity to bypass established ways of interest aggregation, that makes them an attractive choice for individuals.

On the other hand, the organizational specifics of new social movements, their fluidity and lack of role specificity plus their underdefined role in the institutionalized political process, make it extremely unlikely that political parties and interest groups will be replaced as core collective actors in that process. This is because accumulating empirical evidence indicates that social movements are frequently building coalitions with parties, interest groups, administrations and governments. Another reason is that established political parties and interest groups respond, as economic theories of politics would predict, by incorporating into their agendas the issues advocated by the new social movements (Budge, Robertson, and Hearl, 1987). There is also evidence showing that the emergence of social movements depends partially on the degree to which the traditional system of socio-political cleavages is still operating. In an analysis of the factors influencing the development of ecology parties in Western Europe, Alber (1985) shows that such parties are weak where class antagonism is still strong. In addition, Schmitter found that functioning neo-corporatist interest cartels between employers' organizations, trade unions, and governments reduce the level of political protest if the old cleavage lines are still highly politicized (Schmitter, 1981, pp. 311–18). Such cartels frequently are, because of their continuing importance in modern industrial societies, very well equipped to control, channel and even neutralize new political issues, as can be witnessed in Scandinavian countries (for Sweden this is shown by Rubart, 1985).

Schmitter (1981), in his analysis of the impact of pluralist versus neo-corporatist modes of interest intermediation on the governability of democratic regimes, may have over-emphasized the neo-corporatist case because his analysis embraces a historical period when social democratic parties had gained broader respectability and government access. It is also interesting that social movements as one potentially new element in the intermediation process did not catch his eye at all, as is generally true for the chapters in the Berger (1981) volume. What is important to note here, however, is Wilson's claim (in chapter 4 in this volume) that the empirical evidence does not support the notion that the emergence and continuity of new social movements is systematically related to the prevailing mode of interest intermediation.

The question remains whether those expecting a lessening role for social movements in the future, or those predicting a large or even larger one, will be borne out to be true in the future. It is in that speculative vein that this chapter will be concluded.

Epilogue

Contemporary social movements radiate a great deal of intellectual fascination to public and scholarly observers alike. The "new" social movements highlight this fascination because they advocate issues that everybody agrees are of tremendous importance for the future of our planet Earth: the environment, peace, pollution, the threat of nuclear plants, and nuclear weapons. This fascination may also be the reason for a certain lack of analytical distance in much of the scholarly debate on the matter (a good case in point is Roth and Rucht, 1987b).

Probably a more important factor in the ambiguities concerning new social movements is the unsatisfactory theoretical underpinnings of research and, even more so, the lack of reliable and comparable information for almost all of the relevant features of these movements. This makes it an exceptionally difficult task to assess where we are standing right now, let alone to speak of where we are going.

One virtue of the intense discussion on this subject is that it has sharpened the senses of observes to the historical depth of the phenomenon. The work by Charles Tilly (e.g., 1978), but also by Raschke (1985) and Zald and McCarthy (1987), are excellent cases in point. The concrete analysis of contemporary social movements in various countries edited by Brand (1985) is also a step in the right direction, although the lack of a clear theoretical frame and appropriate data do more to weaken than to sharpen the contours of the phenomenon. We learn here that movements in specific countries bear elements of anti-modernism as well as of innovation, sometimes even within one movement. Looking at the reaction of the established political actors certainly conveys the impression that the new social movements have produced substantial changes in programs and outcomes in a dialectical process (Budge, Robertson and Hearl, 1987, p. 405). Even stances of anti-modernism (hostility towards technical innovation; civilization criticism) have led to reactions in the political system and thereby created innovations, often in the sense of taking a fresh look at problems. Japp (1984, p. 322) speaks of the ability of the new social movements "to gain currency in the first place for specific interpretations of given problem structures."

Much can be said in favor of an interpretation which sees the new social movements as some kind of an early warning system. This early warning system could be indicative of the lack of institutional fantasy that Huntington (1974), as well as Barnes, Kaase et al. (1979), believes is discovered in present-day democratic structures. These structures, particularly as Huntington has argued, are reflections of political and social systems and problems of two centuries ago and therefore badly in need of reform. Once they are changed, once old elites are replaced or augmented by new elites, will then the societal need for social movements as early warning systems recede? It was mentioned before that this line of reasoning comes dangerously close to circularity: social movements exist, because problems exist; if there are no social movements, there are no major political problems which are not properly taken care of by the established political process. It is only if all variables in a truly explanatory effort are conceptualized and measured independently of each other and are put into an explanatory (causal) dynamic model that we can speak of a theory and of its empirical test.

There is, of course, still the notion that processes of social differentiation lead to new forms of social integration and types of collective action (for example, as sociospatial networks) which alternate between political activity and periods of non-political action or latency. Here, too, theory-guided empirical observations are necessary before we can justifiably speak of any serious test of a theory.

The theoretical and empirical ambiguities just mentioned should alert social scientists to the possibility that their explanations are reconstructions of social realities which depend on their prior historical knowledge of the developments to be explained. In this respect Japp (1984) argues that neither objective societal states nor cognitive-rational states of actors are acceptable as explanatory variables. This is why the interpretations by researchers are "rationalistic metaexplanations of interpretations which themselves have been produced by the social movements" (Japp, 1984, p. 323). Thus, scholars unknowingly reproduce as theory-guided explanation what in fact is one element of the phenomenon studied itself. His alternate approach is to see social movements as collective actors which in a self-referential system create themselves, an approach which is difficult to conceptualize in such a way that an empirically testable theory will derive. Similarly, Luhmann (1986, pp. 227–36), on whom Japp draws extensively, claims that new social movements should be interpreted as artificial forms of self-observation of social systems.

If one accepts the premise that modern societies are characterized through processes of ongoing social and political differentiation, then it can be argued that the emergence of contemporary social movements is at the same time an element and a consequence of that process. Dominant ideologies (of the Left/Right type) and highly institutionalized conflict structures (cleavages) have for a while disguised the structural limits of political parties and interest groups to aggregate and articulate interests. These limits have now surfaced. Little in the previous discussion suggests that the above-mentioned "established" collective actors will completely disappear in the long run. However, it is equally unlikely in the light of what has been said that contemporary social movements will disappear, although there will be ebbs and flows in their overall visibility (the latency notion!).

A particularly interesting question for future research will be the extent to which these movements will become political instruments represented in everyone's political repertoire (Barnes, Kaase et al., 1979) – like the vote – or will these activities remain biased ideologically (in favor of the Left) and resource-wise (in favor of the better educated)? At various instances we have alluded to already existing links between the established and the "new" collective actors. This finding points to an eventual institutionalization of social movements into the political macro-structure of democratic polities. Although this certainly is a contradiction to the self-definition of new social movement actors, in systemic terms it may eventually result in a value-added intermediary structure and by that virtue represent some of the institutional fantasy that Huntington (1974) deemed so necessary for democracies to survive.

Part III

Networks of Action

6

The West European Peace Movement and the Theory of New Social Movements

THOMAS R. ROCHON

The new social movement school is perhaps ready to be supplanted by a theory of "neo-new social movements." As our experience with new social movements grows, claims made about their potential to promote revolutionary social and political change have been muted. One need only compare the early literature on new social movements (Raschke, 1980; Brand, 1982; Brand, et al., 1986; Offe, 1984; Habermas, 1984) with the more recent literature on the subject to see the contrast. Originally, new social movements were said to be new because they have a participatory, non-hierarchical pattern of organization, and because their activists have or are in the process of developing a far-reaching critique of the existing political, economic, and social systems. But organizational democracy and systematic critiques were also to some extent characteristic of earlier movements, such as the labor movement in the late nineteenth and early twentieth centuries (Webb and Webb, 1920). And studies of the nuclear energy, nuclear weapons, women's and environmental movements have shown that they do not consistently exhibit the traits of new social movement (Nelkin and Pollak, 1981; van der Loo, Snel, and Steenbergen, 1984; Katzenstein and Mueller, 1987; Rochon, 1988; Klandermans et al., 1988). As Offe points out in this volume (chapter 12), new social movements seem to be following the same trajectory of institutionalization that has characterized previous movements. There is considerably more continuity between old and new social movements than had been thought.

If contemporary political movements are not totally new, they nonetheless have wider aspirations than most movements that arose in an earlier period of democratic development. In contrast to the labor and women's movements of the late nineteenth and early twentieth centuries, which in large part sought political incorporation, contemporary movements seek to change social values as well as public policy. These extended ambitions create acute tactical dilemmas for political movements, as they sharpen the trade-offs between the activities that would reach the greatest numbers of people and activities that hold the most promise of influencing the government. This tension overlays the classic dilemma between maintaining ideological purity within a movement and diluting that purity in order to widen the breadth of movement support.

Both dilemmas are clearly illustrated in the tactical choices faced by the recent mobilization against nuclear weapons in Western Europe. The experience of the Dutch Interchurch Peace Council (IKV), discussed in this volume by Klandermans (chapter 7), is entirely typical. The peace movement[1] was enmeshed in a wide network of contacts ranging from political parties of the Left and center to trade unions and churches. These contacts exposed many people to new perspectives on European security and led them to question some of the security doctrines most central to NATO's nuclear strategy. Yet the ability of the peace movement to mobilize hundreds of thousands of people in the early 1980s did not reflect the spread of a fundamental critique of militarism and superpower domination of Europe. On the contrary, as the social breadth of the peace movement expanded, the depth of its ideological critique contracted. Although individual activists frequently followed the path of what Rucht (chapter 9 in this volume) calls ideological generalization and radicalization, the peace movement as a whole moderated its demands as increasing numbers of people became involved in it. By the time the movement reached its peak in 1983, the demand for denuclearization of Europe "from Poland to Portugal" had been lost in the single-minded effort to prevent deployment of Cruise and Pershing II missiles.

The peace movement thus offers an example of the tension between mass mobilization and the more diffuse pattern of social action said to be characteristic of new social movements. Indeed, the example of the peace movement suggests that it is impossible to be a movement of national proportions and still retain new social movement traits. To see why this is so, we must first examine the constraints on movement activity posed by the desire for mass mobilization and for media attention.

The Costs of Mass Mobilization

The first problem faced by any political movement seeking mass support is the limited range of political tactics that are acceptable to the public. Although public acceptance of the repertoire of protest activities has widened in the last generation, disapproval of protest is still a factor in how political movements are perceived. The only unconventional political action that achieves consensus approval in the five nations studied by Barnes, Kaase et al. (1979) is the circulation of petitions (85 percent approval), hardly a sufficient activity to sustain a political movement. Only slightly more than two-thirds of the publics of these nations approve of lawful demonstrations;

boycotts are approved of by only 37 percent; and no other activity is approved of by more than one-fifth of the population.[2]

These are levels of approval for protest activities in the abstract, unconnected to any particular cause. In practice, acts of protest are identified in people's minds with particular causes. Fewer people approve of peaceful demonstrations against deployment of nuclear missiles than approve of peaceful demonstrations in general.[3] In France, where the peace movement suffers particularly low public esteem, only 48 percent of the adult population would be willing even to sign a petition against the arms race (Les français et leur défense, 1982). In Britain there is a clear divergence in public reactions to the Campaign for Nuclear Disarmament (CND), which is associated primarily with large, peaceful protest marches, and to the Greenham Common peace camp, which was associated chiefly with civil disobedience in the form of trespass onto the neighboring cruise missile base.

Public reactions to the Greenham Common peace camp and to CND are indicative of how these contrasting styles are viewed. Most survey questions about the deployment of cruise missiles in Great Britain produced a plurality opposed to them, ranging between 40 and 60 percent.[4] A plurality nearly as large supports "demonstrations against nuclear weapons [that] have recently taken place in Britain and elsewhere in Europe" (Hastings and Hastings, 1984, p. 323). These demonstrations call to mind the kind of marches sponsored by CND, and they suggest widespread approval of such tactics.

The Greenham Common peace camp, on the other hand, was better at gaining media attention than at winning the approval of potential supporters. Only 6 percent of Britons did not know of the camp's existence in 1983, compared to 41 percent who three years earlier did not know that there were nuclear missiles in their country (Hastings and Hastings, 1984, p. 323; Hastings and Hastings, 1982, p. 330). But 54 percent of the public (67 percent of those with an opinion) had an unfavorable view of the Greenham Common protesters. Of those who said that the peace camp had affected their opinion on the anti-missile campaign, one-third said that it made them *less* favorable to the movement. Asked their opinion of the anti-nuclear protesters more generally, twice as many Britons said they disapprove of peace movement tactics as said they approve (58 to 29 percent), and nearly twice as many said that the protesters made peace more difficult to achieve, rather than contributing to peace (46 to 24 percent) (Hastings and Hastings, 1984, p. 323–4).

Although peace campers in Britain, Germany and the Netherlands all stressed the development of positive relations with area residents, the inconvenience posed by their camps and by the influx of thousands of people for protest events generally stood in the way of local acceptance of the camps. At all three bases, unhappiness with the campers supplanted unhappiness about the missiles as the prime issue among local residents. Saul Alinsky's (1971, p. 127) injunction to movement activists to go outside the experience of "the enemy" without going outside the experience of supporters is easier said than done. The dilemma for the peace movement was to find a way to convey to the public its ideas about nuclear weapons without having those ideas associated with protest tactics of which the public does not approve. The peace movement may have had a majority of the West European publics behind its demand that the cruise missiles be removed, but for many in that public there was a divorce between acceptance of the message and acceptance of the medium. The net result is that the peace movement failed to mobilize many potential supporters of nuclear disarmament. To the extent that any movement aspires to mobilize widely, it must limit its tactics to those few unconventional political actions that enjoy public approval.

The Costs of Media Attention

Part of the reason the peace movement failed to mobilize the full complement of its potential support was the way it was portrayed in the mass media. The media generally present images of movement protest without elaboration of the substantive issues involved. Demonstrations are described as large or small, well-behaved or unruly, a cross-section of the populace or composed of fringe elements. But the issues that brought the protesters together are presented in terms of one-line slogans, if at all. The problem is not so much one of political bias as it is a matter of the exacting criteria used by the media to determine what is newsworthy. Size, novelty, and militancy are the chief elements of newsworthiness. Legal demonstrations and petition campaigns are generally not newsworthy unless they are quite large. But because of its confrontational potential, an attempt to blockade the gates of a missile base is newsworthy even if it involves no more than several dozen people.

Though the peace movement was successful in obtaining media coverage, it was less successful in obtaining favorable coverage. One problem was the tendency of the media to play up the sensational aspects of the movement. The Greenham Common peace camp did not get much coverage until almost a year after it began, when the campers occupied a sentry box in August 1982 and impeded sewer construction that October. At first the typical story about the camp tended to favor 'the brave women who showed their concern for peace at the cost of great personal sacrifice.' But the media were also relentless inquisitive about the personal life-styles of a group of women camping out in the open. The focus on issues peripheral to nuclear weapons themselves not only made it difficult for the campers to convey their point of view, but also led to hostile reporting. The idea that the women should be at home caring for their families eventually became the dominant theme of news coverage of the Greenham Common peace camp.

The media's definition of what is newsworthy thus presents movement activists with a dilemma. On the one hand, the media require a departure

Table 6.1 The various activities of the British peace movement, 1980–1986 (percentages)

Type of event	*1980*	*1981*	*1982*	*1983*	*1984*	*1985*	*1986*
Meeting, speech, film	70.2	47.1	41.3	36.5	31.4	31.5	28.6
Forum, exhibit, conference	15.5	24.0	26.9	28.2	19.1	18.0	22.7
Art, music, festival	8.3	13.5	12.8	12.4	26.2	18.0	22.2
Petition	5.9	2.9	1.3	0.0	0.0	1.2	0.0
March, rally, demonstration	0.0	12.5	17.3	16.2	16.0	19.2	23.6
Direct action, blockade	0.0	0.0	0.3	6.7	7.2	12.0	3.0
Total	100.0	100.0	100.0	100.0	100.0	100.0	100.0
(N)	(84)	(104)	(312)	(550)	(582)	(501)	(487)

Source: Sampled from the 'Listings' column of the CND monthly, *Sanity*.

from social convention as an indication of newsworthiness. On the other hand, too much deviation from those norms is liable to be treated unsympathetically.

Tactical Developments during Mass Mobilization

Beginning in 1980, the peace movement found itself seeking media coverage and mass support, as well as trying to cope with the demands of newsworthiness and public approval. The response to these demands was to diversify the activities undertaken by movement organizations. This diversity is illustrated in table 6.1, which shows the development of British peace movement activities between 1980 and 1986, from the period before the mass mobilization to just after its peak.[5]

The most common activity throughout this period was the holding of a meeting, generally featuring either a speaker or a film about nuclear war. Between 1980 and 1986, however, meetings became proportionately less common in the movement, even though their absolute number rose fourfold. In part, the meeting was supplanted by more elaborate gatherings such as forums, exhibits, and conferences.[6] There was also an increase in the proportion of events in which art or music played a central role. But the most important shift between 1980 and 1986 was from information events to action events, such as petitions, demonstrations, and direct action. The number of peace movement actions in the sample increased from 5 in 1980 to 162 in 1985, before falling back to 129 in 1986. Within this category of action events, there was a shift from relatively innocuous forms of action, such as petitions, to more militant types of direct action, such as blockades.

Compared to reports of peace movement activities in the mainstream media,the variety of events listed in table 6.1 is amazing. Even in 1985, activities in which there was a potential confrontation with the police accounted for less than a third of all peace movement activities. An enormous amount of effort was put into events that are inherently non-confrontational in nature. In addition, some events were adaptations of confrontational tactics to non-confrontational purposes. For example, Oxford Mothers for Nuclear Disarmament planned a walk under the theme "Children Need Smiles not Missiles." This was advertised as a walk, not a march or demonstration. It went from a church to a park located in the same town, and concluded with puppets and balloons for the children, a leaflet for the mothers on what they could do about nuclear disarmament, and a picnic. Even the widely accepted CND peace sign was kept off posters advertising the walk, in an attempt to draw the broadest possible participation by women previously uninvolved in the movement (Lavelle, 1983).

Walks, poetry readings, film showings, leafleting actions and other non-confrontational activities are not without disadvantages. They rarely gain much publicity and they do not often attract the attention of the authorities. But the key advantage of non-confrontational tactics is that they are acceptable to a wide range of people. They present the organization, activists and ideas of the peace movement in a setting in which people are at ease.

If the form of the action has an impact on how it will be received, so does the content of the message. Many peace movement activities attempted to make a link in people's minds between the goals of the movement and the values people already held. The peace movement sponsored tribunals on the morality of nuclear weapons in order to capitalize on popular respect for international law. Other actions appealed to national pride or community loyalty. The Dutch movement argued that the Netherlands is well suited to showing moral leadership by rejecting nuclear weapons. The campaign to have municipalities in several countries declare themselves nuclear-free stressed that the interests of foreign powers should not be allowed to threaten the existence of the local community. CND rallies in Scotland were sometimes led by a piper playing "Scotland the Brave." Former CND head Bruce Kent, visiting an Easter week rally in Scotland in 1984, played to local feelings by claiming to "see in Scotland a more healthy skepticism to government statements than in England" (*Guardian*, 23 April 1984). Kent added, "We want foreign bases out of this country."

Table 6.2 gives some indication of the success of the British peace movement in mobilizing the support of different segments of society. The table is based on the same events that are categorized in table 6.1, but this time the events are classified by the nature of the organizing group. As the peace movement expanded, there were some remarkable shifts in the organizational basis of its support. Pacifist groups organized more than a third of all peace movement activities in 1980. In 1985 such groups organized about the same number of meetings, but their efforts were swamped by increased activity among trade unions, professional associations and women's groups. In 1986, as the wave of popular mobilization against the nuclear missiles diminished, the activities of labor unions and professional associations began to subside and the pacifist organizations

Table 6.2 The various sponsors of British peace movement activities, 1980–1986 (percentages)

Sponsor or venue	*1980*	*1981*	*1982*	*1983*	*1984*	*1985*	*1986*
Pacifist group	34.9	26.4	11.2	5.9	6.8	3.9	21.4
Religious organization	28.6	26.4	31.4	39.7	35.1	34.2	30.4
Educational association, school or university	17.5	18.4	12.8	9.5	6.8	5.3	14.3
Trade unions	6.3	12.6	7.9	7.5	12.2	17.1	3.6
Women's group	6.3	6.9	11.6	11.8	14.9	15.8	14.3
Professional associations	6.3	9.3	25.2	25.6	24.3	23.7	16.1
Total	100.0	100.0	100.0	100.0	100.0	100.0	100.0
(Number of events)	(63)	(87)	(242)	(305)	(222)	(228)	(134)

The number of cases in table 2 is smaller than in table 1 because of unidentified sponsors and because events sponsored by CND itself were not counted.
Source: Sampled from the 'Listings' column of the CND monthly, *Sanity*.

regained their prominence. These shifts in the social basis of peace movement activity help account for the trends in type of activity illustrated in table 6.1. Groups such as the Society of Friends and the Peace Pledge Union, mainstays of peace movement activity in quiet times, are most likely to organize meetings with a speaker. Educational and professional associations have the resources necessary to put together a conference, so increased involvement by such groups raised the number of weekend-long gatherings. More than half of the instances of direct action beginning in 1983 were organized by women's groups. Thus, changing styles of peace movement activity were associated with changing bases of organizational support for the movement.

There are still other ways in which the expansion of the peace movement linked it to a variety of social institutions. Religious themes were prominent in peace events sponsored by church organizations. By planting wheat at the Molesworth missile base to be sent to starving people in Eritrea, the peace movement linked itself to humanitarian values. Traditional themes of mothers as nurturers resurfaced in the contemporary peace movement. Women have also found a feminist basis for participation in the movement, becoming active against nuclear weapons as a challenge to "masculine politics" (Schenk, 1982; Women Against the Bomb, 1983; Cambridge Women's Peace Collective, 1984). Like the connections with the unions and various professional organizations, links to the women's movement help the peace movement mobilize more widely than would otherwise be the case.

Based on the experience of the anti-nuclear energy movement, Herbert Kitschelt (1986) has suggested that movements facing a receptive political system will use assimilative strategies such as lobbying, petitioning, electoral campaigning and litigating. Movements without hope of influencing policy from within will resort to demonstrations and civil disobedience. His argument holds for the peace movement only as a general tendency that shows up in the contrast between the French movement and those elsewhere in Western Europe. It is far more accurate to say that each nation's peace movement employed both assimilative and confrontational strategies, and that each strategy was pushed as far as it could be taken. Assimilative strategies (of seeking political support) were pursued to the limits of receptivity of established political institutions. Confrontational strategies were used simultaneously, but were bounded by the public's tolerance for direct action.

Integration into the Party System

The diversity of peace movement activities increased during the period of mass mobilization. This is true both in the types of activities undertaken and in the range of groups that sponsored them. This diversity of tactics enabled the peace movement to penetrate a variety of social institutions, to expand beyond the strictly political arenas in which policy can be influenced, and so to work for the social transformation said to characterize new social movements.

But the decision taken by most peace movement organizations in 1980 to focus on the proposed INF deployment led the movement to take on other traits less compatible with the requirements of a new social movement. As it developed specific policy goals, the peace movement was forced to establish a working relationship with established political authority. The reasons why this is so are obvious. Even Petra Kelly (1984, p. 17), a member of the fundamentalist wing of the German Greens, argued that:

> a movement operating exclusively outside parliament does not have as many opportunities to implement demands, say, for a new attitude to security, as it would if these demands were also put forward in parliament. Despite the great autonomy of the peace and ecology movements, it seems to me that they have no option but to relate to the political system as it is, given the nature of power in our society.

One of the key forms of relationship with the political system is the development of ties with political parties. Not only does a majority party or coalition determine government policy, but even opposition parties can focus attention on nuclear weapons by questioning government ministers, proposing amendments to legislative proposals, and other parliamentary devices. Sympathetic political parties carry out research on alternative security policies that is better publicized and more authoritative in the public mind than that done within the peace movement itself. Party connections also increase the ability of a peace movement organization to mobilize support for its rallies. In many ways, then, the efforts of allied parties helped prevent the cruise and Pershing II missiles from fading away as issues, even after their deployment.

The specific ways in which movement organizations have interacted with political parties vary from country to country, depending not only on the resources and inclinations of the movement organizations themselves, but also on the established patterns of contact between parties and other social groups. In the Netherlands there is a long tradition of formalized contact between parties and leading interest groups, following the corporatist pattern. The best examples of this are found in relations between employers and employees, who advise the government through the Socio-Economic Council and the Foundation of Labor. Thus, when the "Stop the Neutron Bomb" campaign gathered steam in the Netherlands, it seemed entirely natural that there should be a committee of peace groups and party leaders to discuss nuclear weapons in the country. The Consultative Group Against Nuclear Weapons included representatives from all the major Dutch peace groups, from the trade unions, and from political parties ranging from the Communists to the Christian Democrats. The Consultative Group had no statutory authority and in any event its very diversity worked against the possibility of agreement within the group. But the Consultative Group did enable leaders of the peace movement and the political parties to hear each others' points of view. Representatives of the parties and of the peace movement each found the Group to be a useful forum for explaining their positions and for exploring possible compromises (Groeneveld, 1984; Faber, 1985).

In Great Britain, the contact between peace movement and parties occurred not so much through high level discussions as it did in the framework of party organization and electoral politics. The Labour party has official representation on CND's executive council. CND, in turn, has a network of active sympathizers in Labour's constituency organizations. The Parliamentary and Elections Committee of CND supplied elected MPs of all parties with arguments for nuclear disarmament much the way a regular lobbying organization does. This committee also monitored statements on defense policy made by each MP and sent representatives to the member's local political meetings in order to raise questions about specific NATO weapons and policies.

In Germany, too, the peace movement influenced the Social Democratic Party (SPD) by working through the constituency parties. The end of SPD support for the cruise and Pershing II missiles was presaged by the conversion of a number of state party organizations. In 1981, the SPD in Schleswig-Holstein and South Hesse voted to oppose the missiles. In 1982 a larger number of state party groups, including the biggest ones in North-Rhine Westphalia and Bavaria, were in favor of INF only if it was clear that the possibility of negotiation had been pursued to the fullest extent possible. At the special SPD conference on the missiles held in November 1983, just before the INF debate in the Bundestag, a resolution to reject the missiles was carried by a 385–15 vote, despite ex-Chancellor Schmidt's continued advocacy of deployment. Although the coalition realignment that put the SPD in the opposition was an important factor in making this conversion possible, the groundwork had been laid by peace movement pressure on the party at the local and state levels.

Efforts by peace movement activists to influence political parties were a logical consequence of the political involvement of many movement activists. The peace movement was part of the existing political institutions because its activists were. Surveys of movement activists showed that they were more likely than the rest of the population to become involved in party politics. Europeans who in 1984 said they had taken part in peace movement activities were twice as likely as others to describe themselves as very or fairly strong supporters of a political party, by 60 percent to 30 percent.[7] A survey of participants in the 1981 peace demonstration in Amsterdam found that 23 percent of the marchers were also members of a political party (Schennink, Bertrand, and Fun, 1982). Another study of peace movement activists in six Dutch cities found that 47 percent were party members (Kriesi and van Praag, 1987). This compared to 9 percent of the Dutch population who were party members, according to the 1981 Dutch Election Study. As Klandermans has put it elsewhere in this volume (chapter 7), the peace movement was part of a multi-organizational field of alliances.

The integration of peace movement and social democratic parties was extensive at the elite level as well. Several members of the SPD's federal executive were active in the peace movement, with the most prominent example being Erhard Eppler. Sixty British MPs endorsed E. P. Thompson's appeal for European Nuclear Disarmament in April 1980. At the height of the peace movement mobilization, approximately 90 Labour MPs were members of CND, in addition to several Welsh Nationalists, Scottish Nationalists, and MPs from the now-defunct Alliance (Marchant, 1983).

The primary benefit for the peace movement from its close links to political parties of the Left and center was its ability to exert a profound influence on the defense programs of such parties as British Labour and the German Social Democrats. The costs were that it became more difficult for the peace movement to maintain an independent and critical stance of the kind that no mainstream party would accept.

Conflict and Cooperation with the Police

The limits on peace movement protests are visible even in clashes between protesters and police. Such clashes occur because of the basic tension between political protest and public order. But there is within this conflict also a large element of common interest. Although tension and frustration may cause members of either side to lash out at the other, neither police nor activists want a violent confrontation to occur. More than that, both sides have an interest in an orderly process in which there is a set of clearly defined mutual expectations. This does not mean that the activists want the police to know their plans in great detail, for that would make it too easy for the police to neutralize the protest. Nor do the police want the activists to have a precise idea of which protest activities will lead to arrest and which will not: uncertainty is likely to be a deterrent. But at the point of actual contact between activist and police, when the two lines are facing each other and each side must decide what to do next, there is a mutual interest in not being misinterpreted in a way that might provoke an unwanted escalation of the conflict. The best way to avoid misinterpretation is to develop rules of protest agreed to by both sides, and then to play by those rules.

This is one reason for the emergence of a stable repertoire of protest actions. At times, these standard procedures may lead to open displays of courtesy. In January 1960 there was a rally at a Thor missile base in England at which CND marched past the base legally, while the Direct Action Committee held an illegal sit-in. Participants in the two actions became hopelessly intermingled, and the response of the police was to ask each person there if he or she intended to be arrested. The only people arrested were those who replied affirmatively (Minnion and Bolsover, 1983, p. 20).

This level of cooperation between police and activist is uncommon, but the extent of mutual understanding about how to handle protest is not. Time and again in West European capitals in the early 1980s, up to a half million people marched through the streets, held a rally in a central area, and then dispersed. These massive events were arranged ahead of time with the local police, who kept traffic off the streets and kept an eye on the proceedings. When the police did intervene, most commonly because a section of the march stopped or deviated from the official route, designated stewards from the organizations sponsoring the march would hurry to the scene. They generally cleared up the problem without arrests.

Even in the case of sit-ins and blockades, where the action is illegal and is likely to lead to arrest, there was often a great deal of cooperation between peace protesters and police. Large actions, whether legal or illegal, were unlikely to lead to uncontrolled clashes with the police because they were widely publicized in advance and both police and demonstrators were able to prepare for them. Demonstrations were carefully choreographed in advance. Activists who expected to participate in actions of civil disobedience, such as blockades, were usually required to undergo training in passive resistance and non-violence. The police they faced had been trained in crowd control and in dealing with non-violent protest.

Paradoxically, the smaller and more spontaneous actions have the most conflict potential. There are more unknown elements in such protests, and unplanned developments introduce a greater degree of spontaneity in how the demonstrators and police react to each other. The police have on-the-spot latitude in determining what actions will lead to arrest, and they have less information about what actions will be attempted by the protesters. One participant in a 1983 blockade of the Neu Ulm missile base in Germany described the experience as a psychological duel between police and demonstrators, in which the police gained the upper hand when they realized that the protesters had no clear goals and were confused about what to do next. The relationship between activists and police was viewed as a form of psychological warfare, for "even with new action forms we will only have short term successes, until the police psychologists analyze our motivations and the game will begin again" (Bedürfnisanstalt Blockade, 1983).

This is a reminder of the conflictual aspect of the relationship between protesters and police. But the irony of peace movement protest is that the massive demonstrations that achieve worldwide press coverage and that seem so ominous to many in government and among the public are actually more controlled and less conflictual than are the smaller actions involving at most a couple of dozen people. The visual image so frequently broadcast of police carrying a demonstrator off to jail looks like an image of conflict. It is. But it is also an instance of two groups of professionals carrying out their jobs with precision.

Allies in Municipal Government

Between peace movement and police, the relationship may be characterized as limited conflict moderated by a mutual recognition of common interests. In municipal governments, by contrast, many peace movement organizations found an active ally. Municipalities are involved in nuclear weapons politics primarily by their obligation to participate in civil defense planning. Local authorities must, in several West European countries, draw up plans for providing services and maintaining order in the wake of a nuclear attack. They may also be obliged to maintain bomb-proof bunkers to house government officials and other important personnel. These obligations are not always carried out to the specifications of the national government, especially in cities controlled by parties sympathetic to the peace movement. The Greater London Council, which when it still existed was particularly critical of Britain's civil defense program, decided in 1984 to open its bunkers for public inspection. The bunkers remained open for a week, and visitors to them could obtain both official civil defense literature and antinuclear leaflets printed by CND. The GLC also created the Greater London Conversion Council to develop plans to convert arms factories to other forms of production. It declared 1983 to be "Peace Year," and spent $600,000 on projects and information that presented alternatives to the government's military and civil defense program (Byrd, 1985).

The example of London is not unusual. The Derbyshire City Council awarded 23,000 female employees the option of an unpaid week off work to allow them to attend a rally at the Greenham Common peace camp in September 1984. The West Yorkshire County Council sent a peace information bus on a tour of the county on the fortieth anniversary of the nuclear bombings of Hiroshima and Nagasaki. The Strathclyde District Council gave a grant to the Faslane Peace Camp to do peace education work in the area. The campers acquired a bus and printed leaflets to pass out in shopping areas and at residential complexes (Faslane Peace Camp, 1984, pp. 48–50).

In the Netherlands, the local government of Amsterdam made all public transportation in the city free on the day of the large demonstration there in November 1981. The Woensdrecht city council voted against having cruise missiles in their municipality on the same day the Dutch government announced that missiles would be stationed there. Numerous local governments in the Netherlands have adopted the IKV's "Local Peace Program," which commits them to helping peace groups by providing meeting space and printing facilities at minimal cost, to maintaining contacts with East European cities, to encouraging peace education in the local schools, and to conducting a critical review of civil defense plans (Klandermans and Oegema, 1987).

Great Britain's CND and West Germany's Action for Reconciliation urged municipalities to declare themselves nuclear-free, with substantial success in each country.[8] The British movement went the furthest, with 170 nuclear-free cities established by the end of 1983. A national civil defense exercise planned by the British government had to be cancelled when over half of the targeted municipalities refused to cooperate. All over Western Europe, sympathetic local governments gave subsidies to community groups involved in spreading information about the effects of nuclear war. Towns in various West European countries adapted the practice of "twinning," begun to foster cultural exchanges between member states of the European Community, to anti-nuclear politics. New twins were created with towns in Eastern Europe or the Soviet Union. Old twins within Western Europe began to exchange children's peace art and other peace education projects. In West Germany and in Great Britain, movement activists obtained representation on peace committees set up by the local governments (Sevill, 1985, pp. 52–3). In Britain, local governments took out ads in CND's monthly magazine to announce their nuclear-free status and their sponsorship of peace events ranging from conferences to exhibitions of peace art. Local governments thus lent their authority, credibility and resources to the peace movement.

The Peace Movement Faces the State

It is a common presumption that the peace movement represented a challenge to the authority of the state. That contemporary movements seek to undermine the liberal democratic state is the most important claim made by the new social movement theory. But although peace movement activists pursued their goals outside of the state, they did not conduct politics against the state. Contacts between the peace movement and West European governments have proven to be more dense and fruitful than one would expect from such rhetoric as the slogan from the German peace movement, *Kein Frieden mit dem Staat* (No Peace with the State).

Among movement leaders, the relationship with government was frequently one of intensive consultation and negotiations. Most peace movement leaders maintained regular contact with political figures who were willing to help, who could provide information, or who might be persuaded to support the movement. The published diary of IKV secretary Mient Jan Faber (1985) provides the most detailed account available of a peace movement leader's daily telephone calls, meetings and exchanges of notes. His range of contacts included ministers and members of parliament, high civil servants in the ministries of foreign affairs and defense, and staffers at the American embassy in The Hague and at NATO headquarters in Brussels.

Faber's case is without doubt unusual because of the openness of the Dutch political system and the extent to which prominent people tend to know each other in a small country. Faber also represents the moderate wing of the Dutch peace movement; for campers outside of the Woensdrecht missile base, contacts with authority took on a very different character. But portions of the peace movement in all West European countries actively sought the support of politically important individuals and organizations, attempting to exert influence in election campaigns, legislative debates, court cases and bureaucratic hearings.

By accepting the help of some local governments and political parties, peace movement organizations tended to integrate themselves with established political institutions. By working through legislative, judicial and bureaucratic channels of appeal, these organizations undermined the inclination of some of their own activists to declare the illegitimacy of liberal democracy. The desire for mass mobilization led to a narrowing of the movement's policy focus. That narrowed policy focus, in turn, blunted the critique of the political system that was felt so deeply by so many activists. The rise of the peace movement was accompanied by the decline of its new social movement characteristics. It may not be possible to be a mass movement and a new social movement at the same time.

Lessons for New Social Movement Theory

It is not possible literally to test the new social movement theory with respect to the experience of the contemporary West European peace movement. For one thing, forms of political action are not as important to the theory of new social movements as is the claim that such movements champion new kinds of issues and politicize areas of life previously considered to be part of the private domain. As Melucci (1985, p. 809) put it, "movements are not qualified by what they do, but by what they are."

In addition, there is real ambiguity on precisely what the new social movement approach expects to see in the tactics of the peace movement.

An account by Claus Offe (1985, p. 830), for example, refers to new social movements as incapable of engaging in political negotiations, "because they do not have anything to offer in return for any concessions made to their demands." Yet a later section of that same article is devoted to a consideration of possible alliances that new social movements may conclude with other political forces, with comments on the policy compromises that could be expected to occur in different alliances. The peace movement does not resemble a new social movement incapable of engaging in policy negotiations, but it looks very much like the alliance-forming political actor whose dilemmas Offe described so insightfully in the later part of his article.

As a perspective on the meaning of contemporary movements, the new social movement approach is more than a series of testable propositions. It is a theory of the role of political movements in advanced industrial societies. The experience of the West European peace movement adds an important caveat to that theory, namely that a movement cannot do all things at once. To the extent that a movement attempts to achieve specific policy goals, its broader aspect of cultural criticism and transformation is necessarily subordinated.

This does not mean that broad cultural criticism does not continue during a political campaign. Movement activities at the community level can be enhanced by the existence of a specific national policy goal. Petitions are not a particularly effective way of convincing a government to change its policy, and pledge campaigns in which people sign a vow not to participate in a war misunderstand both the causes of war and the psychology of participation in wars. But the petitions and the pledges do bring movement activists and potential supporters together. They are an excuse to talk about nuclear weapons.

The campaign to establish municipal nuclear-free zones had the same benefits. The systematic canvassing and circulation of petitions in connection with nuclear-free zones was such a useful device in generating discussions that local peace organizations typically prolonged this preliminary phase by going to the city council as late as possible. The emphasis on the campaign itself, rather than just the outcome, is nicely summarized by the Action for Reconciliation brochure on the subject: "The campaign for nuclear free zones in regions and communities is a powerful tool to intensify the discussion about the arms race. A sticker alone does not make a nuclear free zone" (Aktion Sühnezeichen/Friedensdienste, 1982).

Even so, the decision to seek media coverage of the anti-INF campaign forced a shift of focus from community action to events far larger in scale. The campaign against INF redirected the peace movement from cultural action to political action. People were brought into the peace movement on the assumption that specific policy reforms were to be achieved. These new recruits concerned themselves with mobilizing trade union support, penetrating party organizations, and influencing electoral campaigns. They spent a great deal of time with municipal council members, and less with community organizations. The policy goals of the movement had a very great effect on its tactics, moving it away from the forms of political action characteristic of a new social movement. By appealing to the values of a diverse array of social groups, the peace movement at least temporarily abandoned its broader critique of militarism.

The experience of the peace movement suggests that the aspirations of new social movements and the achievement of mass mobilization for a specific policy goal are not compatible objectives. What Joseph Gusfield (1981) has called the trait of fluidity in a political movement clashes with the trait of linearity. New social movements are expected to be fluid (oriented to social change), but a political movement that undertakes mass mobilization tends to become linear (oriented to political change). It is perhaps for this reason that Melucci refers to "peace mobilizations" rather than to a peace movement (Melucci, 1985, pp. 801–9). The new social movement approach makes sense only if we remain mindful of the distinction between the cyclical peaks of mobilization, of which the peace movement is Western Europe's largest post-war example, and the continuous but far more restricted current of support for fundamental social change.

This does not mean that the peace movement would have been better advised to avoid the opportunity for mass mobilization presented by the decision to deploy cruise and Pershing II missiles. Most of the leading peace movement organizations had been in existence for some time before the 1979 INF decision, carrying on their activities without publicity or popularity. They developed critical views of nuclear deterrence and of the partition of Europe between spheres of superpower influence, but their ideas received little attention beyond the circle of those already committed. Had they not seized the opportunity for mass mobilization presented by INF, peace movement organizations would have been doomed to continuing marginality. The strategy of mass mobilization limited the ability of the peace movement to preserve its critique of the political system and to maintain its focus on cultural revolution. The dilemma for a new social movement is that the alternative, to remain a counter-cultural sect, presents even dimmer prospects of achieving significant change.

NOTES

1. For simplicity and in deference to its self-labelling, I will refer to the West European movement against Pershing II and cruise missiles as the peace movement. This is not meant to imply that those who supported deployment of these weapons were in favor of war.
2. The other activities, in descending order of approval, are rent strikes, occupying buildings, blocking traffic, unofficial strikes, painting slogans, personal violence, and damaging property. These figures are derived from Barnes, Kaase et al. (1979, pp. 544–6). A panel survey of the United States, West Germany and the Netherlands shows that the levels of protest approval have generally remained stable since 1974 (Kaase, 1990).
3. The slippage is not enormous, however. Fifty percent of the French, 52 percent

of the British, 59 percent of the Germans, and 79 percent of the Dutch declared themselves sympathetic to recent demonstrations for peace in Western Europe in a poll published in *Le nouvel observateur* on 21 November 1981, right after the first international wave of demonstrations, as reported by Everts (1982).

4 The United States Information Agency, using a question wording that maximizes support for the missiles, does find a plurality in favor of deployment in Great Britain, but these results are an outlier. For a report on the British response to a series of questions asked by commercial polling firms, see den Oudsten (1984).

5 The events in table 6.1 are sampled from those listed in CND's monthly magazine, *Sanity*. This source implies certain biases. Events unconnected with CND are less likely to be listed, and activities that require the participation of only a few people are unlikely to be listed since the purpose of the compilation is to publicize the event to potential participants. The effect of these biases is to understate the extent of small scale, non-confrontational activities, particularly when organized by local groups.

6 I have defined the second category in table 6.1 as being those activities that involve at least three speakers or presentations.

7 *Eurobarometer* 21. A survey carried out in April 1982, *Eurobarometer* 17, shows a similar relationship between activity in the peace movement and adherence to a political party. On party involvement among peace activists, see also Bolton (1972); Schmid (1982); and Inglehart's chapter in this volume, chapter 3.

8 The French peace movement has also attempted to create municipal nuclear-free zones, but without the benefit of sympathetic political parties they have had little success. See Pourquoi créer des zones dénucléarisées? (1983).

7

Linking the "Old" and the "New": Movement Networks in the Netherlands

P. BERT KLANDERMANS

Students of new social movements tend to emphasize the distance of contemporary social movements from established socio-political institutions. They define the "newness" of new social movements partly in terms of the presumed marginality of these movements in the socio-political system. Apart from citations from movement leaders and other self-conceptual evidence, little empirical data is available to test these assumptions. In contrast with the new social movement literature, it will be argued here that whatever characteristics define the newness of new social movements, neither marginality nor detachment from the social and political organization of society are very crucial. On the contrary, on theoretical grounds it is stated that new social movement organizations, like every movement organization, are part of a larger network of supporting and opposing organizations. From this viewpoint, it is much more instructive to study a movement organization's linkages to these networks, than to dismiss them by making detachment or marginality defining characteristics of new social movement organizations. By mapping out and identifying their allies, their opponents, and those who are indifferent, we might improve our explanations of a social movement organization's ability to mobilize resources, use opportunities, and exert influence.

It would be less convincing, however, to simply confront new social movement literature with theoretical arguments. Therefore the theoretical argumentation will be accompanied by empirical evidence from studies on the Dutch peace movement.

Social Movement Organizations and Multi-organizational Fields

New social movement literature was not the first to emphasize the marginality of social movements. Long before, classical theories on social movements defined insurgent action and social movements as relatively spontaneous, disorganized forms of action enacted by marginal, non-integrated individuals. It was supposed that integration into socio-political organizations would restrict political protest. In the Anglo-Saxon literature, resource mobilization theory signified a drastic break with this conceptualization of social movements. Resource mobilization theorists supposed that indigenous organizations supported rather than restrained political protest (McAdam, 1982; Morris, 1984). Organization was seen by them as an indispensable resource for effective insurgency, rather than as evidence against the existence of a social movement.

Over the last two decades the significance of support structures for the generation and survival of social movement organizations has become increasingly apparent. Curtis and Zurcher (1973) were among the first to point to the operation of multi-organizational fields in the development and maintenance of protest organizations. According to their definition, a multi-organizational field "is the total possible number of organizations with which the focal organization might establish specific linkages" (p. 53). Linkages can be established on the organizational level "by joint activities, staff, board of directors, target clientele, resources, etc.," and on the individual level "by multiple affiliations of members" (p. 53). Their study of the anti-pornography movement revealed that anti-pornography organizations were enmeshed in a network of other community organizations and that participants in anti-pornography organizations were members of other voluntary associations as well. These linkages were based on common interests, ideologies, audiences, or other shared characteristics.

A few years earlier, Heirich (1968) showed how the Berkeley Free Speech Movement developed out of existing networks of activists and student organizations at Berkeley. More recently, Morris (1984) observed that the existence of internal social institutions and organizations is a basic requirement for the development of a social movement organization. What proved to be true in the case of the anti-pornography movement and the Free Speech Movement turned out to be equally true for the civil rights movement (Morris, 1984): social movement organizations develop out of existing networks of organizations which are necessary both to generate and to maintain them. In the meantime, numerous studies reported the importance of social networks as support structures for a variety of movements ranging from terrorist organizations to neighborhood movements (see Klandermans, 1989b, for an overview).

Although Curtis and Zurcher note that the opponents to the anti-pornography movement also constitute a multi-organizational field, surprisingly little attention has been given to the fact that multi-organizational fields need not necessarily be supportive.

The multi-organizational field of a social movement organization has both supporting *and* opposing sectors. These two sectors can be described as (1) a social movement organization's *alliance system*, consisting of groups and organizations that support the organization, and (2) its *conflict system*, consisting of representatives and supporters of the challenged political system, including counter-movement organizations (Kriesi, 1985). The boundaries between the two systems remain vague and may change in the course of events. Specific organizations that try to remain aloof from the controversy may be forced to take sides. Parts of the political system (political parties, elites, governmental institutions) can coalesce with social movement organizations and join the alliance system. Coalitions can fall apart, and previous allies can become part of the conflict system.

Alliance systems serve to support social movement organizations by providing resources and creating political opportunities; conflict systems serve to drain resources and restrict opportunities. Within conflict systems, sooner or later bargaining relations develop between representatives of social movement organizations and target institutions, at the initiative either of the movement organization or of its opponent.

Different social movement organizations have different but overlapping conflict and alliance systems. The greatest overlap will exist among organizations from the same social movement industry, that is, social movement organizations that belong to the same social movement (organizations from the women's movement, from the environmental movement, etc.). But, movement organizations from different social movement industries may also have overlapping conflict and alliance systems. The cleavage between a social movement organization's alliance and conflict systems may coincide with other cleavages, such as social class, ethnic lines, or Left/Right affiliation. The proportion of the multi-organizational field that becomes engaged in conflict or alliance systems will vary from situation to situation. Moreover, there will be variation over time. The proportion of the multi-organizational field engaged in one of the two systems expands or contracts according to cycles of protest (Tarrow, 1983). At the peaks of protest, almost any organization may be enmeshed in either system; in the valleys most organizations wil belong to neither.

The specific make-up of alliance and conflict systems fluctuates over time, and with the particular movement, situation, and circumstances. For example, Walsh (1981) showed that in the different communities surrounding Three Mile island, after the reactor accident, social movement organizations with different alliance systems developed, depending on existing differences in the multi-organizational field in each community.

This conception of social movement organizations (i.e., as emerging from multi-organizational fields) implies a historical continuity in the generation of social movements. Moreover, it denies that marginality is a distinguishing feature of social movements. By partitioning multi-organizational fields into alliance and conflict systems, one is alerted to look for linkages between social movement organizations and existing organizations and institutions.

Such an approach seems to be at odds with new social movement literature, in which the very use of "new" suggests discontinuity. Indeed, much of the new social movement literature stresses the distinction of new social movement organizations from traditional organizations such as trade unions, political parties, pressure groups, interest organizations, churches, etc. (cf. Brand, 1985; Klandermans, 1986). New social movements are presented as developing in the margins of post-industrial society, detached from and opposed to traditional organizations of interest intermediation. A closer look, however, reveals that this representation is oversimplified. New social movement organizations, like any other social movement organization, have all kinds of links with organizations and institutions in their socio-political environment. Far from being detached from traditional organizations of interest intermediation, they appear to be enmeshed in the socio-political networks of the societies in which they develop.

In this chapter I will demonstrate this to be true for an example of new social movements, the Dutch peace movement. But, before doing so, I will elaborate the concepts of alliance and conflict systems.

ALLIANCE SYSTEMS

Research abounds on the multi-organizational fields of social movement organizations functioning as alliance systems. Major parts of the alliance system consist of other movement organizations, from the same movement industry or from different movement industries (cf. Klandermans, 1989b).

Although coalitions between cognate social movement organizations seem natural, competition is equally likely (Zald and McCarthy, 1980). Wilson (1973) argued that we can expect greater competition among cognate movement organizations appealing to the same mobilization potential than among movement organizations with non-overlapping mobilization potentials. This phenomenon is exactly what Benford (1984) described in his study of the interorganizational dynamics of the Austin peace movement, and what Rucht (1989) reports in his study of the French and German environmental movement.

In general, enduring coalitions between social movement organizations are not very likely. Coalition formation can be threatening for many movement organizations. In Wilson's (1973) words, "Resources, autonomy, and purposes can be jeopardized if the organization must share the credit for victory and the blame for defeat." Wilson goes on to argue that coalition formation is stimulated if resources are significantly shrinking or if it is thought that joining a coalition will increase them significantly. In line with this argument, Staggenborg, in one of the rare empirical studies of coalition formation in social movements, concluded that organizations are most likely to form coalitions "(1) when individual organizations lack the resources needed to take advantage of opportunities or fend off threats, or (2) when coalition work allows movement organizations to conserve resources for tactics other than those engaged in by the coalition" (1986, p. 388). Once formed, coalitions easily break down, because of ideological conflicts and because the organizational maintenance needs of individual organizations compete with the needs of the coalition.

Much more common than enduring coalitions are *ad hoc* alliances among movement organizations: loose, cooperative bonds between two or more organizations. Since *ad hoc* alliances usually develop among organizations that agree on substantial goals, similar alliances return over time and in different circumstances.

Although other movement organizations form a major part of the alliance system of a social movement organization, almost any kind of organization can become engaged in an alliance system: political parties, unions, churches, recreational organizations, youth organizations, student and campus organizations, traditional and new women's organizations, nature conservation organizations, business organizations, consumer organizations, community organizations, and sometimes even governmental institutions. Further, the composition of alliance systems changes in the course of the cycle, as Tarrow's (1989) research on the Italian protest cycle in the 1960s and 1970s shows. Although traditional and institutional actors were absent in the initial phases of this cycle, these actors joined in and took over events at the peak of the cycle, either to channel protest into more moderate directions or to use the accumulated political pressure to advance their own interests. At the downturn of this cycle the alliance system rapidly disintegrated, radical organizations became more dominant, and the alliance system eventually contracted into a network of political radicals (Della Porta and Tarrow, 1986).

Because alliance systems provide a social movement organization with resources and political opportunities, they are indispensable to social movement organizations, if the organizations are to survive and have any impact. The greater resources and the more political opportunities an organization in the multi-organizational field can supply, the more attractive an ally it is.

Resources

The resources furnished to a social movement organization by its allies are both tangible (money, space, equipment, etc.) and intangible (organizational experience, leadership, strategic and tactical know-how, ideological justifications, etc.). More important, alliance systems provide social movement organizations with extended communication and recruitment networks. Non-movement networks turn out to be extremely important for a movement organization in consensus and action mobilization, especially in the earlier stages of mobilization (Klandermans, 1988).

Opportunities

Alliance systems create political opportunities. Interorganizational networks link social movement organizations with the political system, connecting elites and political parties with the social movement sector. Political entrepreneurs occupy strategic positions linking social movement networks with the political system (McCarthy and Zald, 1977). Costain and Costain (1986) and Jenkins (1987) have shown that in the US the social movement sector and the political system have become increasingly enmeshed. Links to political parties and elites are indispensable for political influence. Drawing on Lipsky's and Wilson's models of protest, bargaining, and political culture, Terchek (1974) showed that the political influence of the civil rights movement depended not only on public opinion and the level of insurgency, but also on the extent to which movement organizations could activate allies with substantial political resources. Sophisticated lobbying and ties with the party system and governmental institutions increase a social movement organization's impact on policy-making considerably (Costain and Costain, 1986). In our study of grassroots mobilization and local government in the Netherlands, the inclusion of some of the local political parties in the alliance system of the peace movement turned out to be decisive in passing freeze motions or peace programs in Dutch city councils (Klandermans and Oegema, 1987).

CONFLICT SYSTEMS

Part of the multi-organizational field of a social movement organization consists of organizations and institutions that are opposed to the social movement organization. Central to the conflict system are target organizations and institutions: governmental institutions, employers' organizations, business organizations, elites, political parties, etc. But, just as a social movement organization's alliance system is open to any kind of organizational support, so a social movement organization's conflict system can admit any kind of organizational opponent. Occasionally the actions of the social movement organization itself push organizations and institutions into the conflict system. Protests inevitably have spillover effects that penalize people other than the intended target. Consequently, these people may affiliate themselves with the opponents of the social movement organization.

Within conflict systems, an "us–them" dynamic tends to develop. Mansbridge describes this process, frequently discussed in the literature on intergroup relations, for the proponents and opponents of the Equal Rights Amendment (ERA) in the United States: "Building an organization on belief in a principle, when the world refuses to go along with that principle, produces a deep sense of us against them; when two movements are pitted against each other reality will provide plenty of temptations to see the opposition as evil incarnate" (1986, p. 179). Organizations such as social movement organizations, which rely on volunteers, can easily become engaged in such ingroup–outgroup dynamics because it "requires an exaggerated, black or white vision of events to justify spending time and money on the cause" (1986, p. 6). Especially if a countermovement develops, the conflict system easily becomes dominated by such intergroup dynamics.

A countermovement tends to develop if a social movement organization is successful (Zald and Useem, 1982). Countermovements are often (and sometimes openly) supported by elites and the established order. As a consequence they frequently have extensive financial resources at their disposal. Some examples of countermovements are the anti-abortion movement in Western countries, the anti-ERA movement in the US, the new Christian Right in US churches, and in the Netherlands the Interdenominational Committee for Mutual Disarmament (ICTO), established by conservative church members, many of whom have military backgrounds, in reaction to the successful mobilization against cruise missiles by the Interdenominational Peace Council (IKV).

Draining resources and restricting opportunities

Authorities and countermovements and their allies try to deprive a social movement organization of resources and political opportunities by increasing the costs of participation in the social movement organization, by undermining its organizational strength, declaring specific tactics illegal, abolishing specific opportunities, criminalizing the organization, offering symbolic concessions, and campaigning to turn the public against the social movement organization (Griffin et al., 1986; Henig, 1982; Terchek, 1974). To accomplish these ends, a variety of tactics are used (Griffin et al., 1986; Zald and Useem, 1982); opponents attempt to (1) break the unity of the movement; (2) impair the social movement organization directly (through infiltration, bribery, penalizing and arresting members and leaders, anti-social movement organization legislation, restricting resources); (3) increase the costs of mobilization and collective action (by using the police, strong-arm boys, repression, and by threatening activists); and (4) constrain the political opportunity structure (through anti-propaganda and litigation, and by undermining the moral and political bases of the social movement organization).

Sometimes, however, attempts to impair a social movement organization can actually work to the organization's advantage, fostering internal cohesion, increasing support from the alliance system, persuading individuals and organizations to become engaged in the alliance system, and sometimes forcing governmental institutions to side with the social movement organization (Terchek, 1974; McAdam, 1983; Tarrow, 1988b).

Bargaining

Sooner or later, social movement organizations and their targets have to enter into some bargaining relationship. Bargaining can serve the interests of challengers *and* the polity. As a rule, challengers try to arrive at substantive gains, whereas the polity tries to prevent or put an end to insurgent action. Consequently, the authorities' interests in bargaining increase with the degree of disorder (Morris, 1984; Mueller, 1978; Terchek, 1974). Sometimes challengers and opponents negotiate to reduce the risks associated with collective action (as in agreements between police officials and the organizers of a demonstration to prevent violent confrontations, or negotiations between a strike committee and management on the termination of a complex production process).

In time, more or less elaborate bargaining structures develop as a component of the conflict system. Specialized institutions are created, in which representatives of social movement organizations and target institutions meet. As a result, parties develop a vested interest in maintaining these bargaining structures. Bargaining structures form a bridge between the conflict and alliance systems. The more they expand, the more the social movement organizations run the risk of co-optation, that is, of participating in the structures they were fighting against. No wonder social movement organizations are ambivalent about becoming too engaged in those structures. Little empirical work has been done on collective bargaining in social movements' conflict systems, but students of movement organization conflict systems can learn much from the literature on collective bargaining in industrial relations (see Stephenson and Brotherton, 1979, pp. 155–239).

Movement Networks in the Netherlands

In an attempt to support the arguments made above, the remainder of this chapter presents data from research on the Dutch peace movement. This research is conducted by our own group at the Free University and by political scientists from the University of Amsterdam with Hanspeter Kriesi as a core member. It is clearly not a test of the complete reasoning developed thus far. This would require a comparative, longitudinal framework to compare the multi-organizational fields of individual movements and to assess their impact. It will, however, provide evidence in support of my basic argument, that is, that new social movement organizations, like any other social movement organization, are embedded in a multi-organizational field and that alliances cut across the traditional/non-traditional, institutional/non-institutional, or conventional/non-conventional cleavages in society.

The data presented stem from two different data bases: our own data collected in June 1985 among the 115 members of 10 local IKV-groups. The IKV (Interdenominational Peace Council) is the core group of the Christian tendency within the Dutch peace movement (see Klandermans and Oegema, 1987, and Kriesi and van Praag, 1987, for a discussion of the organization of the Dutch peace movement). Over the last decade the IKV has been one of the key organizations of the Dutch peace movement. Kriesi's data stem from a survey among movement activists and the population in four communities conducted in fall 1985. The data presented here are restricted to the questionnaires completed by the 144 core members of the four local peace movement networks. Apart from IKV-groups, local peace movement networks comprise several other organizations, like chapters of Women for Peace, Stop the Neutron-bomb, union of conscientious objectors (Kriesi and van Praag, 1987). Note that in both cases data were collected at the local level. Altogether in the two studies, data were collected in 14 different communities from different regions of the country. Although figures to assert representativeness are not available, I feel confident that the picture drawn will be valid for the Dutch local peace movement in general.

MULTI-ORGANIZATIONAL FIELDS OF THE DUTCH LOCAL PEACE MOVEMENT

Curtis and Zurcher (1973) made a distinction between linkages among organizations and multiple membership affiliations. Our presentation will make this same distinction: with regard to the former, the formal ties between representatives of local peace movement groups and local organizations are taken into account; with regard to the latter, the memberships of participants in the local peace groups in other local organizations are of relevance. The data on linkages between organizations are restricted to the local IKV-groups. The data on overlapping membership are both from the IKV-groups and the local peace movement networks.

LINKAGES WITH ORGANIZATIONS

Local IKV-groups try to develop cooperative relationships with as many local organizations as possible (cf. Klandermans and Oegema, 1987). Usually there is an agreed-upon division of labor by which some of the group members specialize in contacting specific organizations, as a rule related to their affiliations to other local organizations. IKV-groups are engaged in two types of relationship with local organizations. First, *ad hoc* coalitions in peace platforms are related to the mobilization for collective action, like peace demonstrations and the public petitions. Parties from the Left and center-Left also usually participate in the platforms, together with a changing variety of other local organizations. Second, relatively enduring collaborations are based on some joint undertaking. The data in table 7.1 refer to both types of relationship.

Among the 115 peace activists in our IKV-group sample, 58 percent (66 of 115) had engaged in "frequent" or "very frequent" contacts with other local organizations on behalf of the IKV group during the six months before we conducted our interviews. These 66 individuals were maintaining 109 linkages, or an average of 11 ties for each of the 10 IKV groups in the sample. Table 7.1 presents the distribution of these ties over distinct categories of local organizations. Twenty-four ties exist between IKV groups and *ad hoc* peace platform groups. Since local branches of political parties comprised a large number of the groups participating in these platform groups,

Table 7.1 Ties to other local organizations among IKV-group members, 1985

Relations with	*Number of ties*
Ad hoc relations in peace platforms	24
Churches	22
Unions	15
Other peace movement organizations	13
Political parties	10
Educational organizations (schools)	6
Health organizations	6
Youth organizations	4
Immigrant organizations	4
Neighborhood organizations	3
Women's organizations	2

Table entries are the absolute number of 'frequent' or 'very frequent' contacts between a member of the IKV-group and other organizations, among the 66 respondents having such ties.
Source: Sample of IKV-activists by Free University Research team (Klandermans and Oegema, 1987).

political parties ranked relatively low as separate organizations with which peace groups had contact. Among the 72 linkages outside the platform, 47 were with major established socio-political institutions, churches, unions, or political parties. The remaining 25 linkages were with a variety of local groups dealing with social welfare, neighborhood issues, and specialized interest groups (youth, women, elderly, immigrant). The prominent position of churches in the multi-organizational field of local IKV groups is a consequence of the prominent position of Dutch churches in the peace movement in general, as well as of the specific position of the IKV as the core organization of the Christian tendency within the peace movement. Of interest for our present discussion is the fairly high proportion of linkages with unions, since unions are usually viewed as the epitome of "old" movements. Linkages with political parties are somewhat underestimated, since parties are represented in the platform. It is predominantly parties from the center-Left and Left that groups have contact with.

The picture that emerges from table 7.1 demonstrates that the IKV groups occupy a far from isolated or marginal position in the multi-organizational field of their community. Linkages are both with traditional (like churches, unions, or political parties) and with non-traditional organizations (like neighborhood organizations or women's organizations), institutional (like schools, churches, or health organizations) and non-institutional organizations (like youth organizations and immigrant organizations). Not all these relations are equally supportive, and some are rather bargaining relations with opponents, like the relations with some of the political parties.

Whereas table 7.1 presented the number of relations representatives of

Table 7.2 Affiliations to other organizations among IKV-group members, 1985

Organization	*No. of affiliations*
Unions	13
Church-related organizations	16
Social welfare/neighborhood organizations	11
Other social movement organizations	20
Political parties	20
(N of total affiliations)	(80)

Table entries are the absolute number of formal memberships in other organizations among the 50 peace movement members having such affiliations.
Source: Sample of IKV activists collected by Free University Research team (Klandermans and Oegema, 1987).

IKV groups had with organizations in their community on behalf of their group, table 7.2 summarizes the affiliations of IKV group members to other organizations in their community. Over 40 percent of the IKV-group members were members of one or more local social or political organizations in their community, two-fifths of them even in some managerial position. Together these 50 people were responsible for 80 mutual affiliations among IKV groups and local organizations. A quarter of these overlapping memberships were with other social movement organizations, that is, within the social movement sector of the multi-organizational field. The remaining three-quarters of multiple affiliations were within the non-movement sector: political parties, church-related organizations, unions, social welfare and neighborhood organizations each take a share.

Altogether, the basic argument of this chapter finds clear support: local peace movement groups are embedded in the multi-organizational fields of their community.

OVERLAPPING MEMBERSHIP

In our theoretical introduction it was argued that major parts of the alliance system of social movement organizations consists of other movement organizations. This is clearly illustrated by the figures in table 7.3. The table presents the percentages of respondents of both studies who sympathize with or are active in six contemporary movements. As a standard for comparison, figures from a national sample are also given. The results almost speak for themselves. Compared to the national figures, exceptionally large proportions of the members of IKV-groups and local peace networks sympathize with or are active in other movements. At the attitudinal level, four movements – the anti-nuclear power movement, the environmental movement, the anti-racism/fascism movement, and the third world move-

Table 7.3 Sympathy and activity in other social movements by peace movement activists, 1985 (percentages)

	IKV sample		*Kriesi sample*		*National sample*	
Movement	*Sympathize*	*Active*	*Sympathize*	*Active*	*Sympathize*	*Active*
Anti-nuclear power	–	–	98	50	28	8
Environmental	89	18	99	35	50	9
Women's	60	16	96	33	24	6
Anti-racism/fascism	89	24	–	–	–	–
Third world	100	30	–	–	–	–
Squatters'	–	–	40	10	5	2
(N)	(115)		(144)		(579)	

Dashes indicate no data.
Source: Sample of IKV activists collected by Free University Research team (Klandermans and Oegema, 1987); Kriesi sample of peace activists (Kriesi and van Praag, 1987); national sample from yearly socio-political survey from University of Amsterdam.

ment show an almost complete overlap. Almost every activist in the peace movement sympathizes with each of these movements. In the case of the women's movement, the two samples appear to be somewhat different, but even in the IKV-sample the overlap is considerable. The squatters' movement is more distant, but there we also see a strong overlap compared to the figures from the national sample. Moving from attitudinal to behavioral support, the figures are even more striking. Large proportions of the local peace movement activists are active in other movements as well. Because of differences between the two questionnaires, the data from the two samples are not completely comparable; but for our objective it suffices to conclude that local peace movement organizations are connected with other movement organizations through a tight network of multiple affiliations by their activists.

The same can be said with regard to political party membership (table 7.4). Party membership in the Netherlands must not be confused with party preference or voting behavior. Party members are fee-paying affiliates of the party organization. In a sense, party members are the activists of the political parties. Of a population of 14 million in the Netherlands, 3 percent are members of a political party. Compared to these figures, an astonishing proportion of the peace movement activists are members of a political party: almost one-fifth of the IKV-activists and exactly half of the members of the local peace networks. Clearly, the activists of the local peace movement organization are among those who are politically active above the average.

The figures in table 7.4 are interesting in other respects as well. First, they show a significant but understandable difference between the two

Table 7.4 Party membership among peace movement activists and the Dutch public, 1985 (percentages)

	IKV sample	*Kriesi sample*	*National sample*
Member of party	17	50	3
Party preferences among members			
PvdA (Labour)	25	39	24
PPR (Radicals – Catholic)	50	26	2
CPN (Communist)	0	13	0
PSP (Pacifist socialists)	0	13	2
EVP (Radicals – Protestant)	20	6	0
CDA (Christian democrats)	0	0	35
VVD (Liberals – conservative)	0	0	23
Other party	5	3	13
Total	100	100	99

Source: Sample of IKV activists collected by Free University Research team (Klandermans and Oegema, 1987); Kriesi sample of peace activists (Kriesi and van Praag, 1987); national sample from yearly socio-political survey from University of Amsterdam.

samples: As they belong to the Christian tendency within the peace movement, IKV-group members are primarily affiliated to the two Leftist parties with a Christian origin (PPR and EVP). Members of the local peace movement network, in contrast, are affiliated to Leftist parties with socialist or social democratic credentials, although the radical Christian parties are well represented there as well. Second, overlapping membership with political parties restricts itself completely to the parties from the Left. There are *no* mutual affiliations with the two major parties of the Right (CDA and VVD), although more than half of the population of party members is affiliated to either CDA or VVD. Evidently, as far as overlapping membership with the political parties goes, the alliance system of the peace movement restricted itself to parties on the Left of the Dutch political spectrum.

As they reveal the linkages with the political system, the data on political party membership confirm our observations with regard to the linkages of the local peace movement with local organizations. However, the local peace movement was embedded in a specific part of the political system. Only political parties from the Left belonged to the alliance system of the peace movement organizations. Polarized as Dutch society was on cruise missiles, the parties to the Right obviously belonged to the conflict system of the movement. To a large extent the controversy on cruise missiles was reproducing the Left/Right cleavage in Dutch society. This was confirmed by data on party preference and voting behavior stemming from several national surveys (Kriesi and van Praag, 1987; Schennink, 1988).

Summary and Discussion

Social movement organizations emerge from multi-organizational fields. Through linkages and overlapping membership they are solidly integrated in the organizational networks of their community. I see no reason to assume that new social movement organizations are different from "old" social movement organizations in this regard, although there might be differences in degree. Instead of discontinuity and detachment from the political and social system, I have stressed the inclusion of new social movement organizations in cyclical patterns of expansion and contraction of alliance and conflict systems in multi-organizational fields. In this analysis, alliance and conflict systems have different functions to those of social movement organizations. Whereas alliance systems serve to create the resources and opportunities to be deployed by the social movement organization, conflict systems serve to drain resources and opportunities and to generate bargaining structures. Alliance and conflict systems can absorb a smaller or larger subset of the multi-organizational field, depending on how polarized a community or society is on the issues the social movement organization puts on the agenda. Outcomes depend on resources and opportunities available to each of the two systems.

Neither this analysis nor the empirical evidence on the Dutch peace movement necessarily mean that there is nothing "new" about contemporary social movements. People involved in movements have a background in conventional political groups and traditional social organizations and will exploit these ties on behalf of their movements. It does mean, however, that at least in the case of the Dutch peace movement "what is new about new social movements" cannot be defined in terms of marginality or detachment from existing social and political institutions.

The cleavage between a social movement organization's alliance and conflict systems may coincide with other cleavages in a society, for instance the Left/Right cleavage. This makes it much more difficult to arrive at some compromise. If, as in the case of the Dutch peace movement, the parties dominating the conflict system of a movement organization are in power, it is extremely difficult to exert any influence on policy-making.

During a protest cycle, resources and opportunities created by one movement organization's allies become available to other movement organizations. Overlapping memberships and interlinkages between movement organizations play an important role in this regard. Old social movements, like unions, can, and as shown by our study of the Dutch peace movement, in fact do become part of the alliance system of new social movements. New social movements use the resources and opportunities provided by old movements. Old movements revive in the wake of the protest cycle using resources, innovations and opportunities created by new social movements. As a consequence, the distinction between old and new social movements is blurred.

Although our empirical evidence stems from Dutch findings, there is no reason to assume that these results are exceptional. Admittedly, the Dutch political culture is different from that in other countries. But similar findings are reported on the Italian, Swiss, and French new social movements (Kriesi, 1985; Diani and Lodi, 1988; Rucht, 1989; Tarrow, 1989). To be sure, one may find differences between countries in the make-up of a social movement's alliance and conflict systems, and in the way they divide a multi-organizational field into supportive, opposing, and indifferent sectors. It might even be true that new social movements rely on a different blend of supporting organizations from those of old social movements. But, rather than questioning whether new social movements in countries other than the Netherlands are embedded in multi-organizational fields, it is more instructive to compare the characteristics of the multi-organizational fields they are embedded in, and to find out what difference varying characteristics make to a movement's fate. Despite the differences that inevitably will be found, the principle will remain the same, however: new social movement organizations like every movement organization are part of the social organization of their communities. They develop in multi-organizational fields, wherein supporting and opposing structures help to shape the movement's structure and ideology. By defining contemporary social movements as detached from the social and political organization of society, new social movement literature can easily overlook important aspects of new social movements' reality.

8

Feminism and Political Action

JOYCE GELB

The 1960s and 1970s saw the resurgence of feminism as a social movement in virtually every Western nation. However, in each nation the movement has adapted to the history, culture, and politics of its own society (Bouchier, 1984). This chapter will focus on the ways in which culture and political institutions have shaped feminist movements in Britain, Sweden, and the United States. As Helga Hernes (1983, p. 33) has pointed out, women in the state occupy similar roles regardless of national boundaries: they are citizens, consumers, clients, and employees. But it is also important to add that they are claimants in the arena of public policy, seeking to gain economic and political rewards and contesting for power and access.

This chapter primarily examines the mechanisms and impact of women in their role as political claimants, helping to identify and influence policy with regard to their lives. The study contends that differences in the "political opportunity structure" or institutions, alignments, and ideology of each nation, structure the development, goals and values of feminist activists. In turn, it will be argued that movement structure and systemic differences have affected and constrained opportunities for movement impact within each nation. While the feminist movements in the United States, Sweden and the United Kingdom share many joint objectives, they differ significantly with regard to styles of political activism, leadership orientation and organizational values (Jensen, 1983). These differences interact with contrasting political opportunities to shape the success of feminist claims.[1]

Central to the analysis presented here is the question of how political alternatives are shaped by the role of women activists themselves. We focus on the role and structure of feminism in the nations studied by examining two "faces" of feminism. The first is the Women's Liberation Movement, often characterized as the more radical or "younger" branch of the movement. Such groups in both the United States and England are distinguished by their emphasis on life-style change, provision of alternative services, and decentralization and anti-elitist values (Stacey and Price, 1980, p. 180).

The second segment of the movement operates within the traditional policy-making structure. In the case of Britain, this sector of the movement participates in parties and unions which play a dominant role in the British political system. In the United States, although we examine the role of women in parties and unions for comparative purposes, the emphasis is on women as interest-group activists, given the primacy of pressure groups in the American political arena. These two sets of feminist activists are roughly analogous to what Jo Freeman (1975) has called the "militant" and "reformist" branches of contemporary feminism, although many British feminists would decline the honor of inclusion in the reformist camp. In Sweden, there is only one manifest face of feminism, that represented by women in parties and unions; the minute feminist movement is almost subterranean in character and visibility.

The three countries – Britain, Sweden, the United States – were chosen because they are all post-industrial Western democracies, with policies that are superficially similar with regard to women's rights. Each has legislation related to abortion rights and labor force equality (Sweden was the last to adopt the latter) and each has set up an administrative Equal Opportunities Commission to monitor sex discrimination laws, albeit with very different powers. While all three societies have experienced similar trends, they differ with regard to their significance for feminist politics. For example, while all three nations have experienced the simultaneous mobilization and incorporation of new participants into the political process, and the expansion of the tasks of the modern state, particularly as these relate to women, these processes have proceeded in different ways and with different impacts in each (Hernes, 1982, p. 7).

In order to analyze the role of women's movements in helping to structure policy alternatives and outcomes, we will contrast three different models of women's participation and activism in this analysis: the first is *interest-group feminism* in the United States; the second, *left-wing/ideological feminism* in the United Kingdom; and finally, *state equality* in Sweden.[2]

Interest-group feminism is characterized by a relatively open political system, a focus on equal rights and legal equality (although many demands may go beyond mere reformism) and the creation of lobbying groups which may have a mass membership or be staff-dominated. Networking and inclusiveness typify the approach to different political orientations within the women's movement.

Ideological or left-wing feminism is characterized by insistence on ideological purity and a reluctance to work with groups espousing different viewpoints. This type of feminist politics is decentralized and locally based, largely lacking a national political presence and impact. Fragmentation as well as enthusiastic commitment to sectarian (feminist) views are characteristic of this model.

State equality is characterized by the absence of a visible and influential feminist movement. Instead, women are active via political parties and, to a lesser degree, trade unions (true to a lesser degree in left-wing feminism, as an alternative to local political action). The state has tended to anticipate or coopt women's concerns into public policy, usually without discernible pressure from women's groups. Policies related to women are generally discussed within the framework of "equality" or "family" policy.

While these models may be transferable to other political cultures, the idiosyncratic nature of each movement and its interplay with national politics and ideology prevents us from generalizing beyond the three cases addressed in this article. However, it may be suggested that of the three models discussed here, the one closest to being a "new" social movement is the British, because it alone represents an innovative approach to organization and participation. We will return to analysis of the distinction between "old" and "new" social movements as they relate to the three cases discussed here in the conclusion.

Crucial variables affecting movement emergence and activism are the degree of corporatism-pluralism and political centralization. Schmitter's well-known definition characterizes corporatism as interest representation in which constituent units are organized into a limited number of singular, non-competitive, hierarchically ordered and functionally differentiated categories, recognized or licensed by the state and granted a deliberate representational monopoly within their respective categories (Schmitter, 1984, pp. 85–131). This system is most characteristic of Sweden, while Britain is a more weakly structured corporatist or "tripartite" system. In contrast, in the American system, strong or strongly structured pluralism prevails.

The remainder of this chapter contrasts the organizational styles of the movements in each of the three nations under consideration. We examine the nature of the coalitions and alliances which have been developed in order to gain resources, access, skills, and members. We look briefly at the strategies employed by each movement, drawing some conclusions about the efficacy of autonomous versus integrative approaches in relation to the government, parties, and other institutions. Finally, we discuss the conditions for the success of feminism in these three countries, hoping to illuminate the extent to which an essentially transformational movement may influence its members and the larger society and political system, and answering the question of what is "new" about this new social movement.

Movement Politics in Three Nations

Social movements may be described as movements seeking to achieve social change through collective action. Directed groups – or those with more formal organizational structures – have formal leadership structures, definitive ideology, stated objectives and specific programs, while non-directed movements stress a reshaping of perspectives and values through personal interaction (Hanmer, 1977, p. 92).

To a large extent, the most visible portion of the American feminist movement belongs to the "directed" category, at least in terms of formal leadership structure and definite programs of action. To a greater degree, British feminism is of a more "non-directed" nature, with greater emphasis on personal interaction, expression and articulation of feminist values, and the importance of internal democracy.

Coalitional activities across issue and ideological lines are less well developed in the United Kingdom than in the United States, while opportunities for British feminists as political activists are constrained by the values inherent in a traditional society, a declining economy (until recently) and a highly centralized, closed political system. Hence, interest-group politics are not the primary avenue for feminist involvement in Britain. However, the absence of formal, structured coalitional groups comparable to those in the United States should not obscure the existence of alternative groups actively involved in other aspects of political life.

In both the United States and the United Kingdom, feminist movements operate in an autonomous manner, creating gender-based organizations. At the same time, they interact with government, and political parties and unions. In Sweden, feminist movement activity *per se* is limited and women's organizations tend to gravitate toward political parties. An effort to create all-party gender-based groups is evident in recent years, and *ad hoc* activity (including demonstrations) takes place around such issues as prostitution and pornography and against efforts to tighten abortion restrictions.

THE CASE OF BRITAIN

Movements may be viewed along a continuum from unconventional, almost spontaneous, sometimes illegal activity to movement groups that seek more conventional political and legal change. Social movement organizations are the acting components of the movement, which may help to "turn grievances into program" (Lowi, 1971). While American "women's liberation" groups have moved closer to an accommodation both to the political system and to the more "reformist" wing of the movement, largely because of an emphasis on coalition politics and pragmatic achievements (goals), no such tendencies are evident in the British movement. Rather, there has been little institutional transformation since the movement's inception in Britain, although constant changes, proliferation, and fragmentation occur due to conflicts regarding strategies and ideology (Banks, 1981, p. 227).

The most active part of the British feminist movement emphasizes expressiveness, personal transformation, consciousness, and changed belief systems. It eschews formal structure and hierarchy and is centered in small groups which stress life experience and self-help politics. In Gerlach and Hines' (1970) terms, it is segmented – localized, autonomous, and ever-changing – and decentralized. Nonetheless, it lacks the *reticulate* or networking structure which they see as inherent to movement groupings. Largely as a consequence of ideological conflicts and the consequent failure to coordinate action, as well as permit the sharing of resources, networking efforts, particularly at the national level, are absent. Feminist movement organizations in the United States are far more likely to form coalitions (with other women's groups and more traditional interest-group activists) in order to realize their goals (McCarthy and Zald 1977, pp. 121–4).

It appears accurate, as Zald and Ash (1966) suggest, that movements adopt different forms depending on their goals, with personal change movements adapting decentralized structures with exclusive membership, while institutional change movements are typically centralized and inclusive (also see Curtis and Zurcher, 1974). While goals of grassroots participation, service, and goal transformation are more likely to be realized in decentralized structures, the costs may lie in the failure to influence the larger political system. Bureaucratic structures provide skill and coordination but may sacrifice participatory goals. One trend in the United Kingdom which may effectively interface with the movement structure and values that predominate among women's groups is the growing interest of local council governments in aiding feminist efforts.

LOCAL WOMEN'S COMMITTEES – A NEW DIRECTION IN BRITISH POLITICS?

Recent efforts at the local level of British politics suggest that in some instances it may be possible for feminist women to work closely with Labour councils and local authorities. New developments in numerous British urban areas have interfaced effectively with the decentralized structure of British feminism. Since 1969, a variety of multifaceted action groups have developed in many British urban centers, centered around feminist collectives, rape crisis and battered women's shelters, health clinics, black women's groups and groups of women in law, media and other professions (Bouchier, 1984). Feminists have turned to local government for funding, access and space, in order to maintain their activities; and (fearing hierarchy and male cooption) have tended to prefer local-level dialogues to those at the national level. Aided in part by the increased number of local councillors who are women (18.4 percent) (Equal Opportunities Commission, 1983), several local councils have created women's committees to address the representation of women's interests (Goss, 1984, pp. 109–32). Since 1982 such committees have been established by the Greater London Council (GLC) (since disbanded), numerous London boroughs including Camden, Islington, Southwark, and Hackney, and in 22 other British urban centers (including in Scotland). Virtually all of these constituencies are under Labour party domination. In some communities, support for women's committees came from local councillors (e.g., London), while in others the initiative for establishing a commission emanated from the women's movement itself. The GLC committee sought to involve a wide spectrum of women by holding open meetings and coopting women to represent such groups as lesbians, the disabled, and trade unionists (Flannery and Roelof, 1984). Funding was given to Greenham Common women for child care and health facilities; all to a generally hostile press reception (Goss, 1984). A major thrust of the GLC women's committee was to foster multi-action centers for women, thus bringing some order to the centrifugal politics we have described, perhaps building on the women's action centers which already exist in many British cities (*Spare Rib*, February 1985). At the GLC, efforts were made to emphasize participatory democracy and openness through public meetings. These were met with enthusiastic response and large attendance (interview, Ken Livingstone, 1984). Meetings were open to all interested and tasks rotated, in accordance with feminist theory. Despite media emphasis on funding of "radical" projects, in fact child-care subsidies comprised the bulk of GLC funding efforts.

Other women's committees have been established in the London boroughs of Camden and Islington, with varying degrees of success and funding. In Islington, as early as 1984, the Women's Unit was all but defunded and defunct (interview, Hilary Potter, 1984). The Camden group, with strong support from local Labour councillors, has devoted considerable funding to women's efforts, although in recent years cutbacks have occurred. As of 1988, a Women's Bus which travelled around the community to focus discussions and provide information and advice, circumventing the media by going directly to local women, was curtailed owing to lack of funds (Goss, 1984; Flannery and Roelof, 1984).

The functioning women's committees have been unique in creating new structures which bring women into the political process, seeking to break down traditional hierarchical processes and, through innovative systems of publicity and participation, reaching out to women who are not normally part of the political process. They also attempt to broaden the framework of service provision from a feminist perspective. Fears exist among feminists that such efforts may coopt the movement, especially among those wary of Labour party practices in the past (Goss, 1984, p. 128). In fact, however, these committees may act as a bridge between the movement and labour and governmental structures, if they are successful.

A major drawback to these promising efforts is the attack of the Tory government on both the GLC and Left-dominated local London councils (of which 12 had women's committees) (*Spare Rib*, February 1985). The GLC was dissolved in 1986 by the Thatcher government. After the demise of the GLC, the London Borough Grants Scheme and the City of London Trust have continued to fund numerous women's services, although the close political ties between the movement and local political structures which existed previously have been severed, in some instances irrevocably. While the trend toward local-level support for women's efforts in Britain is still too new to assess fully, particularly in a period of flux, these actions signify efforts to provide linkages between traditional, socialist, and radical feminist groups and to reach out to the vast number of British women who were previously unaffected by the feminist movement.

Clearly, the British movement has succeeded in creating local activities emphasizing consciousness and life-style transformation in numerous (primarily urban) centers throughout the country. British feminism also defies traditional sociological rules regarding the origins and maintenance of social movements. Rather than becoming bureaucratized and less radical, the British movement retains its ideological fervor and commitment and continues to seek new alternative structures (Bouchier, 1984, p. 179). The degree of activism and commitment is impressive, even to the casual observer. Movement groupings and activist outposts created in the 1970s survive. These include Women in the Media, the National Abortion Campaign, the National Woman's Aid Federation, Rights of Women (the voluntary legal arm of British feminism) and the Women's Rights unit of the National Council for Civil Liberties (now renamed Liberty), as well as Women Against Violence Against Women (WAVAW) (Van der Gaag, 1985). As Jensen (1982) observes, the movement's main contribution may be its survival, in contrast to the largely esoteric state of the formerly active French and Italian (as well as other continental) movements. The British movement remains the most vital and important one in Europe, although the numbers involved in some grassroots groups may have declined since the 1970s (Coultas, 1981, p. 36).

While the British movement's origins were in many ways similar to that in the United States (even prompted initially by infusions of Americans who served as catalysts), the British movement has never been characterized by either mass demonstrations (other than the mass effort to defeat the anti-abortion Corrie Bill in 1979) or a strong national presence. Internecine struggles within the movement, largely between sectors of radical and socialist feminism, remain unresolved. One effort in conjunction with the TUC, to fight restrictions in the Abortion Act, both demonstrated the mass potential of the movement and also its internal conflicts. This march (against the Corrie Bill) was probably the largest demonstration around a social issue that Britain has seen (Bassnett, 1986). Nonetheless, it also reveals the chasm between women and the trade union movement and between socialist and radical feminists: the latter resented what they perceived as male usurpation of the movement. A feeling of hostility was thus one result of the day's effort, though not the only one. The Greenham Common peace effort three years later resolved some of these conflicts through a separatist women-only effort. Thirty thousand strong, it helped to bring women from all groups together and gain widespread attention for the issue of nuclear proliferation. A major question for British feminism remains the centrality of "cultural" versus "political" feminism.

In the United Kingdom, parties and unions occupy a major – if declining – political role and a tradition of Left/socialist thought is strong. While activity in party and trade union politics has been viewed by some as equivalent to the American liberal/equal rights movement, little evidence for this perspective exists. This analysis is thus in considerable disagreement with the view expressed by Hewlett, who argues that in both England and

Sweden the most effective women's groups are embedded in parties and unions, not in separate feminist organizations (Hewlett, 1986, p. 170). Rather, it appears that women's participation in established British institutions is marked by marginalization, with women organized into separate advisory groups and limited to a handful of mandated seats on executive committees. The major union force, the Trades Union Congress (TUC) endorses numerous progressive policies on behalf of women, lobbies for them and even demonstrates on their behalf (especially in the case of abortion, referred to above). Nonetheless, where issues of economic and political power are involved, there is greater hesitancy. In addition, tensions exist between socialism and feminism and between the hierarchical unions and parties and feminist ideology, as the march against the Corrie Bill also demonstrates. However, at their most effective, women's groups within parties may serve as forums through which women's demands and concerns may be highlighted. But Jensen's (1982) contention that British feminism has emerged from a decade of political isolation and an exclusive grassroots focus to help unite the Left and the Labour party seems premature, based on the analysis presented here. While we have commented on the energetic activity of women within Labour and the organized Left, the policy or decision-making impact has thus far been limited.

Despite the existence of numerous progressive policies related to women in the United Kingdom, the absence of a feminist movement which can set a policy agenda, speak for itself, and engage in dialogue regarding specific policy initiatives, resulted in serious gaps between policy and implementation. The product is a strangely limited vision of feminist goals and ideals which, in fact, produces less societal change than might be expected.

THE AMERICAN MOVEMENT

In the United States, a tradition of reform, the absence of a strong socialist Left, and the significance of interest groups in decision-making combine to produce a different type of movement. In part reflecting the increased weakness of parties and unions politically, feminists have organized as separatist or gender-based groups outside established structures (Adams and Winston, 1980). This gives them significant autonomy in terms of strategy, as a recent trend in the direction of electoral efforts demonstrates. American feminism is characterized by far greater diversity and inclusivity of different views than its British counterpart; coalition-building and networking are movement watchwords. The American movement has also forged an accommodation between the more "radical" women's liberation movement and the middle-class reformist one (it has also developed strong linkages with traditional women's groups, such as the League of Women Voters and the YWCA).

Because political parties are less central to the political process in the United States than in the United Kingdom, feminist interest groups such as the National Women's Political Caucus (NWPC), the National Organization for Women, and other groups play an important role in recruiting women for political office, providing training and some campaign support, and actively campaigning for key political issues such as abortion rights and the Equal Rights Amendment. As a result of the Democratic party rules change in 1972, women in that party have achieved virtually equal representation with men (even in the Reaganite Republican party, unaffected by rules changes, women comprised some 48 percent of the delegates in 1984) (Freeman, 1988, pp. 215–46). In 1988, 52 percent of the delegates to the Democratic convention were women in comparison to 38 percent at the Republican convention (down significantly from four years earlier). The Democratic delegates reflected continued attention and support for the feminist agenda, despite a diminished visible presence for organized groups at the 1988 Democratic convention. Seven out of ten Democratic delegates said they thought the federal government was not paying enough attention to the needs and problems of women. Seventy-two percent of Democratic delegates (in contrast to their Republican counterparts) expressed support for continued legal abortion (*New York Times*, July 17, 1988, p. 16; August 14, 1988, p. 32).

Particularly striking is the increasing significance of feminist interest groups within the Democratic party over the past decade. Jo Freeman (1985–6, p. 87) has pointed to the different political cultures that dominate the Republican and Democratic parties. In the Republican party power flows from the top down and delegates' relationship to party leaders and loyalty to the party itself tend to be significant. In the Democratic party, in contrast, constituencies are seen as the party's building blocks, and power flows from the bottom up. These distinctions help to explain the important role feminist groups have gained within the Democratic party.

In the Democratic party, because of the different structure and mechanism for representation, feminist groups have a major policy decision-making role. By 1980, over 20 percent of the Democratic delegates were members of the National Organization for Women (NOW) or the National Women's Political Caucus (Freeman, 1987). They won convention approval for a proposal to deny Democratic party funds to any candidate who did not support the Equal Rights Amendment (Minority platform report). Like the National Women's Political Caucus, NARAL (National Abortion Rights Action League) had its own floor operation and lobbied successfully for another minority plank (no. 11) supporting government funding for abortions for poor women (Freeman, 1987).

Therefore, by 1984 feminists had successfully demonstrated their political clout. In that year, NOW met with five of the Democratic presidential candidates and pressed them on their support for women's issues, appointments, and willingness to select a female vice-presidential candidate. NOW Vice-president Mary Jean Collins was appointed to the platform drafting committee and given the leading role on matters of concern to NOW. A coalition of women's groups presented a list of acceptable female vice-presidential candidates to Mondale (who had earlier received the endorsement of NOW and later that of NWPC and NARAL as well). After the nomination of Geraldine Ferraro, a committed feminist, the feminist coalition had little to do at the convention itself.

By 1988 women were "insiders" at the Democratic convention; the fifteen associations that formed the "Women's Central" held a women's caucus every day of the convention. Prior to the convention, and during it as well, alarmed over concern that the Democratic National Committee wanted to soft-pedal issues such as abortion and the Equal Rights Amendment, representatives from groups such as NOW, the NWPC and AAUW (the American Association of University Women) met the committee chair and persuaded him that to omit these issues would be more dangerous to the party than to include them. Subsequently, these women's groups and others met continually to ensure that strong statements on these issues and other key policies, including pay equity and family leave, as well as the full and equal access of women to elective office and party endorsement, were included in the platform (Freeman, 1988, pp. 875–7, provides a full analysis of these strategies).

In contrast to 1984, when NOW had been a prime mover behind the vice-presidential nomination of Geraldine Ferraro, women's groups were less visible in 1988. In part, this was a response to the fear of party leaders, who had been accused of pandering to "special interests", especially women, by Republican campaigners in the past. Thus while the party has opened its inner sanctum to women and listens to feminist groups on a variety of key issues, continuing pressure to ensure support is apparently necessary in coming election years.

Despite an increased role in Democratic party decision-making, the most visible manifestation of American feminism remains the traditional interest group, organized with hierarchical structure and staff dominance. Groups such as NOW have moved in the direction of mass membership, while such feminist groups as WEAL (Women's Equity Action League), Center for Women Policy Studies and the National Women's Political Caucus fit the McCarthy and Zald (1973, p. 11) model of funded social movement organizations which rely on "conscience constituencies" or contributors for resources and staff for day-to-day decision-making and long-term strategizing (also see Handler, 1978, p. 8). As the history of the movement against domestic violence demonstrates, even non-traditional groups with grassroots origins are pulled in the direction of political engagement and greater professionalization. American feminists are eclectic and pragmatic in their use of strategies – from protest to litigation and campaigning. Because in the United States, in contrast to Sweden and the UK, political power is diffused, more opportunities for different types of activism are available.

American feminists have been far less reluctant than their British sisters to engage with political and bureaucratic forces and seek legitimacy; they have emphasized networking, coalition-building, lobbying, and legal change in order to gain resources and reform legal and other procedures.

While protest has never been a central strategy for American feminists, mass rallies such as the 1970 "Women's Strike for Equality" brought together disparate elements of the movement in various cities around the nation, in order to advocate child care, abortion on demand, and equal educational and employment opportunity. More recent demonstrations of militancy include the 1986 march in Washington in support of reproductive freedom; about 85,000 people participated in the march and the April 1989 march for abortion and women's rights which attracted half a million demonstrators. However, protest has often seemed unnecessary, because of the perception that channels of access to the political system were opened to feminist influence with a minimum of difficulty in the 1970s and 1980s.

Because of the unique character of the American judicial system, with its powers of judicial review and the ability to provide remedies for an entire "class", the role of litigation has been far more important for American feminists than their European counterparts (see Gelb and Palley, 1987). Litigation has been seen as a mechanism for developing case law and precedents for an entire class of women and a consciousness-raiser for the public and political system, even if the case is lost. Litigation is also used as a tool to prod the rest of the political system into greater action regarding enforcement.

Through "modified collectives" (now adapted by at least one British feminist group as well, Women's Aid), local-level feminists have sought to balance external imperatives and ideological principles. Often, from local-level efforts, state and then federal coalitions have been created. Coalitions often include traditional women's groups, radical feminists, and equal rights advocates, as well as other supporters in a particular issue area (Gelb and Palley, 1987). Nonetheless, although American feminists have moved much further in the direction of traditional group organizational structure (which most reformist lobbying groups have always utilized in the United States), they have not abandoned feminist policy goals in the interest of organizational maintenance. The movement has successfully altered the agenda of policy, at the state and local level in particular, and has been recognized as a "legitimate" interest group of political significance, while deepening its commitment to key feminist issues and concerns.

THE SWEDISH MOVEMENT

The feminist movement *per se* plays a minor role in Swedish politics, although issues related to "equality" and "family" are given primacy by the Social Democratic party and other parties. Women's federations in four of Sweden's five parties provide a context for some policies related to women and for recruitment of women to party and political office. In all five major Swedish political parties, including the dominant Social Democratic party, the Left Communists (VPK), and the more centrist parties, women have achieved a degree of influence, although party structures differ and women's concerns are not in the forefront of party concerns and/or platforms. While public opinion data suggest that women may be dissatisfied with aspects of efforts to achieve "equality" (78 percent in Sweden, 94 percent UK, 58 percent US) (Hastings and Hastings, 1985, pp. 707–17), at present no mechanism appears to exist to operationalize such discontent and move beyond the egalitarian rhetoric that is such a dominant feature of society to a new stage of political endeavor on behalf of women.

Women in Sweden have gained an unusually high degree of representation in the political system. Women comprised 31 percent of members of parliament in 1985, with a tremendous increase in electoral representation occurring in the 1970s; about one-third of those elected at the county and municipal levels are women (Eduards et al., 1986, p. 6). It may be only somewhat cynical to note that women have achieved legislative representation in a system in which power is now centered in the corporatist system and bureaucracy, sectors in which women have not made comparable gains. In any event, women are absent from most leadership roles and their public visibility is often limited as well (Adams and Winston, 1980, p. 141). Women office-holders tend to be given responsibility largely in areas related to traditional 'female' concerns such as education and social welfare policy; even in these areas they usually have peripheral as opposed to central decision-making positions (Eduards et al., 1986, pp. 6–7). Also, women in government have generally supported their partisan loyalties rather than gender concerns once in office, in keeping with the structure outlined above. Female party leaders' attitudes resemble party dictates more than feminist views; they diverge on attitudes related to women's concerns as much or more than men (Kelman, 1984). While women have gradually increased their role in union offices over the past two decades, in the powerful LO, which has strong links to the dominant Social Democratic party, there is little evidence of changed attitudes towards power and policy for women.

Swedish women's organizations have had little impact in relation to public authorities and minimal impact on public policy *per se* (Eduards et al., 1986, p. 15). Gender is not seen as an explicitly conscious, deliberate, or legitimate dimension of politics in Scandinavia (Scott, 1982).

In contrast to the US and Britain, a women's liberation movement has existed in only a minimal way in Sweden. Although the militant and active Group 8 was created in the 1960s, it never developed into a strong and coherent movement with major influence in Swedish politics. Most feminists were concerned with class rather than gender (Eduards, 1981) and the struggle for women's concerns was conducted primarily within the political parties and other political institutions. Militant feminism was unacceptable in Sweden's consensus-oriented society. Neither consciousness-raising nor alternative political structures, nor theoretical and cultural feminism, ever became strong forces within the Swedish political framework, except among a few activists. Nor did new approaches to service delivery (e.g., rape crisis activity or domestic violence shelters) gain a large number of adherents.

Thus, despite Sweden's progressive policies regarding women, there is little room for experimentation and grassroots efforts in a society in which virtually all concerns are encompassed by the state. Issues of male power and violence have only begun to be raised in this state so devoted to the establishment of equality in a sex-neutral fashion (see Gelb, 1989). In 1973, Rita Liljjestrom wrote that if women are to achieve collective liberation, they "need to rally around a community of values, around a program which roots them in shared experiences, and which gives them political identity for "sisterhood" and an alternative value system to keep them from being devoured by equality under the terms set by the male value system" (Hernes, 1984, pp. 32–3). This has not occurred in Sweden, nor has there been a public dialogue on issues related to women conducted by women on *their* terms. As Baude (1979) and others point out, "women's liberation and/or feminism" are viewed as confrontational and "anti-Swedish." Another writer aptly suggests that:

> When conflicts have arisen in Sweden and demands are put forth by different groups, they must be swallowed quickly by the all-encompassing Social Democracy. Social Democracy must show without delay that it is also capable of coping with the new demands emanating from women. In other words, the special interests of women have been subordinated to the general good, the definition of which still lies in the hands of men . . . Reforms can also be considered as a preventive measure – one which is necessary to divert more radical demands and conflicts which could lead to a more serious split. (Register, as quoted in Eduards et al., 1986, pp. 148–9)

In this type of system, significant issues of male–female relations and power are never seriously discussed. Autonomous women's organizations are viewed as suspect and women have difficulty shaping policy on their own terms.

Patterns of Action

The analysis presented here leads to several conclusions relating to the role of feminist movements in comparative perspective. First, our findings underscore the significance of systemic factors, or the "political opportunity structure", in encouraging the particular style and significance of movement mobilization and development. With Klandermans (chapter 7 in this volume), we stress the degree to which linkages to other networks help to create opportunities for political access and resource acquisition for movement activists in the three nations examined. Such linkages seem most advanced in the United States and Sweden. In the case of the United States, links to other feminist groups, on single issues such as abortion, as well as to traditional women's groups and other group activists, have been created. Coalitions take either an *ad hoc* or a more permanent form depending on the issue. In the United States, the movement's dynamic was restored as a result of threats from the Right in the aftermath of Reagan's victory and the necessity to coalesce around the Equal Rights Amendment (and then the need to regroup after its failure to gain ratification). In Sweden, programs related to women's concerns, primarily in terms of labor force policy, were advanced first by the dominant Social Democratic party, but in recent years the centrist parties have also adopted policies favorable to women and advanced women's political representation. In Britain, history and ideology lead feminists toward energetic activity within the Labour party; such efforts have achieved their greatest success at the local council level. Unions are still reluctant allies of feminists in all three nations, although they provide auxiliary support on some policies, as the TUC's commitment to continued abortion rights demonstrates in Britain. A particularly good example of feminist penetration into existing institutions of power is evident in the relationship between feminist interest groups in the United States and the Democratic party in the period after 1972.

We may suggest that support for feminist demands regarding representation and policy commitments has occurred in large measure because the relatively unstable nature of traditional electoral coalitions in the post-war era means that women voters are perceived as a new source of constituent strength. Similarly, although blue-collar unions are generally hostile to feminist demands and claims, the declining relationship between the traditional working class and unions produces new interests in women workers. As potential recruits, women workers may be crucial to the future role of trade unions, especially in white-collar sectors in which women comprise the major (and often only) new base for mobilization. This has produced responsiveness to feminist demands for increased representation, special education, and training and child care in both Britain and the United States. In Sweden and the United States, electoral instability in the 1970s and 1980s increased responsiveness to feminist and/or women's demands. As we have seen, the Democratic party in the United States and the Swedish Social Democratic and centrist parties deal with women's demands for greater influence and policy commitments. In the case of Britain, the Social Democratic Party/Liberal Party Alliance, as a new third party seeking new sources of electoral support, made the most promises to women activists, especially in terms of increased representation. Their efforts (to date unsuccessful) have nonetheless persuaded the Labour party to take greater account of their active feminist constituency, although few actual gains have resulted.

Analysis of the relationship between Left politics, which by virtue of ideology is thought to have a special relationship to feminism and other new social movements, suggests several difficulties inherent in the supposed "natural" relationship, despite mutual attraction. Because of their paramount interest in class, as opposed to gender issues, feminism has had an uneasy relationship with Left parties. In Sweden, the Social Democrats, in part due to their close connection to the LO, the dominant blue-collar trade union, were unwilling to support the concept of equal opportunity legislation (the LO has preferred to retain its primacy *vis-à-vis* retention of the collective bargaining system as opposed to statutory enactment). Only when a centrist coalition attained electoral dominance from 1976 to 1982 was such legislation enacted; in part because women's votes were seen as a key element in the government's electoral success. In Britain, while the Left has supplied ample rhetoric related to feminist issues, its traditional base of support (largely in the trade unions) constrains meaningful commitment to feminist concerns. In Britain, only at the local level has a congruence developed between Left pronouncements and feminist aspirations. Our analysis suggests that new social movement organizations are often capable of engaging in political negotiations, because they have votes and members to offer in return for policy commitments. We find that feminists in particular and women in general are able to bargain successfully, and have the potential to do even better in the future, with political institutions which are resource-rich, but constituency-poor in the electoral arena.

The contention that new social movements are following the same trajectory in institutionalization that has characterized previous movements may be worthy of some qualification based on our analysis (see Offe, chapter 12 in this volume, for a different treatment and analysis of this point). The feminist movement was founded around two different organizational styles: the equal rights sector, which tends to be organized around traditional interest-group lines and to participate in traditional politics; and the women's liberation sector, which is organized in a more decentralized participatory style and which seeks to develop political alternatives, rather than pressure-group strategies. The mainstream movement in the United States follows the first model, has become relatively institutionalized, and has coopted the second to a degree; in Britain, the reverse is true. To a degree unique in the annals of movement literature and even in comparison to other British groups which began in a similar fashion (the Child Poverty Action Group, for example), the British feminist movement has remained remarkably faithful to its initial mobilizing precepts. The movement continues to be decentralized, localized and non-hierarchical even at the cost of not increasing political influence and not creating a national political presence.

While most analysts see protest as central to the activities of social movements, the analysis presented here suggests that protest has never been employed as a central tool by most feminists. Rather, a mix of confrontational and assimilative techniques is employed (see Rochon as well as Rucht in this volume, chapters 6 and 9), with most emphasis on the assimilative and integrative. However, protest is utilized in each nation to demonstrate support for key feminist issues, particularly when these issues are challenged by opposition groups. Examples are the demonstrations held in Britain to oppose the Corrie Bill's attempt to restrict abortion rights, and similar protests in the United States and Sweden on abortion, pornography, and rape. In general, however, feminists in all three nations lobby, petition, and engage in electoral campaigning to gain support for their demands; they rely on protest in order to demonstrate their ability to mobilize mass constituencies, and seek to coerce elites to make concessions and bargain accordingly, especially when previous efforts have not produced change.

Political Outcomes

An important area of analysis for our inquiry is that of impact and outcomes. Because feminism is a movement as well as an ideology, its impact must be judged both in terms of specific reforms and in terms of the development of a collective consciousness. With regard to specific reforms, feminists have had a degree of success in all three nations. Issues on the feminist agenda – abortion, equal rights, domestic violence, and the like – have been incorporated into legislation, albeit with varying degrees of feminist input and degrees of commitment to vigorous implementation. In each nation, the state has created new machinery to deal with issues related to equal rights (usually in the form of an equal opportunities commission), again with varying powers and sanctions at the disposal of the new apparatus. Women's access to electoral and political office has increased – in the case of Sweden and the United States more often through recruitment of women leaders from women's organizations than is the case in Britain. In Britain and the United States, new alternative forms of political organization have been created and sustained over several decades. The existence of two wings of movement activism are seen by this author as a strength; this simplistic dichotomy masks a complex structure of organizations, initiatives, and ideas, and movement specialization helps to create possibilities for diverse strategies and approaches (Lovenduski, 1986, pp. 62–5). This research strongly suggests that in neo-corporatist systems such as Sweden, policy gains are likely to be of an economic sort, providing benefits to women which are largely labor-related. In the more pluralist American setting, policy gains tend more to accrue to groups and/or individuals in terms of economic mobility and political power reflecting a weak central state and government sector, but strong opportunities for access to new roles. The weakly corporatist British system lies somewhere in between.

In his analysis of social movements and policy change, Tarrow (1983, p. 8) suggests that "the range and flexibility of its tactical repertory is often a good predictor of movement success." We contend that the American movement has enjoyed the greatest autonomy in terms of choice of tactics, because it is not constrained by the necessity to work through the intermediation of existing groups, such as parties and unions. This permits the utilization of a variety of tactics from protest to participation in existing power groupings. British feminist participation is constrained by the rigidity of the political system and the ambivalence of movement activists about engaging in national-level politics in a coordinated fashion. Swedish women are, in the main, limited to behaving as participants in existing institutions.

A final measure of movement impact lies in the area of the creation of a new "collective consciousness" among supporters, allies, and/or the general public. Such a "collective consciousness" refers to a transforming set of ideas related to new norms, roles, institutions, and/or redistribution of resources

(Mueller, 1987, pp. 89–110). Among the indices we may use are attitudes to work for women with children.

In all three nations, attitudinal change relating to women's role is evident, although least so in Britain. However, even in Britain, female attitudes toward work have undergone considerable transformation: among younger women, a vast majority say they wish to return to work after having children. Still, more British women say that women should quit their jobs when they have children and return to work only after their children are grown up (53 percent compared to 21 percent of Americans and 35 percent of Swedes) (Hastings and Hastings, 1985).

In contrast, in the United States, a majority of women indicate that they would want to work outside the home in the future, even if they had enough money to live comfortably without working. Considerable support exists for a new concept of marriage in which husband and wife share work and domestic responsibilities (Roper Organization, 1980).

In Sweden, a dichotomy has emerged between male and female views in relation to a prohibition on pornography. Women from 1979 to 1985 have consistently supported a ban in far greater numbers than men (1982: 51 percent women, 28 percent men; 1985: 68 percent women, 43 percent men).[3] This appears to demonstrate a growing awareness around this issue, which was highlighted by an all-women's party demonstration in Stockholm. Eduards (1981) notes that majorities of women (but not men) also favor the use of quotas to gain election to the Riksdag for women and oppose jobs for men first. It should be noted that some data also suggests a "gender gap" on the issue of abortion, with more women against! (Eduards et al., 1986; Sainsbury, 1985). Whether these trends augur a new feminist consciousness which may be translated into group activism is questionable given the continued prevalence of norms related to Swedish political consensus within the corporate state; what is worth noting, however, is the apparent emergence of separate female consciousness around several key items on the feminist "agenda." Other surveys, in all three nations, dealing with such issues as equal sharing of household tasks, increasing numbers of women in business, and consciousness of continued discrimination against women, reveal a significant degree of consciousness around feminist concerns. In turn, we suggest that one of the most significant outcomes of feminist movement activism has been the creation of a new consciousness relating to women's roles, expectations, and possibilities for change. Further mobilization and change appear to be predicated upon the existence of such "consciousness."

Conclusion: Feminism as a "New" Social Movement

Tarrow (1983) has suggested that political "reforms may be more reversible when they are substantive only, creating no new vested rights of participation or veto in the population groups newly enjoying them." This hypothesis argues, then, for the idea that reforms producing institutionalized participation may be the most significant of all, largely because they help to create resources for future mobilization and ensure continued attention to substantive policy concerns. Seen from this perspective, the American movement, now recognized as a legitimate group in the pluralist system, may be in the strongest position to endure as a significant political force.

This analysis suggests that there are distinctive features of the "new" feminist movements, although these have tended to vary with national factors addressed above. The case of British feminism may present the greatest support for the view that "new" social movements may be distinguished by a more decentralized and democratic internal structure, dedicated to the goals of collective good and personal transformation. The particular political "style" represented by the British movement may reflect a conscious rejection of corporatism and British political institutions in general, as well as the realization that these institutions are, in the main, closed to active participation by "promotional" groups which exist beyond the perceived "legitimate" pale of representation. The absence of an autonomous active feminist movement in Sweden suggests the problems of generalizing beyond specific examples to a "European" model. The movement in the US may serve to demonstrate continuities with the past, in the relationship between feminist and more traditional, historically based women's groups, and the more "reformist" orientation characteristic of American activists.

The movements in all three nations, despite their disparate structures and goals, are distinctly "new" in at least one aspect. This is the manner in which feminist politics has created a new agenda for public policy. The effort, often successful, to abolish distinctions between "public" and "private" spheres of society has resulted in a restructuring of grievances addressed by governments everywhere. The claims made by feminists on public policy have been transformational in bringing new issues to the fore politically. Finally, despite the national differences highlighted in this chapter, the feminist movement is in many ways unique in its international character; national boundaries have often been transcended by the new values and "consciousness" as well as the demands represented by movement goals.

NOTES

1 The material presented in this essay is the result of three visits to the United Kingdom; in spring 1979 and again in the summers of 1982 and 1984. The research in Sweden was conducted in summer 1986. The research related to American feminism is based on research conducted for an earlier book, *Women and Public Policies*, and an article, "The Politics of Wife Abuse," undertaken from 1980 to 1986. In each country, research was based in large measure on interviews with feminist scholars, journalists, and activists, as well as women active in political parties, unions, and civic groups, and holders of elective and appointive

political positions. The period studied runs from the 1960s to the end of the 1980s.

2 The term "state equality" has been deliberately chosen to contrast it with "state feminism," a term employed by Helga Hernes and other Nordic observers of feminism and public policy. While, as Hernes observes, the concept "state feminism" may represent an amalgam of policy from above and mobilization from below, it represents the former more than the latter, at least in Sweden in the 1970s and 1980s. As this writer argues that political mobilization of women from below has had limited influence in the Swedish case, and the term "feminism" is viewed as suspect and rarely discussed publicly in this nation which emphasizes "gender neutrality," "state feminism" appears to be something of a misnomer. Because we should not confuse systems which are, in Hernes's terms, "women friendly" regarding positive policy outcomes for women, with feminist analysis and input, I prefer the model of "state equality." (See Hernes, 1987, pp. 9–29, for a discussion of "state feminism.")

3 Data cited here were made available by Soren Holmberg, University of Gothenburg, Sweden.

9

The Strategies and Action Repertoires of New Movements

DIETER RUCHT

Confronted by opposition, any organization or collective actor seeks to develop adequate strategies and actions in order to pursue its aims successfully. Not surprisingly, strategic issues played a crucial role in past social movements, such as the labor movement or the "old" women's movement, provoking intense debates within the movements. Today, in a society which has achieved a high degree of self-monitoring and self-reflection, and whose features are considered to be not merely given elements but rather matters of deliberate choice, the importance of strategic and tactical decisions has apparently increased.

This holds true for present-day social movements as well. And since scholars no longer perceive social movements merely as a kind of irrational and unstructured "mass behavior," they evaluate aspects of strategy and action repertoire, especially in relation to issues of organization and mobilization. Thus the period since the 1970s has seen the rise of a considerable body of literature on social movements, ranging from political and theoretical reflections on the labor movement in the nineteenth and early twentieth centuries to presentations using the resource-mobilization approach. Nevertheless, our knowledge about the conditions that influence movements' strategies and actions is still relatively poor. To be sure, there are both comparative and/or systematic assessments which focus particularly on this topic (Turner, 1970; Gamson, 1975; Piven and Cloward, 1977; Freeman, 1979, 1983; Kitschelt, 1986), as well as numerous case studies (e.g., Barkan, 1979; Lawson, 1983). Comprehensive treatises on social movements also usually cover problems of strategy and forms of action. But they do this only in a very general, brief, and often ahistorical way. Finally, there is a large body of literature on individual movements and conflicts, dealing more or less implicitly with the determinants of strategies, tactics, and/or specific forms of action (e.g., mass mobilization, lobbying, rioting, civil disobedience). But this literature hardly provides an answer to the questions that need to be answered:

- Which groups/organizations/movements tend to use which kind of strategies/ forms of action under which circumstances? "Circumstances" is used here in the broadest sense, covering not only accidental factors but also parameters which can be located on various structural levels.

- Are there strategies and forms of action which are typical of the so-called new social movements?

- What is the relationship between certain movement issues and stages of conflict development on the one hand, and strategies and forms of action on the other?

With respect to these questions, only a tentative analysis can be offered in this chapter. Within this context, I will try to do three things. First, I will identify some characteristics of the actions of new social movements and formulate some hypotheses about why their repertoire differs from that of old social movements. Second, I will discuss the determinants of the strategies and actions of various new movements in terms of logics of action and societal structures. Third, I will refer to the conflicts engaged in by environmental movements in order to illustrate the complex impacts of structural and accidental factors on strategies and actions. It is not my intention to provide the reader with a stringent empirical analysis, but rather to lay some groundwork which may serve as a guide for further and more empirically oriented (comparative) studies.

The Political Action of New Movements: An Historical Perspective

In the European context, the concept "new social movements" (NSMs) has enjoyed increasingly widespread usage. In contrast, American scholars and movement activists are much less familiar with this concept, or are even baffled when confronted with it. In the United States, both the absence of a powerful and well-organized labor movement and the existence of radical liberal traditions – from the eighteenth-century revolution to left-wing populism, civil rights movements, New Left and student movements, and finally to various contemporary movements such as the women's, ecology, and nuclear freeze movements – make it relatively difficult to contrast the latter movements with the former. While in Europe the concept of NSMs usually refers to a distinct, although broad, political and ideological spectrum, "new" movements in the United States tend to be associated simply with "contemporary" movements, including salient countermovements such as those of the New Right.

But even in Europe, where there seem to be better reasons to distinguish the NSMs from the "old" movements,[1] different answers are given to the question of what is "new" about NSMs. Leaving aside this ongoing debate and neglecting potentially distinctive criteria such as ideologies, targets, social bases of recruitment, and organizational structure, here I will give particular attention to forms of action and their underlying strategies.

Before taking a closer look at the actions of NSMs, we should realize that the dominant forms of action of social movements may vary significantly over long periods. Only a historical perspective can determine if there is anything which is specifically characteristic of the actions of new movements.

Based on a general survey of the prevailing forms of collective action during the past centuries, Charles Tilly (1977) stated that Europeans in the twentieth century, unlike in earlier periods, are concerned with *proactive* actions, that is, struggles for demands that had not been expressed before. *Reactive* collective actions, such as food riots, tax revolts, local protests against military recruitment, and rivalries between local authorities, have declined significantly for two reasons. First of all, the emergence of national states deprived local groups of resources and the means of power. Secondly, organizational changes related to urbanization, industrialization, and the expansion of capitalism decreased the significance of communal-based groups. New organizational forms (e.g., associations, political parties, trade unions) and new forms of actions (e.g., strikes) emerged. Politics became more and more nationalized and the national state became an interested participant in all forms of collective violence (Tilly, 1977, p. 158).

In a recent article, Tilly (1986) sharpens his contrast between the eighteenth and nineteenth century action repertoires. He maintains that the scope of action shifted from the local to the national sphere, and, with respect to the orientation to powerholders, "patronized" forms of action were increasingly replaced by "autonomous" actions. The action repertoire of the nineteenth century, exemplified by election rallies, the invading of assemblies, public meetings, strikes, demonstrations, and social movements, looks very familiar to us. According to Tilly, this is the repertoire "with which we still live" (1986, p. 173). Even if this general statement should be true, I maintain that a closer look reveals forms of action characteristic to and preferred particularly by the NSMs.

In comparing the actions of social movements in the first half of the century to the actions of NSMs, I would like to point out some differences that deserve explanation. In contrast to the "old" movements and in particular to the labor movement or the fascist movements in Europe, groups and organizations within the NSMs seem to act more independently from each other. They reject any hegemonic position or concept, in particular the Leninist model of the "avantgarde." Even protest parties claiming to represent movement issues in the field of national parliamentary politics and with a strong emphasis on locally based activities are met with strong reservations by significant parts of the NSMs. As a result of this quest for autonomy (von Beyme, 1986, p. 37) – and in contrast to Tilly's assertion – contemporary movements' activities do not necessarily focus on the national level. They emphasize the role of independent and small groups and the importance of local activities, and they promote grassroots politics

(Gundelach, 1984; Rucht, 1984b).[2] Of course, these tendencies can also be found in "old" movements, particularly in their early stages of formation. Today, however, the NSMs seem to choose the form of loosely coupled networks deliberately, and not necessarily because of a lack of resources and organizational skills.

With respect to conventional participation, the significance of parliamentary representation (and thus the meaning of elections as well) seems to have decreased, whereas other conventional forms of participation, including court action, have apparently become increasingly relevant.

Some historical movements, such as the "old" women's movement or the "reformist" labor movement in the twentieth century, expected to achieve societial change via parliamentary majority. The role of the legislative was largely overestimated not only by these movements, but also by the wider public. Today a more realistic perception seems to prevail. Contemporary movements are less confident of achieving social change not only by revolution, but also by legislation. Moreover, voting for a party or a political leader today depends much more on situational considerations ("issue-voting") than on an overall world-view. People may vote for a party at the same time that they publicly oppose some of its particular policies. This would have been very difficult for a working-class voter at the turn of the century.

Litigation by social movements has also increased because possibilities to intervene in specific policies by using administrative and constitutional courts previously did not exist at all, or only to a very limited degree.[3] This is also true for procedures such as public hearings and inquiries. Moreover, in many Western democracies the judicial system presently has a more independent position in relation to the holders of political and administrative power than it did in the nineteenth and early twentieth centuries. In addition, courts are more sensitive to certain issues and claims not propagated by strong interest groups. Finally, just as respect for authority seems to be lower than in earlier decades, the willingness to challenge legal decisions has grown. This is also due to the political schisms within societal elites on most issues raised by NSMs, be it abortion, biotechnology, nuclear power, or military defense.

With respect to unconventional behavior, acts of violence, and particularly the use of arms, have generally diminished in importance over the last few centuries, whereas acts of civil disobedience are becoming more and more relevant. The relative infrequence of armed violence is mainly the result of the institutionalization and extension of formal procedures for the expression of social and political discontent. Moreover, minority rights seem to be respected both socially and juridically to a greater extent than in earlier periods. Finally, the power of the modern state is so great as to discourage the use of violence by social movements, at least when conflicts are carried out in the public domain and do not escalate to the dimension of a civil war.

Still, all the institutionalized means of articulating interests and expressing discontent are powerless to prevent the relative deprivation of certain population groups and social strata, the deterioration of public goods such as the quality of water and air, or even the possible extermination of humankind. Consequently, there are still reasons to protest, and new ones are likely to emerge. In general, the concerns of social movements in the last century, even if these movements started as single-issue movements, tended to be embedded into broad and sharply contrasting ideological world-views. By contrast, the concerns of the NSMs are conceived much more as single issues, not to be solved by a redistribution of the means of production and wealth within a wholely new political system. This trend is a major reason why there is no longer *one* salient movement representing *the* subjected class, but a plurality of coexisting and cooperating movements whose significance can hardly be described in terms of class antagonisms.

In so far as institutionalized channels for expressing discontent are blocked (for example, by a broad coalition of the established parties, by a neocorporatist cartel of elites, or by high procedural thresholds) or are unsuited to the nature of the problems (for example, sexist role behavior), and in so far as a policy or social behavior as a whole cannot be changed via other means, acts of civil disobedience may be an adequate answer for protest groups. These acts allow the venting of specific and intense criticism without necessarily aiming at a general transformation of society.

Moreover, civil disobedience as an act with a highly symbolic and expressive component requires an audience to whom the actor can appeal. Unlike revolutionary or terrorist acts, which are directed specifically at the enemy and which do not necessarily involve or require the support of third parties, civil disobedience must take account of public opinion – and the dependence of power-holders on that opinion – as a crucial variable for its failure or success. Civil disobedience can only flourish when there is a large public which is not *a priori* partisan to the conflict and whose opinions cannot be directly controlled or easily manipulated by power-holders. Obviously, the presence of modern mass media enhances the efficacy of civil disobedience. The disobedient act can become audible and visible to millions of people; further, to a much greater degree than before, such acts are transferred immediately to the large public, without passing a long chain of transfer stations. Even if news is necessarily interpreted and filtered by journalists, competition between news outlets, professionals standards, and the ideological heterogeneity of the press combine to make it difficult for the media to ignore protest or to deny movement representatives a hearing.

A final difference between old and new movements is that at least particular parts of the former tended to rely on a narrow action repertoire, and they rarely shifted quickly from one kind of action to another. Today, cooperation and alliances between groups using different forms of action seem to be more common. Also the parallel use of different forms of action by the same group, or the shift from one form to another, seems more likely than in older movements. In consequence, a broad variety of forms of action, extending from moderate expressions of discontent to overt violence, is likely to be practiced at the same time and within a single movement. This observation is strongly supported by various case studies. Moreover, surveys of contemporary forms of political participation show that people tend to use both conventional and unconventional forms of action (Barnes, Kaase et al., 1979; Fuchs, 1984).

These features may also be partially explained by a kind of reasoning which is determined less by general ideologies than by situational cost–benefit assessments. A still more relevant argument to account for the variety and flexibility of new movements' actions is the autonomous status of groups and organizations within these movements. For instance, in contrast to the German labor movement, there are no hegemonic organizations and elites that have enough authority to tell the rank and file what should be done, whether it is allowed to protest, or whether the time is ripe for radical action. In addition, the loose networks of many groups and activists as well as their overlapping issues and positions within the NSMs allow for a broad dissemination of ideas, experiences, and action forms.[4] There is room for the parallel use of different forms of action by different groups in the same conflict, without forcing each group to legitimize every form of action that is used. Finally, one should take into account factors such as the enormous spatial mobility of people in modern societies, the involvement of many professionals in the NSMs, and the ensuing broad spectrum of knowledge, techniques, and other resources available to contemporary movements. Strikingly, even conservative groups such as those of the New Right tend to use a range of actions (including disruptive tactics) similar to that of their opposite numbers on the Left.

Summarizing this section, we have argued that the prevailing action repertoire of the new movements is not that of earlier movements. While in the next section I will show how various structural factors account for these differences, it is also necessary to understand that movement actions are generally adopted within the context of particular strategies, which are themselves strongly influenced by a range of social structures.

Determinants of Strategies and Forms of Action

In order to examine empirically the determinants of strategies and forms of actions, it is necessary to first discuss them as analytic concepts. *Strategy*, a term which can be differentiated from tactics,[5] refers to a conscious, long-range, planned, and integrated general conception of an actor's conflict behavior based on the overall context (including third parties and potential allies), and with special emphasis on the inherent strengths and weaknesses of the major opponent.

The strategies of social movements have often been described from a variety of perspectives (for different approaches to the classification of strategies see Gamson (1975); Jenkins (1981); Freeman (1983); Kitschelt (1986); and Zald and Useem (1987)), but without any explicit theoretical or conceptual basis. One of the few exceptions is Ralph Turner's presentation of a systematic general typology of social movement strategies by categorizing them as being of the persuasion, bargaining, or coercion type (Turner, 1970). This is a useful starting point for a classification. In my view, however, these categories do not necessarily describe movement strategies but rather fundamental modes of interaction. They may also characterize activities on an individual level. Moreover, individual or collective actors may rely on all these options concomitantly or consecutively, without changing their general strategy. For instance – supposedly because of its low costs and risks – persuasion will usually accompany most competitive and conflictual political activities in democratic societies. Finally, Turner tells us little about any systematic relationship between the character of the movement and the strategic options.

Because most movement studies have a limited focus (the restriction to a certain movement, to certain policies, or to the selected variables) or have not separated distinct movement strategies, they are not adequately suited to the tasks of this essay, which requires a general typology of social movements and to relate different kinds of movements to different strategies.

Based on a distinction between conflicts over "systemic control" on the one hand and those over the patterns of the "life world" on the other (see Habermas, 1981), two logics of action can be analytically identified; these correspond to two ideal types of movement strategies (Rucht, 1988). The *instrumental* logic of action implies a *power-oriented* strategy, which is concerned with the outcomes of political decision-making and/or with the distribution of political power. Successful performance in struggles over power requires, above all, instrumental reasoning. On the other hand, the *expressive* logic of action corresponds to the *identity-oriented* strategy, which focuses on cultural codes, role behavior, self-fulfillment, personal identity, authenticity, etc. Such a strategy relies mainly on expressive behavior, trying to change cultural codes by alternative life-styles.

There is nothing especially original or highly sophisticated in suggesting these dualistic strategies.[6] To be sure, a movement may have certain phases, ideological tendencies, and issues in which its overall strategy is relativized or even ruled out. Moreover, for basic reasons, no movement can restrict itself to following only one of these strategies exclusively. Deviant cultural practices are likely to be attacked and hindered by political power-holders, and therefore identity-oriented groups and movements need to become involved in power struggles to defend them. Similarly, power-oriented groups and movements need to maintain their collective identity by adopting cultural symbols, and at least to respond partly to the sentiments and demands of their adherents. Hence, when classifying the labor movement as power-oriented, one should not forget that there once was an important network of labor groups emphasizing cultural issues (e.g., choirs, sport clubs, educational and hobby associations, etc.).

Depending on the circumstances, movements that strive for political power may rely on a *specific strategy*, or on a combination of different

Table 9.1 Typology of social movements' strategies

Logic of action	*General strategy*	*Specific strategies*
Instrumental	Power-oriented	Political participation Bargaining Pressure Political confrontation
Expressive	Identity-oriented	Reformist divergence Subcultural retreatment Countercultural challenge

ones. Here I would distinguish between political participation, bargaining, pressure, and confrontation.

Identity-oriented movements based primarily on expressive behavior seem to have no strategy, if the term "expressive" is taken literally. In fact, however, these movements are not purely spontaneous; they cannot avoid the question of what course to follow and how to deal with opponents. However, unlike power-oriented movements, identity-oriented movements primarily seek to change personal behavior by espousing and adhering to deviant cultural practices. The strategies they use can be reformist divergence,[7] subcultural retreat, or countercultural challenge. Table 9.1 gives an overview of these various strategies.

The conceptual reflections presented above assume that the "logic" of the movement has consequences for its strategies (both general and specific). Moreover, certain strategies may require specific organizational forms and may be unattractive except to people from specific social strata and/or with specific beliefs. Because the organizational structure of a given movement and the social profile of its supporters will hardly change within a short period, not every strategy is available at a given time. But it would be false to assume that in all cases there are necessary and direct correlations between a given movement, its organizational structure, and the chosen strategies. For instance, a decentralized network of autonomous groups may be beneficial to a peaceful effort to change a specific policy as well as to the revolutionary use of violence by terrorist groups. Conversely, there may be groups with similar objectives and strategies, but based on totally different organizational structures.

Depending on the type of society and the available resources and institutional channels of action, each of these strategies may be pursued through specific forms of action. For instance, in democratic societies, struggles over power usually involve attempts to create a political party or to exercise influence on parties as well as parliamentary and administrative organs by direct intervention or by the election of leaders.

Looking at social movements in general and over long periods, we can find that a broad variety of collective actions have been used. However, if we investigate specific types of movement or movements within shorter time periods, this potentially available set of actions turns out to be narrower. Moving further to specific currents, organizations, and groups within a given movement, the scope of available actions becomes quite narrow. It is on this level that the category *action repertoire* should be located. It denotes the range of specific kinds of action carried out by a given collective actor in a cycle of conflict, usually lasting from some years to some decades. Within such a period, the actor becomes acquainted with this set of actions. It experiences the corresponding costs and benefits, and it gets an idea of which action has to be chosen under which circumstances. This is in line with the statement that the notion of repertoire "incorporates a sense of regularity, order, and deliberate choice into conflict" (Tilly, 1986, p. 174).

It should be mentioned that "action repertoire," as used here, covers only part of the actions of any social movement. When we look at social movements' actions, our attention is usually captured by more or less spectacular protest events such as petitions and campaigns, mass meetings and rallies, acts of civil disobedience and violence. These activities are outward-directed in the sense that they are addressed to an opponent or to a third party.[8] Usually in the shadow of these visible activities we can detect not only non-spectacular protests embedded in everyday life (Lüdtke, 1984), but also many inward-directed activities. Collecting money from supporters, creating specific facilities and operating tools, training activists, assessing costs and risks, making decisions, evaluating the result of actions – these are movement activities which may be directly connected to outward-directed actions, but are not, or are only in part, visible to an external observer. In fact, a great deal of every movement's energy is not directly devoted to public actions. In the long run, organizations have to be created and maintained, members and adherents must be informed and motivated, newcomers have to learn the use and meaning of group symbols, internal conflicts must be solved, etc. In short, every movement has a visible and a latent reality (Melucci, 1984a).[9]

STRATEGIES AND ACTIONS OF NEW MOVEMENTS

Leaving aside the inward-directed actions mentioned above and focusing exclusively on public actions, one could ask which particular movements within the new movements tend toward specific strategies and respective forms of action.

I argue that the NSMs in general are not exclusively oriented towards either questions of power or questions of identity. Typically, both logics are represented, even if each may have a different weight at different periods and within different currents. But if we take a closer look at various issue-oriented movements within the NSMs (see table 9.2), some of them can be classified relatively easily with respect to their overall orientation and their

Table 9.2 Prevailing outward-directed strategies and actions of various new movements

Movement	*General strategy*	*Specific strategies*	*Forms of action*
Student movement/ New Left	Ambivalent	Extraparliamentary pressure	Congresses, teach/go/sit-ins, demonstrations
		Confrontation, countercultural challenge	Blockades, happenings, communes
Women's movement	Identity-oriented	Reformist divergence, subcultural retreat	Practising different gender roles, consciousness raising,
		Countercultural challenge, pressure	provocations, demonstrations litigation (USA) lobbying (USA)
Anti-nuclear/ecology movement	Power-oriented	Extraparliamentary and parliamentary pressure	Campaigns, rallies, petitions, referenda, litigation, hearings, lobbying, party formation, demonstrations, civil disobedience,
		Confrontation	Sabotage, violence
Alternative movement	Ambivalent	Subcultural retreat, countercultural challenge	Self-run community, living/working projects
		Bargaining	Negotiations with local bureaucracies
Peace movement	Power-oriented	Extraparliamentary pressure	Campaigns, mass meetings, rallies, marches, petitions, civil disobedience

corresponding strategies and forms of action, whereas other movements typically combine power- and identity-orientation without any clear preference for either aspect.

An example of the combination of power- and identity-orientation is the so-called alternative movement, whereas the anti-nuclear movement, as it focuses on a specific policy implemented by large firms and state authorities, clearly is predominantly oriented towards the political decision-making process, and thus can be described as power-oriented. The end of the nuclear power program would no doubt cause the end of the movement.

In contrast, the new women's movement, at least in Europe, focuses less on state policies and much more on cultural patterns, particularly on gender roles. Obviously this movement also works for specific policies to liberalize abortion, to guarantee equal rights, etc. These were central concerns of the "old" women's movement, and the more these aims were realized, the more that movement lost its *raison d'être*. Since the transformation of deeply rooted gender roles can hardly be accomplished by means of policy changes alone, it is improbable that even outstanding victories on the political level would put an end to the new women's movement.

Admittedly, there are remarkable differences between women's movements in Europe and the United States (see Gelb in this volume, chapter 8). Because of an opportunity structure much more favorable to forms of action characteristic of pressure group politics and lobbying, the women's movement in the United States is better represented in the realm of established politics than its European counterparts. This example demonstrates that the movement's logic implies only probabilities in terms of strategies and action, but does not entirely determine them. It is in this sense that table 9.2 should be understood.

STRUCTURAL AND ACCIDENTAL FACTORS FOR STRATEGIC CHOICES

Strategies and actions depend not only upon a movement's logic, but also on various other factors, among which are the basic structural patterns of a given society. Some of them, in their turn, may influence the movement's logic. Because of this interplay of factors, one should be careful about suggesting a linear causal model to explain particular strategies, that is, with a movement's logic of action as an independent variable and an additional set of intervening structural variables. Even if the status of these structural variables still has to be clarified, for analytical purposes it seems helpful to differentiate these factors as follows.

First, on a macrostructural level, one can locate relatively stable historical conditions, such as the economic "regime", the pattern of social stratification, state structures, or basic values. These factors are given to collective actors, and hardly change in the medium term (e.g., a single generation). The underlying theme and the overall strategy of social movements are largely shaped by these structures, which form a kind of corridor that limits the range of specific strategies, the types of organization and the actions which can be chosen. The movement cannot go beyond the structurally defined limits. Only in the long run, and by the aggregate effects of many minor changes, can these "big structures" be altered and transformed (see Tilly, 1984). It would be false, however, to perceive structures only as restricting social action, as they also provide a constitutive basis for collective action. For instance, party politics could only flourish after the establishment of modern parliaments. Similarly, contemporary movements may profit from public hearings and mass media, which help put their issues on the political agenda. Similarly, while the liberal state limits protests by channeling interests into parliamentary and interest-group channels, it also offers important opportunities for social movements, e.g. through public hearings, mass media, or the right to demonstrate. Hence we must evaluate the "duality of structure" (Giddens, 1976) both as a basis for and a constraint on the actions of social movements.

Second, the choice of specific strategies and forms of action is restricted and facilitated to a high degree by what could be called "mesostructural" factors.[10] A movement or a movement organization has to deal with conditions that are clearly structural but not stable in epochal terms. The movement itself may influence these structures to a certain degree, or powerful institutions may change them within a period of few years or less. It is on this level of analysis that one finds factors such as pre-existing organizations (both of the movement and its opponents), party constellations, distinct rules and policies, the state of public opinion, factions internal to the movement, the lessons derived by core activists from their experiences in the course of the movement's earlier conflicts, or from similar conflicts.

Third, if the focus of analysis shifts to a still more specific level, we enter the microcosm of groups, be they independent or part of a formal organization. Even on this level, where social phenomena become relatively fluid and unstable, we find structural parameters such as informal hierarchies, patterns of internal and external interaction, and the distribution of resources. These may account for a specific action repertoire, and on an aggregate level they may shape the movement on the whole, as was the case with the spread of consciousness-raising groups within the new women's movement.

Finally, which action is chosen in a given structural context also depends on accidental factors, such as precipitating or neutralizing incidents. The financial demise of a movement's organization, the retirement of a charismatic leader, or the death of a demonstrator at the hands of the police may have a considerable impact on the activities of a movement. Clearly, some of these factors are purely accidental, and their weight can only be assessed *ex post facto*. Other factors appear to be contingent as well, but a closer look reveals that they are influenced, or even caused, by structural factors. For instance, the sudden breakdown of an organization may be due to a long causal chain: inadequate collective perception of the historical situation – struggles on false terrain – inappropriate strategy – internal split – decreases in membership – organizational breakdown. Looking from this perspective, we may find that a single event may have a considerable impact on the micro- and mesostructural level. For example, in West Germany, the Chernobyl disaster gave rise to a new branch of the anti-nuclear movement – the so-called Becquerel movement – devoted to specific forms of action, and thus challenging the tactics and specific strategies of the old campaigners of the anti-nuclear struggle. Partly through this rejuvenation of the anti-nuclear movement, the Chernobyl incident may help change or even put an end to the national nuclear program in the long run.

To sum up: in assessing the factors that influence social movements' strategies and actions one has to take into account the movement's logic, which depends basically on the macrostructural setting in a given historical period. The NSMs in general combine two logics, even if some specific movements tend to be either power-oriented or identity-oriented. Moving further to specific currents, organizations, and groups within those movements during relatively short periods, a variety of more specific structural and accidental factors come into play. Further, because a movement or parts of it are not simply a product of structural conditions, but may in turn influence at least micro- and mesostructures, linear causal explanations risk being too shortsighted.

The Case of Environmental Movements

A systematic analysis of the strategies and actions of individual movements within the NSMs has not yet been made. It would be very interesting to find out when and why various forms of action were introduced, or when and why they were adopted and altered by other movements. For instance, one could trace the spread of civil disobedience or some of its features (such as affinity groups) from Indian activists inspired by Gandhi to both Spanish anarchists and the American civil rights movement, and from them to non-violent groups in the ecology and the new peace movements in many countries. Here too, not only trans-national networking but also mass media play an important role in disseminating specific forms of action (Gitlin, 1980; Kielbowicz and Scherer, 1986).

Another tempting research task would be to investigate which strategies and actions dominate in which kind and in which phases of conflict. However, answering these questions would require a complex comparative research design, since many factors and variables must be controlled.

On a far more modest basis I would like to make use of some general categories and to illustrate some hypotheses that have been introduced above with regard to environmental movements. Here I will draw mainly on my knowledge of the French and German environmental movements and on various case studies of environmental conflicts. Of course, these experiences reflect developments in specific areas and periods; they may or may not hold true in general.

STRATEGIES IN DIFFERENT HISTORICAL PERIODS

In the first section of this chapter I argued that during the twentieth century a remarkable shift in the prevailing strategies of social movements has occurred. Some empirical evidence for this statement can be found in analyzing various environmental conflicts in different historical periods. A comparison of two sets of conflicts – over the implementation of two large and technically ambitious hydroelectric power stations in the first decades of this century in Germany and over two major nuclear projects in post-war West Germany – show striking differences in the prevailing strategies and action repertoires of each period (Linse et al., 1988). In the "old" conflicts, the opponents of these projects basically relied on moderate forms of action (e.g. petitions, collection of signatures, articles in journals). Litigation was already used, but it remained of minor importance. In contrast, actors in contemporary conflicts not only used "conventional" means of expressing dissent, but also became engaged in mass rallies, disruptive forms of action (blockades, occupations), and even violent actions. Moreover, disputes between experts and court actions played a significant role in recent conflicts. Thus the range of strategies and action has become much broader and – in an overall view – the emphasis has shifted towards confrontational strategies and more offensive forms of action.

Contrary to what one might expect, these differences cannot be explained primarily by the higher risk of nuclear technology compared to hydroelectric power stations, but rather are mainly due to factors on the macro- and mesostructural level, for example, features of political culture. Particularly in Germany, the political culture underwent a remarkable shift from the authoritarian state with a predominant integrationist ideology towards a more participatory and confrontational political behavior.

These significant changes in strategies and actions in environmental conflicts during the twentieth century are not limited to Germany. Without going into details about the cases, similar developments can also be found in countries such as the United States, France, Switzerland, Spain, and Norway.

GROUP-SPECIFIC STRATEGIES

The contemporary environmental movement has been characterized as power-oriented. Undoubtedly there are certain groups within this movement that do focus primarily not on state politics but rather on personal change (life-style, consumer behavior etc.). But these groups do not play a central role. Even if it is justified to ascribe just one general strategy to the environmental movement, we should not forget that this movement is probably the one among the NSMs that utilizes the broadest range of actions. The reasons accounting for this seem to lie in the vagueness of "ecology" and the non-class-specific nature of many environmental concerns. Hence there is space for a great variety of ideological currents and political issues; there is a large popular basis for recruitment, to which groups with very different organizational patterns and preferred forms of action can appeal. This variety is shaped mainly by factors located on the mesostructural level. Consequently, it is to this level in particular that attention should be paid in order to explain the selection of action forms within this movement.

Mesostructural factors also mainly account for the coherence or schisms between various currents, tendencies, and networks of the environmental movement in a given country. Within such a movement one can usually find groupings of traditional conservationists, of "environmentalists," and of "political ecologists." In West Germany, these currents are relatively closely interrelated and sometimes manage to unite for common campaigns. In France, however, the salience of the Left/Right cleavage contributes toward the gap between political ecologists on the one hand and conservationists and environmentalists on the other (Rucht, 1989). Both camps act largely independently from each other and thus seem to represent at least two environmental movements, each one preferring its own organizations, strategies, and forms of action. In general, conservationist organizations tend to view themselves as non-political, using moderate and non-conflictual forms of action (e.g. buying land to protect animals and plants, clearing forests of trash by volunteer work, etc.), while organizations of environmentalists and ecologists are deliberately engaged in the political arena. The former focus more on activities such as influencing media, petitioning, litigation and lobbying (Dalton et al., 1986), whereas the latter also rely on disruptive means of action, and in some cases even on violence.

Obviously, there is also a correlation between the ideological orientation and the strategic preferences on the one hand and the organizational structure on the other. Conservationist organizations usually are more hierarchical, whereas organizations of political ecologists rely more on grass roots organizations. But there are exceptions to this rule. Greenpeace, for instance, which certainly does not belong to the conservationist tendency, is based on an extremely rigid hierarchy, whereas "Robin Wood," a German offshoot from Greenpeace, pursues similar goals and forms of action on the basis of a grassroots organization. Despite such cases, however, I would maintain that the logic of a movement correlates with distinct strategies, organizational patterns, and forms of action.

LOCALLY SPECIFIC STRATEGIES AND ACTION REPERTOIRES

Comparing local resistance against various nuclear power plants or airport construction projects in the 1970s and the early 1980s in West Germany, one finds enormous differences in the intensity of conflict, the spectrum of activated groups, the number of people involved, and the action repertoire that was used (Kitschelt, 1980; Rucht, 1980, 1984a). At some sites, opposition remained totally insignificant. At others, many people were mobilized and came to use very different action repertoires, whether predominantly moderate and conventional, disruptive and even violent, or mixtures of the two. Moreover, the intensity of the conflict did not necessarily correspond to the inherent burdens and risks of the technology at stake. For instance, despite the fact that a major nuclear accident may kill thousands and even millions of people, the struggle over a "conventional" technical project such as the construction of the Narita airport near Tokyo led to much more violent confrontations (Apter and Sawa, 1984) than have happened at any nuclear construction site in the world. Similarly, a relatively small project such as the expansion of the Frankfurt airport provoked more mass mobilization and more violence than many nuclear projects in West Germany. On the other hand, the construction of a large canal (the Rhine–Main–Danube Canal) which, among its other effects, is destroying the beauty of an idyllic river valley in Bavaria, has provoked only moderate protest.

This confirms the conventional wisdom that the intensity and extent of protest cannot be related directly to the "objective" significance and effects of any problem. One can therefore conclude that there are local and regional factors, such as the social composition of the population around a given site, the established political constellation, the specific strategies used by the proponents of the project, or the attitude of pre-existing organizations engaged in the conflict, etc., which mainly account for the strategies and action repertoires of local and regional protest group.

STRATEGIES IN DIFFERENT STAGES OF CONFLICT

Looking at the course of a given conflict over time, different strategies and forms of action may be used not only by a movement or during an overall movement cycle (Tarrow, 1988a, 1989), but also by specific groups and organizations. With respect to conflicts over large-scale industrial projects, two points can be made.

First, there is a tendency which one could call the "principle of generalization." This means that in the early stages of conflict the project is usually criticized only in some aspects, and often with regard to rather secondary issues. Thus, for instance, protesters criticize the choice of location or the size of the project, or they attempt to reduce side effects without questioning the project *per se*. After a time, if there has been no real success on such secondary issues, the movement's arguments become broader and deeper. People ask about the necessity of the project in principle; they protest against the manner in which they are excluded from information and participation; they relate the project to other policies; they look for substantial alternatives; ultimately they may challenge the prevailing conception of economics, technology, and politics, and even struggle for a new social order. This is precisely what happened in some outstanding conflicts over the construction of nuclear power plants or airport facilities in West Germany (Wyhl, Brokdorf, Gorleben, Wackersdorf, Startbahn West), and also in France (Malville, Ploggoff) and Switzerland (Kaiseraugst, Graben).[11]

Often these changes over time cannot be seen purely as a result of local or regional factors. Behind them lies the development of an entire movement, whose experiences, organizations and strategies influence, or even overshadow, local struggles. Sometimes local groups have successfully warded off the direct intervention of external groups, for example in the conflict over the construction of a nuclear power plant in Wyhl. Sometimes protest groups were happy to gain external support and to form broad alliances; this happened in the conflict over a nuclear processing plant in Gorleben and in the conflict over the extension of the Frankfurt airport. Sometimes, for instance in the conflict over a nuclear reprocessing plant in Wackersdorf, locally based and external groups acted side by side in a kind of implicit division of labor (Kretschmer and Rucht, 1987).

Thus, in order to understand the determinants of the action repertoire, we must weigh these factors, and moreover we must assess the indirect influence of non-local factors on local conflicts, for instance, mass media reports about similar conflicts. A comparative analysis should also evaluate the degree to which intervening variables can reduce or eliminate the generalization tendency. In this respect, I would speculate that the possibility and degree of generalization basically depends on the length of the conflict and the social composition of the involved protest groups. The longer the conflict, the younger and better-educated the activists, and the closer protest groups and organizations are interlinked, the more likely it is that generalization may occur. Similarly, generalization may occur rapidly in local groups which are joined by actors from a movement with a fully developed ideology and a lot of accumulated conflict-experience. This, for instance, was the case with the resistance against the nuclear reprocessing plant in Wackersdorf, Bavaria. Prior to the choice of this locality as the plant site, the local population was neither engaged in the nuclear power debate nor experienced in this kind of local struggle. But after the decision, they rather quickly moved towards a rejection of nuclear power in general, a move which had previously taken more than a decade for the anti-nuclear movement in general (Kretschmer, 1988).

Second, I want to draw attention to another protest phenomenon which I call the "radicalization tendency." This concept implies that activities in the early stages of the conflict are usually moderate and conventional. But if they fail, the action repertoire becomes enlarged, and thus disruptive and militant activities may also be pursued.[12] Of course this tendency can be closely linked to the generalization tendency. If protest groups consider a project to be unnecessary, to be extremely dangerous, or to serve only the interests of capitalists and/or political elites, they will be inclined to revert to more radical forms of conflict (see also Rochon's analysis of the British peace movement, chapter 6 in this volume). Here too, the move toward greater radicalism depends on non-local factors. If there are already radical tendencies within the broader movement supporting the local actors, it is likely that they will try to influence the course of the conflict and to convince local actors that radical actions are best suited to stop the project or to combat the general policy and power structure which underlies it. Of course, the local groups' choice of strategies depends not only on their allies but to a large extent on their opponents, whose conflictual and sometimes accommodative behavior has a great impact on strengthening, weakening, or overruling the radicalization tendency. With respect to these strategies, it is hard to identify any general correlations. Both repression or symbolic integration may or may not reduce the conflict. Partial concessions may encourage further demands or cause splits between protest groups. But as a general rule, the use of a flexible strategy by movement opponents, ranging from calculated repression to symbolic integration and partial concessions, is likely to reduce the radicalism of protest. I do not mean to imply, however, that such cases permit assumptions about the specific outcome of the conflict. Case studies show that both moderate and radical forms of action may or may not be successful.

Conclusion

This chapter has dealt with a broad variety of strategies and actions used by new social movements in public conflicts. Contrary to assertions made by some social scientists, however, substantive differences between the strategies and actions of new movements and those of former movements do exist. Two characteristic features of the new movements' activities are the enrichment of the action repertoire, and the parallel and flexible use of both conventional and unconventional actions. Litigation and other procedural channels to express discontent are becoming quite relevant. With respect to disruptive and militant forms, the most striking feature is the increase of civil disobedience and the relative infrequence of severe violence compared to the periods of civil war and severe class struggle in previous centuries. These differences can be explained basically by macrostructural changes.

A closer look at the more specific strategies and actions of individual movements reveals some differences which can be explained only in part by the movement's logic and its general strategy, both of which are caused mainly by macrostructural factors. Other explanatory variables can be located at a mesostructural and a microstructural level, which in turn, may interplay with accidental factors. The impact of these factors on different levels was illustrated with regard to various struggles of environmental movements.

To compensate for our scanty knowledge of the parameters influencing different movement strategies and forms of actions, a major effort should be made to carry out systematic comparative analysis, at the level both of individual movements in given periods and areas, and of in-depth case studies (also see Lowe and Rüdig, 1986). This requires the creation of categories and hypotheses, and thus a theoretical effort. Unfortunately, many theories on new movements are either implicit generalizations of individual movements or have no empirical basis at all. The emphasis of this paper was to present a conceptual approach that acknowledges the characteristics of new movements' strategies and actions in general without neglecting the specificity of particular movements.

NOTES

I wish to thank the editors of this book, the anonymous reviewers, Sidney Tarrow, and Roger Karapin for their helpful comments on earlier versions of this chapter.

1 In the second half of the nineteenth century, the category "social movement" – in the singular form – was often used synonymously with "labor movement."
2 On the other hand, a trend in the opposite direction toward aggregating interests at a high level and shifting to national or even international conflict arenas cannot be denied, e.g., on issues such as air pollution, nuclear energy, or abortion. In such cases, small and locally based groups often establish large organizations or even Left-libertarian parties on a national level (see Kitschelt in this volume, chapter 10), without, however, giving up their strongly anti-hierarchical base.
3 The crucial role of the administrative courts for the conflict over nuclear power, in particular for the West German case, has been stressed by Nelkin and Pollak (1981).
4 Klandermans (in this volume, chapter 7) documents the organizational overlappings and alliances with respect to the Dutch peace movement.
5 The difference between strategy and tactics seems to be stressed in Europe more than in the United States. Tactics can be understood as a specific concept of conflict behavior based on a situational assessment of available resources as well as of the benefits and costs of various forms of action, both for the actor and its opponent(s). Tactics may rely on the surprise effect of a sudden behavioral change, on the actor's anonymity, on techniques of accusing or ridiculing an opponent, on deception, etc. These various techniques may change from one situation to another, and they are not necessarily embedded in a general strategic concept. "Since tactics pertain to immediate situational contingencies, little of general value can be said about these decisions. This is not true, however, for strategy" (Jenkins, 1981, p. 135).
6 For example, Gordon and Babchuk (1959) distinguish between expressive and instrumental organizations. Similarly, Zald and Ash (1966, p. 329) differentiate between organizations with solidarity incentives and those with purposive incentives. See also Raschke's dichotomy of culture- and power-oriented movements (Raschke, 1985, pp. 396ff), Curtis and Zurcher's (1974) contrast between movements with expressive, diffuse goals and movements with instrumental, specific goals, or Pizzorno's distinction between expressive, non-institutional collective action and instrumental action (Pizzorno, 1978). Turner and Killian's trinity of power-oriented, value-oriented, and participation-oriented movements (Turner and Killian, 1972) seem to me less convincing in so far as (a) striving for participation is always a matter of the distribution of power and (b) participation can be a value in itself.
7 The category denotes moderate attempts at a gradual change of dominant cultural patterns in supporting or practicing deviant forms of behavior which are still considered to be "tolerable" by the mainstream culture.

8 According to Gordon and Babchuk (1959, p. 22), instrumental organizations spend more energy on outward-directed activities than expressive organizations.

9 The idea of visibility and latency is still a crude concept, depending on the standpoint of the observer. The anonymity of terrorist group members, to take an extreme case, exists only for those outside the group. Insofar as social movements typically do not have clear borderlines and membership criteria but can be described more adequately as a set of concentric circles (Lang and Lang, 1961, p. 526), reaching from core activists to very marginal groups (such as mere sympathizers), the duality of inside and outside, and thus of latency and visibility, has to be qualified.

10 The concept of opportunity structure, as it has been introduced by Eisinger (1973), elaborated by Tarrow (1983), and adopted by others (Brand, 1985; Kitschelt, 1985), involves both macro- and mesostructural parameters.

11 To be sure, not all groups and organizations follow such an ever-expanding pathway from the beginning to the end. Some hardly move. Others go a long way in this direction, while, however, experiencing substantial changes in their membership and/or social composition.

12 Kriesi (1982, p. 267) stresses this trend with respect to a conflict over a nuclear plant in Switzerland. In assessing shifts in the activities of peace movements over time, Rochon (chapter 6 in this volume) finds a similar trend.

Part IV

New Movements and Political Parties

10

New Social Movements and the Decline of Party Organization

HERBERT KITSCHELT

Parties, interest groups, and social movements are elements in a closely woven fabric of linkages between civil society and political institutions of the democratic state. Each element can be fully understood only as part of an overall pattern of interest intermediation. Theories of interest group, party, and movement behavior, however, have often violated this theoretical premise and have run into predictable difficulties. For instance, the literature on neo-corporatism almost exclusively focuses on the interaction between capital, labor, and state administration, without considering the constitutive role played by centralized mass parties.[1] Similarly, the theoretical literature on parties and party systems often defines the role of parties exclusively in terms of vote maximization.[2] Yet parties are also organizations that recruit political elites, represent and mediate the claims of competing interest groups, and develop policy programs. The literature on social movements has given little consideration to the institutional fabric in which movements are enmeshed. There has been a general agreement that contemporary social movements in advanced capitalist democracies mobilize in response to the growing intervention of market and states in all spheres of social life and the inability of the existing agencies of interest articulation to transmit movement demands into the political process. But the literature on the role of movement mobilization in networks of interest intermediation is still sparse.[3]

My first task in this chapter is to place the rise of what I call "Left-libertarian" political parties in a political setting characterized by a specific pattern of interaction between interest groups, political parties, social movements, and state agencies. Particularly where the institutional fabric furthers neo-corporatist bargaining among organized producers (business and labor), activists in new social movements are likely to create or at least vote for emerging Left-libertarian parties. Although Left-libertarian parties and new social movements operate in different institutional domains, they both seek to disrupt established relations between state and civil society from the perspective of similar visions and objectives. They are "Left" because they share with traditional socialism a mistrust of the marketplace, of private investment, and of the achievement ethic, and a commitment to egalitarian redistribution. They are "libertarian" because they reject the authority of private or public bureaucracies to regulate individual and collective conduct.[4] They instead favor participatory democracy and the autonomy of groups and individuals to define their economic, political, and cultural institutions unencumbered by market or bureaucratic dictates. In other words, Left-libertarians accept important issues on the socialist agenda, but reject traditional socialism's paternalist-bureaucratic solutions (centralized state planning) as well as the primacy of economic growth over "intangible" social gratifications. As a consequence, they evaluate the welfare state, one centerpiece of modern state activity, ambiguously. They endorse the security and protection the welfare state provides from the vagaries of the labor market. At the same time, they disapprove its hierarchical-bureaucratic organization and call for more decentralization and consumer control of social services (education, health, welfare).

My second task is to explore whether the relationship of Left-libertarian parties to established networks of interest intermediation shapes their internal structure and strategy formation. If Left-libertarian parties represent a "backlash" against existing institutional rules and actors, such as political parties, do they also represent a new organizational form and internal style of collective decision-making that anticipates the open, participatory democracy with which they would displace neo-corporatist elite bargaining in all spheres of politics? Or, will the exigencies of vote-getting compel Left-libertarian parties to adopt organizational structures and strategies similar to those developed by the electorally successful conventional conservative, socialist, and liberal parties? If the latter prevails, Left-libertarian parties will follow a *logic of electoral competition.* If they diverge from conventional models of party organization, they will be inspired by the organizational practices and ideologies of Left-libertarian movements and pursue a *logic of constituency representation*, allowing ideology and policy rather than the constraints of vote maximization to shape the form of political parties.

I will show that Left-libertarian parties are developing a new mode of participation and collective decision-making in parties, but one that frequently operates in ways unintended and unexpected by those who endorse a Left-libertarian ideology. These "perverse effects" (Boudon, 1977) give rise to an *organizational dealignment* of parties and their societal constituencies even though party activists try to realize a logic of constituency representation. In one respect, however, the organizational dealignment of parties and constituencies is consistent with Left-libertarian beliefs. In contrast to the prevailing view in conventional parties, Left-libertarians believe political parties to be only one of many actors that ought to participate in governing and shaping social institutions, but not have a dominant focal position. In this respect, organizational dealignment symbolizes efforts to lower the centrality of political parties in the networks of interest intermediation and enhance the autonomous self-organization of citizens in social movements to gain access to the state.

Parties that, even if inadvertently, engage in a process of organizational dealignment must eventually come to terms with the constraints of electoral competition. Unlike the social movements, parties not only express substantive political demands, but also strive to accumulate a generalized power resource – votes. In order to accomplish this, they immerse themselves in a unique competitive setting with other players pursuing the same objective. A logic of constituency representation and organizational dealignment may not be the most promising of strategies to be successful in electoral competition. The future of Left-libertarian politics may hinge on the partie's ability to render the ideology and aspirations of their core constituencies compatible with the imperatives of gaining sufficient electoral support to influence policy-making.

Rather than focusing on the particular conditions of Left-libertarian politics in individual countries, I will try to extract common experiences from a number of studies in the still limited literature on the Left-libertarian parties: (1) a comparative study of the Belgian and West German ecology parties based on 134 semi-structured elite interviews with municipal councillors, party executives, and parliamentarians belonging to these parties (Kitschelt, 1988b, and 1989); (2) a survey among activists in the Belgian ecology parties Agalev and Ecolo (Kitschelt and Hellemans, 1990); (3) studies about the turn of Left-socialist parties to Left-libertarian politics in France, Denmark, and several other cases (Baumgarten, 1982; Hauss, 1978; Logue, 1982); and (4) studies of Left-libertarian American political activists in the US Democratic Party (especially Wilson, 1962, and Kirkpatrick, 1976). These American activists exhibit many of the same ideological and behavioral predispositions as their European counterparts, but have not founded their own party.

My own research focus has been on Left-libertarianism in Belgium and West Germany because both countries, while being advanced industrial welfare states, have sufficiently different party systems, patterns of interest intermediation, and histories of Left-libertarian protest mobilization to make it possible to explore how the variation of institutions and political events influences the articulation of Left-libertarian parties. It is then possible to develop hypotheses about attributes all Left-libertarian parties are likely to share regardless of national political configurations and those that vary across political settings. In this vein, the Belgian ecologists have been viewed as exponents of a "moderate" Left-libertarianism, whereas the West German Greens are regarded as "radical" Left-libertarians (see Müller-Rommel, 1985b).

Electoral Realignment and Left-libertarian Parties

The rise of new social movements has recently been explained by broad "pull" and "push" theories of social development (see Brand, 1982; Offe,

1985, and chapter 12 in this volume). Various studies hold either new grievances caused by economic and political rationalization or rising expectations and changing value preferences in affluent societies responsible for the surge of protest activism. While a combination of these hypotheses goes a long way toward explaining the *content* of new social, political, or cultural demands and the activists' call for decentralized, participatory social institutions, they cannot sufficiently account for the *organization* and *strategy* of Left-libertarian politics (Melucci, 1985, p. 792). Why would Left-libertarian demands necessitate the formation of protest movements outside the existing channels of political participation? And why would these unconventional demands trigger electoral realignments in which existing or new parties appeal to a Left-libertarian clientele consisting of younger, well-educated voters, primarily working in or aspiring to jobs in the personal service sector?

Although economic affluence is a necessary condition for the rise of Left-libertarian demands, not all affluent countries have experienced an escalation of Left-libertarian protest activism. Even among countries with new social movements, Left-libertarian parties have won a substantial electoral following only in some. Specific institutional arrangements and power relations in a country's networks of interest intermediation affect the probability that Left-libertarian parties will garner a substantial following.

Table 10.1 reports the electoral strength of Left-libertarian parties in Western democracies. One group of countries had electorally significant Left-libertarian parties since the early 1980s. Most of these countries are characterized by comprehensive welfare states, corporatist tripartite bargaining between business, labor, and state, and a long-standing participation of socialist parties in governments.[5]

Corporatist welfare states provide economic security through social insurances and public sector employment that encourage the more affluent and protected groups to shift their political attention away from concerns directly related to material affluence and economic growth. Furthermore, with rising levels of mass education, sensitivity to losses of individual self-determination and the sophistication of social criticism are increasing. But while corporatist welfare states are especially conducive to the rise of Left-libertarian preferences, their channels of interest intermediation offer few opportunities for communicating these new demands into the political process. Centralized parties, interest groups, and bureaucracies rigidly resist the Left-libertarian demands articulated by loosely federated, decentralized grassroots initiatives because of substantive and procedural concerns. Left-libertarians attack the collusion of organized labor and capital which, in their view, furthers the bureaucratization and commodification of modern life and disregards intangible collective goods, such as environmental protection or the autonomy of citizens. Moreover, Left-libertarian efforts to institute participatory democracy may pose a direct threat to the organizational coherence and power of centralized parties and interest associations. In

Table 10.1 Left-libertarian parties in Western democracies, 1980–1988

Country	*Party*	*Percentage of vote*
Countries with electorally significant Left-libertarian parties		
Austria	Greens	4.4 (1986)
Belgium	Agalev and Ecolo	7.1 (1987)
Denmark	Socialist People's Party	14.5 (1987)
Iceland	Women's Party	5.0 (1983)
Luxembourg	The Green Alternative	5.2 (1984)
Netherlands	Parties of the Green Progressive Accord	5.7 (1982)
Norway	Socialist People's Party	5.4 (1985)
Sweden	Left Communist Party	5.6 (1988)
	Environmental Party	5.0 (1988)
Switzerland	Greens	4.8 (1987)
	Progressive Organizations	3.8 (1987)
West Germany	Greens	8.3 (1987)
Borderline cases with Left-libertarian proto-parties		
Finland	Greens	4.0 (1987)
Italy	Radical Party	2.6 (1987)
	Green Lists	2.9 (1987)
Countries without relevant Left-libertarian parties		
Australia	no party	–
Canada	Green Party of Canada	nc
France	Ecologists	1.2 (1986)
Greece	no party	–
Ireland	Comhaontas Glas/Greens	nc
Japan	Green Party	nc
New Zealand	Values Party	0.2 (1984)
Portugal	no party	–
Spain	Green Party	–[a]
United Kingdom	Ecology/Green Party	1.1[b]
United States	Citizen's Party	0.1[c]

Table entries are party's best performance in national parliamentary elections, 1980–8. nc = elections not contested.
[a]unsuccessful participation in leftist alliance.
[b]in districts contested in 1983.
[c]presidential election in 1984.

many ways, Left-libertarians represent "consumer" interests against industrial and bureaucratic "producers."

Left-libertarians develop new vehicles of political mobilization when the existing parties and interest groups in corporatist welfare states do not respond to their demands. Social movements represent a first step in this direction. Founding new parties or trying to redefine the mission of smaller centrist and leftist parties towards Left-libertarian objectives (as attempted in the Netherlands or Sweden) constitutes a second step in placing Left-libertarian demands on a country's political agenda. Empirically, the link between strong corporatist arrangements and Left-libertarian party formation is indicated by a very substantial association between low strike activity, commonly viewed as one consequence of corporatist arrangements (see Hibbs, 1976), and the emergence of the new parties.

Aside from the countries with high corporatism, Left-libertarian parties in the late 1980s have been approaching the threshold of significant electoral support in Finland and Italy, two countries not obviously covered by my explanation. With the decline of the Communist party and an increasingly corporatist type of interest intermediation, Finland begins to resemble the other countries with Left-libertarian parties. Italian Left-libertarianism, in contrast, clearly cannot be explained by corporatist domestic institutions. Yet the Italian "Green lists" are still too new and politically amorphous to determine whether their success in the 1987 election signals a long-term trend or a brief interlude.

Democracies without successful Left-libertarian parties all share relatively pluralist patterns of interest intermediation and have been dominated by non-socialist political forces. In several of them, however, socialist and labor parties have recently gained momentum, an observation I will return to in the conclusion to this chapter. Most of the democracies without Left-libertarian politics have significantly lower living standards than countries where these parties are present. Exceptions are Canada, France, and the United States. Plurality voting systems constitute another impediment to Left-libertarian politics in the Anglo-Saxon democracies, but is certainly not the most powerful explanation of Left-libertarian party performance. Countries with plurality voting systems also have seen new challenging third parties emerge. The French case is of special interest, demonstrating that voting systems by themselves are not a satisfactory explanation for the presence or absence of Left-libertarian parties. The special rules of French plurality voting have actually encouraged a multi-party system, yet this system has not benefited Left-libertarian parties, who received disappointing electoral support. Moreover, the electoral performance of Left-libertarian parties did not improve in the 1986 French parliamentary election, in which seats were allocated according to a system of proportional representation.

The conclusion that the established networks of interest intermediation explain the transition from Left-libertarian movements to political parties thus appears to be on firm ground. Since Left-libertarian parties oppose centralized economic producer associations and their party allies, we must next examine whether they create a credible counter-model of political organization. From an analytical point of view, little can be learned from comparing a group's "ideals" with its actual accomplishments. Instead, we will examine the extent to which the actual mobilization of Left-libertarian parties differs from that of conventional parties in corporatist welfare states.

The Decline of Party Organization in Left-libertarian Politics

Left-libertarian movements reject centralized bureaucratic organization and engage in a participatory, fluid, decentralized, and horizontally coordinated mobilization of activists. They rely on small local organized cores, surrounded by loosely affiliated sympathizers, and on weak national umbrella organizations that are called upon to coordinate a limited number of central protest events or political campaigns.[6] In addition to the anti-corporatist thrust of Left-libertarian movements, a number of factors explain the movement's loosely coupled structure. Movement objectives are non-fungible collective goods which mobilize a relatively limited number of people and make it difficult to attract support through selective incentives (Hardin, 1982, especially chs V-VII). Moreover, in contrast to the highly organized labor movement, Left-libertarians are not concerned with one central struggle (such as industrial conflict about wages and property rights), but a multiplicity of issues and mobilized constituencies in different spheres of society inhibiting a broad collective political mobilization. Finally, Left-libertarian activists cannot easily tap such "natural" social solidarities as class, regional, and cultural identities to build political organizations. Although most activists belong to the educated salaried social stratum, the same applies to their most ardent opponents in the bureaucratic and business elites. Class positions are characterized by cross-cutting interests and make it difficult to predict citizen's political orientation. Members of the educated salaried groups routinely find their "interests" shift or even conflict, depending on whether actors are concerned with their roles in production (as bureaucrats, managers, scientists, and professionals) or in the consumptive spheres of social life (as clients of personal services, urban residents, users of recreation facilities). Class and education as such are therefore weak predictors of Left-libertarian activism compared to the more specific social, political, and occupational experiences that predispose individuals to opt for Left-libertarian politics.[7]

Based on the dynamics of new social movements, I will explore five dimensions along which Left-libertarian parties may diverge from conventional political parties:

- the socio-economic profile of party activists;
- the enrollment of Left-libertarian and conventional parties;

- the commitment mechanisms that bind activists to political parties;
- the structure of party organization in conventional and Left-libertarian parties
- the linkages between parties and external constituency groups.

Thus my discussion will move from the micro-level of individual activists and their party affiliation to the aggregate patterns of interaction and organization inside Left-libertarian parties and between parties and their external clienteles. My key argument is that the degree of organization in Left-libertarian parties is much lower than in conventional parties in the same party systems. I do not identify this decline of organization in terms of formal party statutes elaborating rules of representation and the rights and obligations connected with formal party positions. I am concerned with the organization as an empirical system of patterned activity (i.e., the militants), the processes of involvement, decision-making, and strategic action. Even though formal statutory provisions in Left-libertarian parties are often similar to those of conventional parties, they have become empty shells not filled with the life of a mass party.

THE SOCIAL BACKGROUND OF ACTIVISTS

Survey data confirm that most Left-libertarian party activists belong to the group of young intellectuals and professionals holding salaried positions in education, health care, and cultural services or being educated to fill jobs in these areas. Belgian and West German ecology party activists have a much higher level of education than the average population of their own age cohort (see table 10.2). Many activists have studied social sciences and liberal arts, while administrative and blue-collar jobs are under-represented. Management, liberal professions, and business poll at or above the level representative of the overall national occupational profile (table 10.3).[8]

A similar picture emerges from studies of other Left-libertarian parties. The French Unified Socialist party and the Danish Socialist People's party experienced a strong influx of young highly educated professionals when they moved away from their blue-collar clientele and committed themselves to Left-libertarian causes in the early 1970s (cf. Hauss, 1978, ch. IV; Logue, 1982, ch. 7). A similar sociological profile characterizes Left-libertarian reformers in the American Democratic party. In the Democratic Club movement of the late 1950s, young professionals were dominant (Wilson, 1962, pp. 258–88) and at the 1972 Democratic party convention, delegates committed to George McGovern were younger, more urban, more secular, and more likely to hold Ph.D.s (but less likely to have law degrees) than other delegations (Kirkpatrick, 1976, pp. 68–90). McGovern's campaign also attracted "symbol specialists", such as teachers, academics, or journalists.[9]

Table 10.2 Educational levels of Left-libertarian party activists, party sympathizers, and the general public in Belgium and West Germany (percentages)

Educational level	*Agalev, Ecolo, and Green party elites*	*Agalev and Ecolo party conference participants*	*Agalev, Ecolo and Green party sympathizers*	*Belgian and West German public*
Basic Education				
General secondary schooling and occupational training	7	14		
Completed education by age 16			16	48
Intermediate Education				
Advanced technical training or college	12	42		
Completed eduction by age 21			36	32
Higher education				
Attended university or university degree	81	44		
Completed education after age 21 or still in school			47	20
Total	100	100	100	100
(Number of cases)	(134)	(252)	(409)	(5,830)

Sources: The party elites sample is based on interviews with municipal councillors, members of the state and national party executives, and state and national parliamentarians in the Greens, Agalev and Ecolo (see Kitschelt, 1988b). The party conference sample is based on a survey of party activists taken at the October 1985 Agalev and Ecolo national party conferences (see Kitschelt and Hellemans, 1990). The data for party sympathizers and the general public are pooled from *Eurobarometers* 17 (1982), 21 (1984), and 25 (1986).

Although there are still many more male activists, Left-libertarian parties recruit an unusually large proportion of women. In the Belgian ecology parties Agalev and Ecolo, one quarter of the activists in 1985 were women, and in the West German Greens about a third of the activists are female (Kitschelt, 1989, ch. 4). Case studies illustrate that women and the women's movement have considerable influence on Left-libertarian parties (see Baumgarten, 1982).[10]

Although most Left-libertarian militants belong to the educated middle class, their personal income is not significantly higher than that of the

Table 10.3 Occupations of Left-libertarian party activists, party sympathizers, and the general population in Belgium and West Germany (percentage)

Occupational level	*Agalev, Ecolo, and Green party elites*	*Agalev and Ecolo party conference participants*	*Agalev, Ecolo and Green party sympathizers*	*Belgian and West German public*
Technical and liberal professions, management and business	31	12	6	10
Education, personal services, media, pol. associations	27	40 }		
White collar-employment	–	– }	24	21
Administrative-technical occupations	7	19 }		
Students, military, and alternative service	9	8	36	14
Blue-collar employment	4	3	12	18
Not active in formal labor markets; homemakers; retirees; unemployed	21	13	23	38
No response, insufficient data	1	5	–	–
Total	100	100	101	101
(Number of Cases)	(134)	(256)	(399)	(5,687)

The *Eurobarometer* survey included only the general category of white-collar employment while the party elite studies differentiate between administrative-technical and other white-collar occupations.
Sources: as table 10.2.

average salary-earner.[11] In part, this is due to their youth; in part, this reflects their preference for "symbol specializing" occupations which yield lower incomes than business or administrative pursuits.

Compared to Left-libertarian parties, non-socialist parties of the Right have always been dominated by middle-class businessmen, lawyers, public officials, and farmers with far above average personal incomes.[12] Even socialist, social democratic, and communist parties are increasingly drawing on middle-class activists, particularly state officials. The class background of party members usually rises with their position in the party. With the exception of political parties in Scandinavia and Britain, women still constitute very small minorities in mainstream party politics (Lovenduski, 1986).

Two contrasts between conventional and Left-libertarian parties emerge. First, Left-libertarian parties attract more "symbol specialists" and women than conventional parties. Second, the sociological disparity between voters and party militants is greater in the case of conventional parties than in that of Left-libertarian parties. Conventional parties, including contemporary socialist parties, are upper middle-class, but receive substantial support from lower and lower middle-class voters. Left-libertarian activists, on the other hand, represent average income levels and are sociologically less removed from their voters because the parties draw over-proportionally on young, educated, urban voters.[13]

MASS OR FRAMEWORK PARTIES?

Left-libertarian parties attract only a minute fraction of their electorate as party members. All available estimates indicate a degree of organization not exceeding 2 percent of the voters. The Danish Socialist People's partly (SPP), for instance, had 6,900 members in 1974, but received close to 300,000 votes in the 1973 national election (Logue, 1982, p. 207). The French PSU received up to 900,000 votes in the late 1960s, but had about 14,000 party members or a 1.7 percent level of organization (Hauss, 1978, p. 77). The Norwegian Socialist People's party had 6,500 members in 1965, but 122,700 voters (2.1 percent level of organization). In West Germany, the Greens organized about 1.5 percent of their voters in the 1983 and the 1987 national parliamentary elections. In the Belgian parties Agalev and Ecolo, only 0.5 percent of the voters in the 1985 and 1987 national elections were organized.

Membership turnover in Left-libertarian parties appears to be exceptionally high. In the Belgian and West German Greens, estimates are in the neighborhood of 15 to 25 percent of members leaving or joining every year (cf. Kitschelt, 1989, ch. 3). The French PSU (Hauss, 1978, p. 77) and the Dutch Pacifist Socialist party (Gerretsen and van der Linden, 1982, p. 98) experience similar levels of turnover. Although many Left-libertarian parties are relatively young and have gained electoral support over the past ten years, they have been much less successful in increasing their membership. New members are usually offset by losses. Low membership enrollment and high turnover rates are consequences of the parties' inability to provide strong commitment mechanisms to party activism, as I will discuss below.

In contrast to Left-libertarian parties, most conventional parties, whether they are conservative or socialist, have created a much higher degree of organization or "organizational encapsulation" of their voters (Wellhofer, 1985). This is particularly true for socialist parties, which have levels of organization ranging from 5 to 7 percent – such as in the German Social Democratic party – to close to 30 percent of their electorate as in Sweden, although in most countries organization levels have stagnated or slightly declined since World War II (Bartolini, 1983). At the other end of the political spectrum parties of the Right have improved their organization rate in many European countries since the 1960s.[14]

COMMITMENT MECHANISMS IN LEFT-LIBERTARIAN POLITICS

Activism in political parties depends on the mutual reinforcement of personal motivations to become active and organizational incentives that satisfy the individual's expectations. Following Kanter (1972), I will call the pairs of motivations and corresponding incentives *commitment mechanisms*. Membership and activism in Left-libertarian parties remain limited because the scope and intensity of commitment mechanisms is highly constrained. I will illustrate this for purposive, solidary, and material commitment mechanisms.

All studies of Left-libertarian politics show that party activists are moved by strong purposive motivations. Many activists had been involved in Left-libertarian causes before they joined the parties. In a sample of 134 Belgian and West German ecology party activists and leaders, 55 percent had participated in social movement organizations, 39 percent had been involved in the student movement and New Left groups and 17 percent had been in another party before they joined the ecologists. Only 18 percent had no history of political activism before entry into the party (Kitschelt, 1989, ch. 4). In regions where Left-libertarian politics has led to many protest events and social movements, such as metropolitan areas with a high proportion of salaried professionals and universities, party activists tend to have been more involved in a variety of political experiences than in the "peripheries" of the conflict, such as rural or working-class areas. These political careers show a strong substantive commitment to Left-libertarian politics preceding the entry into the parties.

Left-libertarian purposive motivations, however, do not necessarily translate into sustained party activism. Since Left-libertarian ideologies turn against formal organizations and hierarchies, activists generally develop little sense of "organizational patriotism" and loyalty to their party. In fact, many look upon party membership with disdain. Militants frequently express a certain ambivalence, if not outright cynicism, about the value of party life and party activism. Since many activists are intellectuals with highly individualistic and ideological notions of politics, it is difficult to maintain the organizational viability of Left-libertarian parties. Activists report that intellectuals particularly are unwilling to assume organizational responsibilities. A considerable share of party members subscribes to an anarchist radicalism hostile to any form of organizational work. Others who come from social movements are more concerned with the specific issues they wish to promote both in movement and party, not with the party's organization and progam at large. Feminists, in particular, tend to dislike party organization and reject any obligation to be "generally interested in politics" beyond their specific concerns.

Similar tendencies have emerged in other Left-libertarian parties. Activists in the French Unified Socialist party, too, have little sense of organizational loyalty (Hauss, 1978, p. 176) and feel very uncomfortable with recruiting members (Hauss, 1978, p. 58), a reluctance shared by Belgian and West German ecologists. In the Danish Socialist People's party, ideology and strategy are the most important motivations (Logue, 1982, p. 220), but activists generally lack the self-sacrificing ideological fervor typical of communist party militants (Logue, 1982, p. 146). McGovern's delegates in the American Democratic party also had little organizational commitment (Kirkpatrick, 1976, pp. 122, 144–7), and emphasized individual policy issues rather than the party's electoral fortunes (Kirkpatrick p. 101). A similar anti-organizational bias characterized the Democratic Club movement of the late 1950s (Wilson, 1962, ch. VI).

The importance of purposive motivations linked to an anti-organizational ideology has four negative consequences for commitment mechanisms in Left-libertarian parties. First, few sympathizers actually join and contribute to the parties. Second, many who join become disaffected with the party quickly, accounting for the parties' high turnover. An evolutionary process of selection and retention takes place: only those with greater strategic and organizational commitments survive in the party and gradually also the percentage of new members with organizational commitments increases. German and Belgian ecologists confirm this pattern and observe that, as time has passed, more people have joined because of organizational rather than issue-related reasons. In the words of a Green state parliamentarian, "many join the party now not as a representative of particular issues (*Betroffenheiten*), but as an end in itself." A Belgian activist deplores that "what is left in Agalev are political functionaries; the cultural and social aspects of our work have been dropped."

The importance militants attribute to purposive commitments inevitably leads to a certain sense of disappointment with the parties' accomplishments. Purposively motivated activists want to change the world, but are easily discouraged by the cumbersome process of intra-party collective decision-making, let alone the party's impact on policy formation. The often-heard frustrations with party conferences illustrate this problem well (Kitschelt, 1989, ch. 6).

Finally, the nature of Left-libertarian purposive commitments undercut other commitment mechanisms, primarily those based on social and material incentives, a point which deserves somewhat greater elaboration. Social incentives (atmosphere, friendship, social events) may be relatively unimportant in joining parties, but constitute key "secondary incentives" reinforcing activism (see Wilson, 1962, p. 171). The social life of parties is especially important for activists from lower social status groups with less education (Becker and Hombach, 1983, p. 92). Left-liberatarian activists are mostly well-educated intellectuals with strong purposive motivations, but little concern for social atmosphere and friendship in the party. Consequently, friendship bonds within the party are usually limited to political cliques which usually originated outside the parties. Because many activists ignore the value of organizational loyalty and social atmosphere, intra-party debates often lead to hostile disputes in which accusations and insults disregard any sense of basic solidarity and personal respect among the militants. Characteristically, the organizational culture of Left-libertarian parties is most conflictual in party sections with a high proportion of young intellectuals. These are usually located in the "centers" of the Left-libertarian cleavage, located in large metropolitan areas with many protest events.

Activists in the Belgian and West German ecology parties now often see a loss of the playfulness, utopia, and social atmosphere that characterized their parties initially. The founder of the Flemish party Agalev, the Jesuit Versteylen, warned in 1981 when the party turned towards electoral politics and attracted more ideologically committed activists that it would become a party "without women, without youth, without inspiration" (Versteylen, 1981, p. 60). A Green militant in Frankfurt reports that young people enter, but are soon repelled "by our crazy organizational structures and political style." And for another "many who have not been politically active before they came to the Greens feel that politics gains too much the upper hand in the party" (interview materials employed in Kitschelt, 1989).

The withering of social incentives is not confined to the Belgian or West German ecologists. In the French PSU, young, male, anti-clerical intellectuals with a severe, divisive, and polemic style drove out the older, female, and Catholic working-class activists who constituted the initial core of the party (Hauss, 1978, p. 93). In Kirkpatrick's (1976, p. 104) study of American party delegates, George McGovern's Left-libertarians put the least emphasis on social atmosphere as an incentive to joining.

While the weakness of solidarity incentives is an unintended, perverse consequence of the parties' dynamic, Left-libertarians quite consciously try to undercut material incentives. People should join parties to serve collective purposes, not private interests. In order to remove material incentives, most Left-libertarian parties have rules requiring the frequent turnover of leadership positions, low pay for party employees and elected mandataries, and continuous supervision of the party elite by the rank and file. Given the small number of jobs and mandates Left-libertarian parties are able to fill, it is unlikely that many militants are attracted by material incentives. Aside from its intended effects, the material reward structure in Left-libertarian politics has also some unwanted side effects. Militants generally avoid time-consuming organizational work yielding little political satisfaction. Consequently, it is very difficult to fill service positions, even more so when they are not paid. Other organizations allocate intrinsically unrewarding tasks by material incentives or the promise of promotion. But for ideological reasons Left-libertarian parties cannot make use of such material incentives and thus generate very fragile organizational structures.

To sum up this survey, Left-libertarian parties build on highly fragile, precarious commitment mechanisms. They provide few effective incentives for reinforcing the motivations and aspirations that attract people to the parties. This state of affairs differs greatly from what we know about commitment mechanisms in conventional parties. Although the intrinsic limitations of motivational research make it difficult to determine commitment mechanisms, some evidence suggests that party activism (as against pure membership) is increasingly inspired by material and purely pragmatic considerations in most conventional mass parties. In most corporatist welfare states, parties have developed strong powers of patronage and use them not only to stimulate party adherence, but simultaneously to control policy development and implementation through well-placed supporters in the bureaucracy. Both parties of the Right and the conventional Left have witnessed an influx of government officials and public servants on to the party rolls. Often these groups are over-represented in the higher party echelons. In the German Christian Democratic party, for instance, 12.4 percent of the mostly inactive membership in 1982 were state officials, but 22.8 percent of all party functionaries and elected officers and 35.1 percent of the national parliamentarians and party leaders were state officials (Schönbohm, 1985, pp. 239, 243). Similarly, in Belgium, Austria, and other organized democracies, the "Proporz" system of bureaucratic recruitment generates a strong bond between state bureaucracy and party membership. Left-libertarian parties clearly diverge from this pattern, but intentionally and inadvertently do not provide other mechanisms to bind people to party involvement.[15]

THE POWER STRUCTURE OF LEFT-LIBERTARIAN PARTIES

Left-libertarians reject hierarchical and formal party organization. Instead, they call for decentralized, democratic procedures, broad participation of all activists in the collective decision-making process, and continuous supervision of the parties' leaders by the rank and file. Party statutes usually define individual rights and collective rules in ways to promote these ends: offices are rotated, all meetings are open and publicized, decision-making about policy, personnel, and party finances is decentralized. Yet party organization can achieve participatory procedures only when individual *rights* and organizational *rules* are complemented by sufficient collective *resources* (time for mass meetings, finance, and personnel to maintain an effective organizational communications system) and an egalitarian distribution of individual *capabilities* (time, money, information, political skills) to participate in politics. Because Left-libertarian parties are unable to match rights, rules, resources, and capabilities, participatory statutes generate unwanted, perverse effects. On the one hand, they remove the incentive to run for formal offices that have no substantive power and thus weaken coordination and communication within the parties. On the other hand, as Weber (1978, p. 949) noted, where democratic rules encounter an inegalitarian distribution of resources and capabilities, notables or informal elites will undermine direct democracy.

Left-libertarian party statutes create a loose "stratarchal" organization which relies on small cores of activists at the local, regional, or national party levels.[16] Although party militants are invited to participate at all levels of party organization, limited resources, capabilities, and spans of attention actually lead to highly stratified, decentralized party organizations. Specialists for local and national party affairs emerge who rarely interact or communicate with each other. These centrifugal tendencies also fuel political factionalism because more pragmatic activists have a tendency to focus on local politics while "ideologues" put more emphasis on national politics (cf. Kitschelt and Hellemans, 1990, ch. 7). At each separate organizational level, informal elites who possess the greatest personal commitment and resources tend to dominate the parties.

Individual resource constraints explain why participation remains relatively limited even among Left-libertarian party members. Municipal councillors and national party leaders in the Belgian and West German ecology parties estimate that about 30 to 40 percent of their members regularly attend the monthly or biweekly local party meetings. At those regional or national party conferences that are open to all party members, no more than 10 to 25 percent of the membership can be mobilized (cf. Kitschelt, 1989, chs V, VI). About 20 percent of the Danish Socialist People's party membership participate in a party meeting at least once per month (Logue, 1982, p. 207). Less than 50 percent of the French PSU members are involved in party life (Hauss, 1978, p. 61).

Overall, the effective mobilization may be somewhat higher in Left-libertarian than in conventional parties. In the latter, estimates of participation rates range from as low as 5 to 10 percent to a maximum of one-third of the membership.[17] Given, however, that conventional parties organize a much higher proportion of their voters, their activist/voter ratios are still higher than those of left-libertarian parties. In the Belgian ecology parties, for instance, a 35 percent level of activism translates into 0.2 percent of the parties' voters, whereas a 10 percent level of activism in the Belgian Socialists still indicates that 1 percent of the party's voters are involved in its internal deliberations.

As a consequence of organizational rules and widespread "culture of mistrust," party executives carry little or no formal political weight. Especially in areas with highly mobilized Left-libertarian social movements and many party intellectuals, a conspiratorial interpretation of Michels's theory of oligarchy is popular as the party's folk wisdom and colors the militants' zeal to prevent the emergence of bureaucratic organization and centralized authority. The easiest targets for these anti-organizational sentiments are the party executives who are expected to shoulder the day-to-day party administration, but face fierce resentment if they make independent political statements about substantive issues.

A crucial source of political power is the ability to communicate symbols and messages. Because most party executives have little access to the mass media or internal authority in the party, few people are willing to assume the unrewarding burden of party office. Both in the German Greens as well as the Flemish Agalev, activists generally agree that militants shy away from party office because they involve much administrative work and little political reward. Consequently, politically weaker or less representative personalities populate the parties' executive committees. Some exceptions confirm the rule. In the national executive of Ecolo in Wallonia, for instance, a strong informal clique dominates the party and its small parliamentary delegation. The West German Green national executive has gained access to TV debates with the leaders of the main parties during and after electoral campaigns, and thus represents the party to a very large audience.

It is more typical, however, that politically weak party executives promote the centrifugal, stratarchal fragmentation of Left-libertarian parties. On the one hand, they lack resources and capabilities to organize the communication between local party sections and the regional or national party level. On the other hand, they have no capacity to supervise the parliamentary party groups. Left-libertarian parliamentarians, in contrast, have the natural advantages of being paid full-time politicians with considerably more access to the mass media than party executives. For this reason, elected national office is much more sought-after in Left-libertarian parties than party office. Moreover, party executives have little opportunity to shape and control the parliamentarians' political agenda. The latter become the parties' effective spokespeople, unless the parliamentary groups are very small and exceptionally strong personalities hold executive party office.

It is mistaken, however, to assume that the Left-libertarian parliamentary party groups are cohesive, integrated centers of power controlling the parties' rank and file. They constitute aggregates of competing personalities and political entrepreneurs with very individualistic political outlooks and loyalties to diverse constituencies and party factions. Unlike most conventional parties, Left-libertarian parties usually do not enforce party discipline on legislative votes.

In this environment of fluid rules, but constrained resources and capabilities, the dominant character in Left-libertarian parties is not the career politician who continuously lives "off" politics, nor the independently wealthy notable who lives exclusively "for" politics. It is rather a postmaterialist "political entrepreneur" who has devoted his or her life to politics, but neither practices an independent profession nor derives a regular income from political office. These postmaterialist entrepreneurs rely on occasional jobs, party employment, electoral office, and the state's welfare system for an income.

The irrelevance of formal party office applies not just to the Belgian and West German ecologists, but to all Left-libertarian parties. American Democratic Club members also encountered great difficulties in coordinating their activities (Wilson, 1962, pp. 232, 251–7). Hauss (1978, pp. 7, 56, 148–9) describes the French PSU as a poor, centrifugal, decentralized party with a stratarchal separation of organizational levels and a weak executive. In the 1970s, militants saw their party more as a fluid movement, lacking fixed structure and ideology, than as a real party organization (Kergoat, 1982, p. 119). The Scandinavian left-libertarian parties also have more the character of loose "electoral alliances" than of party organizations (Lorenz, 1982, p. 42; Lund, 1982, pp. 74–5). The Danish Socialist People's party, for instance, was originally dominated by its parliamentary leaders, but their authority suffered with the party's libertarian transformation (Lund, 1982, pp. 81–2). Also the Dutch Pacifist Socialist party exhibits all the attributes of a loosely coupled, stratarchal party organization. Local groups have complete autonomy, militants take little interest in the party organization, executive positions are difficult to fill, and the parliamentary party group is the party's only central representative. Gerretsen and van der - Linden (1982, p. 101) conclude: "Altogether, the PSP is a conglomerate of largely independent partial structures (parliamentary party, party executive, and local sections)."

The organizational structure of Left-libertarian parties contrasts with that of conventional parties in parliamentary democracies. Modern mass parties professionalize the party apparatus and formalize communication across organizational levels. Political recruitment reflects this organizational design: militants win party and elected office through long careers in the "sweatshop" of the party organization where they must first learn the ropes before they are permitted to move into positions of authority.

Although mass parties are not tightly integrated as modern corporations or public bureaucracies, they centralize political leadership around highly visible personalities who usually combine party and electoral office. The personal political authority of the party leaders, combined with a formal, professional party apparatus, increase a party's strategic capacity in negotiations and compromises with competitors. Leaders enjoy sufficient authority and control to pursue a political course even when confronted with intra-party opposition. For governing parties, centralized control provides a reliable parliamentary base of support.[18] Left-libertarian parties diverge from these patterns, partly by intentional design and partly by the "perverse" consequences of weak commitment mechanisms and coordinating organizational structures.

PARTY–CONSTITUENCY LINKAGES

Even though many Left-libertarian parties have grown out of social movement struggles, they typically lack organized linkages to movement constituencies. While most party activists have participated in new social movements, the reverse is not true: only a very small fraction of those who engage in protest events and movements ever become party members, for reasons we have already explained. Conversely, those who contribute to Left-libertarian parties must often sacrifice their movement work to meet party obligations.

In this sense, Left-libertarian parties are the "exponents" rather than the "representatives" of new social movements.

Since Left-libertarian activists explicitly reject the Leninist vanguard model of hierarchical party/constituency linkage or the corporatist concertation between modern mass parties and bureaucratic interest groups, they search for a "cultural interpenetration" which will unite parties and social movements around common, shared discourses, symbols, and political objectives. In theory at least, channels of communication within the Left-libertarian subculture are expected to maintain the congruence of movement and party demands and practices. Given that most supporters of Left-libertarian politics, however, belong to no clearly identifiable and socially cohesive milieu, this cultural interpretation through common protest activities, communication media, and political visions remains relatively limited.

Many activists in the German and Belgian ecology parties report that social movements have an instrumental approach to the parties, tempered with mistrust and a desire to keep the parties at arm's length. Movements encourage the parties to assist them in pursuing limited tactical objectives (gaining access to the mass media, public financial support, etc.) and policy changes, but insist on their organizational autonomy. If regular communication between movements and parties develops at all, it remains confined to the parties' elected representatives, those best able to publicize movement demands effectively. This pattern of party/movement linkage strengthens the focal position of party militants in elected office at all party levels and contributes to the parties' stratarchal decentralization of political control.

Party activists generally see movements as particularistic agents with narrow issue concerns. Feminists and environmentalists, squatters and homosexuals, participate in party debates when their personal concerns are on the agenda, but leave when the debates moves on to other issues. But parties cannot simply aggregate the particular demands of a host of disparate movements and produce some kind of "encyclopedic compromise" between them (Wiesenthal, 1987). Because many party activists wish to synthesize constituency demands in minimally consistent, integrated policy programs, they are not always willing to subscribe to movement demands. Hence, conflicts between "movementist" and more pragmatic or ideological activists with broader concerns periodically erupt in Left-libertarian parties (Lund, 1982, p. 71). At the same time, party activists deliberate about policy issues which, while of general importance, mobilize no movements at all. In those areas, Left-libertarian politicians feel they are operating in a political vacuum without advice from constituency groups.

The relationship between Left-libertarian movements and parties is more tenuous than that between established interest groups and conventional parties. While shared subcultural milieux (working-class suburbs, Catholic agrarian regions) are on the decline, mass parties are nevertheless able to preserve a dense net of organizational bonds between the elites of parties and centralized interest groups. In Belgium, for instance, utilitarian, instrumental clientelism based on interlocking organizations fills the void left by the crumbling religious and ideological communities, and preserves the linkage between parties and interest groups (Billiet, 1984). These networks improve the parties' capacity to arrange policy compromises and coordinate actors across the entire system of interest intermediation. The density and political form of the ties between party and constituency group thus constitutes another contrast between Left-libertarian and conventional parties.

POST-INDUSTRIAL FRAMEWORK PARTIES

The actual features of Left-libertarian parties – weak commitment mechanisms, a limited scope of participation, loosely coupled, stratarchal and fluid organization, relatively unaccountable and disjointed groups of political leaders – fall short of the expectations of a tightly coupled, grassroots-controlled organization with high levels of militancy and participation, solidarity and community spirit most activists probably held when they first joined the parties. In retrospect, these expectations were based on faulty assumptions about what sustains individual commitment to politics, the trade-offs between organizational decentralization and coordination, and the individual and collective transaction costs of grassroots democracy (time, money, information, skill). But rather than comparing "ideals" to "realities" of collective political action, it is more important to emphasize the parties' constructive accomplishments.[19] Although they have not implemented a pure "logic of constituency representation," they have introduced a form a political organization to corporatist welfare states, which Raschke (1983) quite appropriately labels "post-industrial framework parties."

The devolution of party organization is by far the most unusual feature of Left-libertarian politics. Other parties in countries with Left-libertarian challenges have either sustained historically high levels of organization or improved them. Many conservative parties, for instance, have experienced spectacular growth and administrative rationalization. Socialist and social democratic parties have levelled off, but at the high level of organization reached in the middle decades of this century (cf. Bartolini, 1983; Paterson and Thomas, 1986). Overall, there are few signs that strong centralized party organizations are about to disappear. Mintzel (1984) has called the dominant parties in corporatist welfare states "mass apparatus parties." They rationalize party organization in order to combine the tasks of voter mobilization and effective policy-making in highly centralized systems of interest intermediation. They are transformed from pure representatives of voters and social constituencies into agencies of executive policy formation.

It is the irony of modern "state parties" that their electoral success and focal role in a centralized political process simultaneously contributes to the emergence of new movements and parties that call for a decisive break with these patterns of politics. Faced with Left-libertarian parties, mass apparatus parties must cope with this dilemma: in order to maintain their electoral dominance and role in the political process, they must sacrifice those voters who oppose the centralized political institutions and call for a libertarian-participatory change. But giving up these voters may cost parties decisive percentage points in the electoral competition. Conversely, if mass apparatus parties try to win back Left-libertarian voters, they risk their effectiveness in the policy-making process, due to the substantive and procedural reforms to which they would have to submit themselves. Mass apparatus parties, however, are fortunate in that their Left-libertarian competitors face a similar dilemma, to which I will turn next.

Left-libertarian Party Strategy and Electoral Competition

The organizational form and posture of Left-libertarian parties appeals to their core supporters in the new social movements and Left-libertarian political subcultures. The parties draw their strength from this new pattern of interest representation and introduce plasticity and openness in political systems that have solidified around highly organized cleavages. Yet Left-libertarian parties face electoral constraints that force them to reconsider their divergence from the style and strategy of mass apparatus parties. Because Left-libertarian parties find it difficult to choose between radical strategies of fundamental opposition to the established parties and moderate strategies of compromise and alliance aiming at incremental policy change, they risk alienating a large proportion of their actual or potential electorate whenever these issues arise. If they adhere to a radical strategy and a pure logic of constituency representation, they keep their core voters, but fail to attract a broad spectrum of moderate sympathizers. If they "rationalize" their internal organization according to a logic of party competition and engage in a moderate strategy, they risk losing the support of their core electorate or being swallowed up by competing parties.

In all competitive democracies, the ability to influence public policy and thus to provide collective benefits for a party's electorate constitutes one important measure of party success. Left-libertarian parties are especially sensitive to this criterion because the organizational dealignment of voters from the parties rules out substitutes for policy success that other parties employ to maintain a loyal following. For instance, Left-libertarian parties produce few secondary benefits which would be exclusively extended to a broad segment of their electorate. Conventional religious and socialist parties, for instance, employed ideological brotherhood and a network of cultural organizations involving all the faithful in a distinctive life-style to that end.

The imperative to maintain a sufficient electorate through policy success and the need to preserve the parties' uniqueness to please their core constituencies exert contradictory pressures on Left-libertarian activists: should the parties compromise with other political forces or pursue pure Left-libertarian objectives? Given the internal fluidity of organizational rules and the weakness of intra-party agents that could produce unity and cohesiveness (e.g. party executives and parliamentary party groups), Left-libertarian parties are often embroiled in severe internal factionalism. Kirkpatrick (1976, p. 122) has summarized the functional link between party structure and strategy succinctly: 'The looser the organization, the less continuous the interactions, the faster the turnover, the more diverse the recruitment practices, the greater the opportunity for development of divergent role perceptions and for institutional change.'

In the West German Greens, at least four different groups vie for control of the party's politics, its organizational structure, and its external strategy (see Kitschelt, 1990, ch. IX). The key source of contention is whether the party should seek government alliances with the Social Democrats or remain in opposition. A similar struggle split the Dutch Pacifist Socialist party in the early 1970s, forcing the moderate faction to leave when the radical group pulled out of an alliance with the Dutch Labour party (Gerretsen and van der Linden, 1982, p. 98). The Danish Socialist People's party spun off a radical faction in the early 1970s when it entered a temporary alliance with the Danish Social Democrats that broke apart quickly. Within the newly formed Danish Left Socialists, factionalism proliferated further and led to a decline of membership and electoral support, because the party could not define an agreed-upon program or plan for action (Lund, 1982). In the 1970s, the French PSU was also racked by intense factionalism, deplored by many activists, and quickening the party's decline (Hauss, 1978, p. 150). The moderates around Michel Rocard eventually left the party and joined the Socialists in 1974.

Factionalism and organizational dealignment also influence the size and composition of the parties' electorate. The core supporters in Left-libertarian subcultures will be unmoved by conflicts and indecision within Left-libertarian parties. But this core is small because few Left-libertarian sympathizers are situated within the parties' organizational and cultural orbit and many Left-libertarian voters tend to be sophisticated, strategic voters without fixed loyalties. They are well educated and better informed about specific policy issues than the average voter. The absence of organizational bonds to Left-libertarian parties makes their voters especially prone to vote-switching in order to reward or punish parties for their actions.

In parliamentary systems, rational voters expect parties to behave as unitary actors who outline their plans before elections and make good on their promises. If intra-party factions are able to overturn the program and strategy a party promised before an election, rational voters cannot establish any link between their preferences and the act of voting. In other words, rational voters demand some certainty about the future behavior of parties or a vote cannot have the intended impact on power and policy in a democracy. Although no party ever approaches the voters' ideals of calculability and complete agreement between individual and party preferences, mass apparatus parties produce certainty about their policies through a variety of organizational mechanisms, the most important of which are internal hierarchy, bureaucratization, and organizational linkages to external constituencies.

Parties characterized by organizational dealignment and devolution cannot provide this degree of certainty. Consequently, their voter support remains highly volatile, as numerous examples suggest. Support for the Danish, Norwegian, or Dutch Left-libertarian parties has varied by more than 50 percent from one election to the next. The same applies to the West German Greens when one compares electoral support in state elections. A typical "electoral cycle" in Left-libertarian parties has three phases. First, a party wins votes and considers an alliance with more moderate Social Democratic and centrist political forces. Second, it fails to enter the alliance or makes an agreement that quickly breaks down. Third, voter support in the subsequent election drops dramatically. Such cycles of electoral support can be observed in Denmark (1966/68 and 1971/73), the Netherlands (1970s) and in West Germany (Hamburg 1982 and 1986/87). When radical strategies prevail, marginal voters will abandon Left-libertarian parties in favor of more moderate parties and return to socialist or centrist alternatives. Because of the trade-off between voter support and the parties' policy stances, strategy formation in Left-libertarian parties has remained volatile. Although they have occasionally joined the government alliances, they have been natural parties of opposition, with both the advantages (clarity of policy principles) and the disadvantages (low appeal to marginal voters) that accompany this position. Weak organizational structures promote strategic volatility and unreliability *vis-à-vis* potential coalition partners. Where Left-libertarian parties have pursued coalitions, such as the West German Greens in Hesse and West Berlin, they have therefore attempted to strengthen the organizational leadership of the party, without, however, altering most other features of a post-industrial framework party (see Kitschelt 1989, ch. IX).

To explain electoral realignment and dealignment, we must thus examine both the demand for new political alternatives within the electorate as well as the supply of new parties. Studies of electoral demand patterns (Dalton et al., 1984) or of the "sociological" characteristics of Left-libertarian voters (Bürklin, 1987a) are valuable and interesting, but they are not sufficient to reveal why parties rise and fall. The fortunes of parties depend just as much on the "supply side" of competitive politics – the organization, program, and strategy of political parties – as electoral demand for public policies.

Conclusion

Left-libertarian parties have introduced a new structural differentiation and polarity into the party systems of European corporatist welfare states. They are post-industrial framework parties opposing both the institutional form as well as the political-economic substance of the post-World War II class compromise, crystallized around elite negotiations among the leaders of mass apparatus parties and the centralized associations of economic producers.

Challenging Left-libertarian parties adopt innovative organizational structures and strategies in order to express a radical critique of collectivist politics.

The structural polarization in party systems develops most clearly between Left-libertarian and conservative parties. While the former dismantle formal party organization, the latter have expanded their formal structures and encapsulation of voters in many democracies. Although it is always difficult to draw a direct link between a party's formal organization and its electoral performance,[20] organizational rationalization has helped conservative parties recapture some of the ground lost in elections throughout the 1960s and 1970s or to make new inroads into the electorate of other parties.

Traditional challenging parties of the Left – social democrats, socialists, and communists – have always relied on the hierarchical structure of mass parties, but, with few exceptions, have been unable to improve or expand their organization in the past twenty or thirty years. Faced with the organizational polarization between conservatives and Left-libertarians, social democratic parties have assumed an ambiguous position. In one sense, conventional Left parties are a functional element of the corporatist welfare state and subscribe to the mass party model. In another sense, they are challenging parties themselves trying to redress the class balance in advanced capitalism. As such, they may not ignore the emergence of the new Left-libertarian forces pressing for institutional change, and will try to incorporate some of their demands in socialist party programs and organization.

In fact, socialist parties in Austria, the Netherlands, Switzerland, and West Germany experienced tendencies towards "organizational devolution" when Left-libertarian activists joined them in the early 1970s. But the entry of Left-libertarians usually reduced socialist electoral support because moderate voters were repelled by the new political agenda and the ensuing internal struggles in the parties. Eventually, these tendencies have reduced the socialist parties' ability to govern and often forced them into the opposition role. The electoral losses suffered by the Belgian, Danish, Dutch, German, Swiss, and the Austrian and Finish social democrats are noticeable examples. In some respects, even conflicts within the British Labour party are affected by arguments about Left-libertarian policies and organizational change.[21] And yet, in countries where the Left-libertarian cleavage is less mobilized, as in the mediterranean democracies, socialist parties have not faced the same problems (Hine, 1986).

Given the new polarization of party systems in corporatist welfare states, what are the prospects for Left-libertarian politics? Putting the question into historical perspective illustrates how speculative any answer to this question currently must be. Most Left-libertarian parties are less than ten years old. Their oldest precursors, such as the Danish Socialist People's party, the French Unified Socialist party, and the Dutch Pacifist Socialist party, emerged less than thirty years ago, but became true Left-libertarian parties only in the 1970s.

Predicting the prospects for Left-libertarian politics, then, is equivalent to anticipating the future of socialist parties from the perspective of a social scientist writing in the 1880s, when most socialist parties were still extremely small, receiving less electoral support than today's Left-libertarian parties.[22] Moreover, at that time liberal parties began to compete vigorously for the working-class electorate, casting doubt on whether socialists could attract a loyal constituency. Cross-cutting cleavages in the electorate and factional struggles about party strategy limited the electoral success of working-class parties. Just as some authors claim today that Left-libertarian parties are merely a consequence of economic deprivation hitting the educated young,[23] a social scientist writing in the 1880s might have predicted that the specter of socialism would vanish as Western economies moved out of the Great Depression of the 1880s. In retrospect, we know that a very large number of forces impinged upon the success of socialist parties: the development of the industrial labor force, the structure and strategy of labor unionism, the extension of suffrage and the nature of the emerging democratic institutions, the competitive structure of party systems, and certainly the political skills and visions of socialist party activists and organizations.

A social scientist writing in the 1880s probably could not have anticipated the complex interaction of factors that would shape the trajectory of socialist parties. But he or she might have identified some of the important variables. In a similar vein, it is easier to determine the variables affecting the future Left-libertarian politics than to map out its course over the coming decades. I have argued that the external socio-economic, institutional, and political factors as well as the internal capacities and strategic orientations in Left-libertarian parties affect their future. Depending on the value of each of these independent variables, we can envision three different outcomes for left-libertarian politics:

1 Left-libertarian politics does not survive. The Left-libertarian cleavage will not be institutionalized in party systems; neither an electoral realignment nor an organizational dealignment will take place.

2 Left-libertarian politics wil be represented by one or several established political parties who will modify their political program to appeal to the majority of Left-libertarian voters. This scenario implies electoral management, but no organizational dealignment.

3 Left-libertarian politics will produce well-entrenched, specialized new Left-libertarian parties that permanently reduce electoral support for the established parties. In this instance, electoral realignment and organizational dealignment converge.

Table 10.4 Conditions determining realignment and organizational crystallization around Left-libertarian parties

Determining conditions			
Economic affluence	Decrease	Increase	Increase
Welfare state provisions	Decrease	Increase	Increase
Level of corporatism	Decrease	Stabile	Stabile
Left-libertarian governments	Infrequent	Infrequent	Frequent
Left-libertarian issues/movements	Decrease	Increase	Increase
Left-libertarian party strategies	Factionalism and radical strategy	Factionalism and radical strategy	Cohesion and moderate strategy
Outcome	Left-libertarian politics will vanish	Left-libertarian politics institutionalized by conventional parties	Left-libertarian politics organized by distinct parties

Table 10.4 represents the values on six critical variables affecting these three outcomes. Left-libertarian politics is likely to disappear if the socio-economic and institutional substructure, which has promoted the growth of Left-libertarian political preferences, erodes. Conventional parties, primarily the traditional parties of the Left, will capture the Left-libertarian electorate, but preserve the latters' agenda, if this substructure remains in place and several other conditions are met: Left-libertarian parties are divided by factionalism, adhere to a pure logic of constituency representation, and subscribe to a radical strategy of non-alliance with other parties, while socialist parties become effective representatives of Left-libertarian demands. Conversely, unique left-libertarian parties will survive and grow only if the institutional substructure of Left-libertarian demands flourishes, socialist parties prove incapable of assimilating the Left-libertarian electorates, and Left-libertarian parties can find a balance between the logics of constituency representation and party competition that enables them to engage in moderate strategies and join government alliances.

Over the past fifteen years, conditions have been generally conducive to the growth of specialized Left-libertarian parties. If we examine long-term trends of party performance in all ten countries with significant or borderline Left-libertarian parties, we find not a single case in which conventional Left parties have won votes back from Left-libertarian parties (table 10.5).[24] This applies to cases where Left-libertarian parties have already existed for a long time (Denmark, Norway, the Netherlands) as well as to countries where Left-libertarian politics is of recent vintage (West Germany, Switzerland). In five countries, there is a direct trade-off between votes for the

Table 10.5 Levels of electoral support for groups of political parties, 1970s–1980s (in percentage of votes received)

	Far Right parties	*Bourgeois parties*[a]	*Other parties*	*Working-class parties*[b]	*Left-libertarian parties*
Austria	0.0	50.0 (+1.2)	0.5 (+0.4)	46.1 (−5.0)	3.4 (+3.4)
Belgium	2.8 (−1.1)	48.2 (−1.1)	15.4 (−3.5)[c]	29.1 (−0.3)	6.0 (+6.0)
Denmark	5.8 (−4.4)	42.4 (+5.2)	6.0 (−2.3)	31.5 (−4.1)	14.3 (+5.6)
Finland	7.4 (+0.4)	45.7 (−0.2)	4.7 (−0.3)	39.4 (−2.7)	2.8 (+2.8)
Italy	6.4 (−0.3)	40.5 (−3.5)	1.6 (−0.2)	47.8 (+1.8)	3.7 (+2.2)
Netherlands	0.0	61.3 (+1.2)	3.0 (−1.0)	30.7 (−1.2)	5.0 (+1.0)
Norway	4.1 (+4.1)	48.2 (+5.3)	3.2 (−3.6)	39.3 (+0.2)	5.2 (−2.5)
Sweden	0.0	47.4 (−1.7)	0.3 (−2.2)	45.2 (+1.9)	7.1 (+2.0)
Switzerland	3.2 (+0.5)	61.4 (+1.6)	6.6 (−2.5)	21.4 (−5.0)	7.4 (+6.8)
W. Germany	0.0	54.8 (−0.1)	0.3 (−0.3)	39.7 (−4.8)	5.2 (+5.2)

Figures represent average party votes shares in 1980s and change from 1970s in parentheses.
[a]Conservative, liberal, agrarian, centrist, and religious parties.
[b]Socialist, social democratic, communist and proletarian parties, except Sweden's Left Communists.
[c]Belgium's residual of 'other' parties includes predominantly the ethno-linguistic parties which account for 11.3 percent of the total vote.

traditional and the new parties (Austria, Denmark, West Germany, Switzerland, and Finland). Also in Italy, where the Left made some aggregate gains in the 1980s, the Communists lost heavily to the emerging Italian Green party in the 1987 election, although they had tried to cater to Left-libertarian demands in the areas of ecology and feminism. In Belgium, Left-libertarians won mostly at the expense of the ethno-linguistic parties, which had begun appealing to Left-libertarian sentiments in the early 1970s (Lijphart, 1977).

The rise of Left-libertarian parties has meant the stagnation or weakening of the traditional Left in almost all ten party systems. In no single case were conventional Left parties able to gain more than 2 percent in electoral support in the 1980s as compared to their 1970s average. Yet socialist parties in five democracies without Left-libertarian parties managed to improve their electoral share by at least that much (Australia, France, Greece, New Zealand, Spain).

In the light of these developments, there is presently little reason to believe that Left-libertarian parties will disappear soon. Overall, they have entrenched themselves in national electorates, although their support remains volatile. Their sympathizers do not consist solely of free-floating protest voters, but have a base in identifiable social groups and political activists who cast votes in strategic ways. At the same time, socialist and communist parties encounter great difficulties in making the demands of their traditional blue-collar and lower-middle-classs electorates compatible with those of Left-libertarian constituencies. These problems surface in intra-party factionalism and programmatic and strategic paralysis. Until 1988, at least, no Left party has been able to develop a successful formula bringing these groups together in a new coalition capable of defeating the non-socialist parties. The case coming closest to this objective is Sweden. But even here, the hegemonic Social Democratic party relies on the tacit or active support of a slowly growing Left-libertarian political sector.

The changes in modern parties and party systems I have sketched force us to reconsider a number of popular theorems about the transformation of parties in contemporary democracies. In the 1950s, Duverger (1954) speculated that a "contagion from the left" would lead all parties to adopt the structural properties of socialist mass parties in order to remain electorally competitive. In the 1960s, Epstein (1967, pp. 234–60) proposed precisely the opposite that a "contagion from the right" would dissolve mass party organization under the impact of changed campaign styles, the growing role of the mass media, and new techniques of party financing. This argument, in turn, fed into the popular "decline of parties" literature of the 1970s, which maintained that issue-related political mobilization would undermine the cohesiveness of political parties and bring about a "pluralization" of political interest intermediation through parties, pressure groups, and social movements (Olsen, 1983).[25]

Duverger was half-correct in predicting an increasing scope, articulation, and cohesiveness of bourgeois party organization, but did not anticipate counter-tendencies of organizational dissolution on the libertarian Left. Nor did Epstein predict the persistence of party organization on the Right. And both Epstein and Duverger, just like Kirchheimer (1966), were too concerned with the imperatives of electoral competition to realize that broader institutional patterns of political interest intermediation shape modern party systems. Here the circle closes. I began with my claim that parties must be analyzed as more than elements in a system of interest intermediation, not merely as vehicles of party competition. Left-libertarian politics illustrates that forces other than a pure logic of electoral competition shape the dynamics of political parties and determine their internal structure and external strategy.

NOTES

1 Exceptions to this rule are the early contribution by Rokkan (1966) as well as Lehmbruch's (1977) work.

2 Examples of sophisticated, but "reductionist" views of political parties are Wellhofer and Hennessey (1974), Robertson (1976), and Schlesinger (1984). As a consequence, literature on the linkages between parties, interest groups,

and social movements has remained sparse. I have criticized the dominant theoretical literature on parties in Kitschelt (1989, ch. 2).

3 As examples of such efforts, see especially the excellent articles by Eisinger (1973) and Jensen (1982), and Kitschelt (1985, 1986).

4 I use the adjective "Left-libertarian" rather than "postmaterialist" to describe the new political phenomena because prefixes such as "post" or "new" are too vague. As the debate about Inglehart's analysis of postmaterialism has shown, Left-libertarians represent only *one distinct component* among supporters of postmaterialist politics (see also the debate between Inglehart, 1987a, and Flanagan, 1987).

5 The argument is empirically substantiated in Kitschelt (1988a). See also Müller-Rommel's (1985a, chapter 11 in this volume) work on the rise of Left-libertarian parties as well as Wilson's chapter in this volume, ch. 4.

6 Organizational patterns of Left-libertarian movements have been examined by Oberschall (1980) and Donati (1984).

7 I am building on Beck's (1983) argument that social consciousness is governed to a decreasing extent by broad structural patterns of class and inequality, but by an individualization of life-styles and world-views.

8 Tables 10.2 and 10.3 present data at a high level of aggregation, combining party militants, sympathizers, and the general adult population from two countries at different points in time (*Eurobarometers* 1982–6). While the aggregation masks some interesting cross-national differences between ecology parties, more disaggregated data would not have affected the basic message of my interpretation.

9 As a true conservative, Kirkpatrick (1976) interprets this phenomenon in a psychological and materialist "class self-interest" framework: "A symbol specialist may be said to have a vested interest in the intellectual and moral aspects of politics because he is expert in articulating, analyzing, critizing, and moralizing . . . Unlike wealth, status, knowledge, and health – rectitude can, for all practical purposes, be had for the claiming" (pp. 253–4). Her study does not provide, however, any empirical data to back up these attributions of motivations.

10 See for instance Gerretsen and van der Linden (1982) on the Dutch Socialist Pacifist party, Kergoat (1982) on the French PSU, and Lorenz (1982) on the Norwegian Socialist People's party. Also Kirkpatrick (1976) found more women among McGovern's delegates than those of other candidates. And in the mid-1980s the Dutch Communist party was taken over by radical feminists.

11 Compare Kitschelt and Hellemans (1990, ch. 5) on income in the Belgian parties Agalev and Ecolo and Logue (1982, pp. 180) on the Danish Socialist People's party.

12 See, e.g. Falke (1982) on the German Christian Democrats, Rappoport et al. (1986) on American convention delegates or Galli and Prandi (1970) on Italian parties.

13 For analyses of left-libertarian voting patterns see Logue (1982, chs. VI–VIII) on Scandinavia, Boy (1981) on France, Bürklin (1985, 1987a) on West Germany, and Deschouwer and Stouthuysen (1984) on Belgium.

14 Data on these developments are provided by Falke (1982), Wellhofer (1985) and Kuhnle et al. (1986) on Scandinavia, and Wilson (1979) on France.

15 Of course, left-libertarian parties also tend to attract civil servants. Yet these recruits rarely work at higher echelons of the central administrative apparatus, but primarily in the social services agencies.

16 The concept of "stratarchy" is of course Eldersveld's (1964) and was developed in an empirical application of pluralist political theory to the internal dynamic of US parties. At least in this organizational respect, Left-libertarian parties represent a definite "Americanization" of European party politics.
17 For West Germany, see Becker and Hombach (1983, p. 86). For Belgium, a party representative (in Ceulers, 1981, p. 165) reports a 10 percent participation rate as an informal estimate. Galli and Prandi (1970, pp. 126–8) suggest similar levels in the Italian Communist party in the late 1950s and 1960s and lower levels in the Italian Christian Democrats.
18 A classic study of party structure emphasizing the role of leadership is McKenzie (1955). For a more recent treatments of the same subject, emphasizing the organizational rationalization of parties, see De Winter (1981) on Belgium, Kuhnle et al. (1986) on Norway, Schönbohm (1985) on West Germany, and Wilson (1979) on France.
19 Most literature on Left-libertarian parties, e.g. Fogt (1984) suffers from a fixation on the contrast between the normative ideal and the reality of party structures.
20 For a discussion of this issue, compare Bartolini (1983, p. 194), Schönbohm (1985, pp. 244–53), Wellhofer (1985), and Kuhnle et al. (1986).
21 In the British case, however, it would be more appropriate to speak of a combined attack of the Old Left and the New Left-libertarian forces on the moderate party establishment.
22 Bürklin (1988a) maintains that a comparison between the early socialists and today's Left-libertarians is misleading, because few countries had universal suffrage before World War I. In a number of countries, however, such as Belgium, Britain, Denmark, France, Germany, and the Netherlands, almost universal suffrage was introduced several decades before World War I.
23 See Bürklin (1987a), my critique (Kitschelt, 1988c) and his rejoinder (Bürklin, 1988a).
24 Even Norway does not count because the reversion of Left-libertarian electoral support in the 1980s is more apparent than real. The share of the Norwegian Socialist People's party has been stable over the last several elections. In the 1970s, Left-libertarians won more votes, however, when they formed an alliance with other forces opposed to joining the European Community in the 1973 election (11.2 percent). In the Netherlands, the Social Democratic party won back votes from the small parties of the New Left in 1980s, yet could not help to prevent the resurgence of the moderate borderline Left-libertarian Democrats '66.
25 Representatives of this argument are Berger (1979) and Sjoeblom (1983).

11

New Political Movements and "New Politics" Parties in Western Europe

FERDINAND MÜLLER-ROMMEL

The remarkable extraparliamentary actions of new movements in Western Europe and the electoral success of small left-wing and green parties in recent years have attracted considerable attention in the journalistic and academic worlds. The questions of why those small parties have become an important and powerful force in Western European politics, and of how their political influence might develop in the future cannot be answered without examining their roots in the student movements of the 1960s, the subsequent flowering of the environmental movement, and, more recently, of the anti-nuclear and the peace movements.

When the student movements of the 1960s disappeared, other political movements rapidly developed in most European countries. Most of these emerged spontaneously on the local level as politically independent citizen-initiative groups. These groups were interested in single issues such as the provision of parks; they protested against urban renewal, new highways, or the construction of nuclear power plants. The citizen initiatives employed a variety of methods in seeking to influence policy decisions. They mobilized public opinion via the unconventional political behavior characteristic of the earlier student movements, such as demonstrations and occasional sit-ins, information campaigns, and similar tactics. They also utilized local and national laws to obtain public access to urban renewal and construction plans, and to force compulsory hearings for those directly concerned with local environmental issues. This led to the establishment of federal umbrella organizations in order to strengthen the political impact of the environmental movement nationally. For instance, the Bundesverband Bürgerinitiativen Umweltschutz (BBU) was founded in Germany in 1972, the Amis de la Terre in France in 1971, the Swedish Miljvardsgruppernas Riförbund (MIGRI) in 1971, and the Dutch Vereiniging Milieudefensie (VDM) in 1972.

In the mid-1970s, one particular issue became dominant in several European countries: nuclear energy. Heavily influenced by the oil crisis, most European governments decided to expand their nuclear program. It was precisely the nuclear power issue, however, that demonstrated the need for organizing political movements at the national level, since energy problems could not be resolved at the local level. More and more local action groups in various countries formed nationally organized "anti-nuclear power" organizations such as the Organization for Information on Nuclear Power (OOA) in Denmark, founded in 1974, the Committee for the Co-ordination of Regional Anti-nuclear Power Initiatives (LEK) in the Netherlands (1973), the Environmental Federation (*Miljöverbund*) in Sweden (1976), the Initiative of Anti-Nuclear Power Plants (IAG) in Austria (1976), and the Action against Nuclear Power (AMA) in Norway (1974).

In the late 1970s, the environmental issue was joined by debate over the NATO dual-track decision on intermediate nuclear forces and the eventual stationing of Cruise and Pershing II missiles in Western Europe. This political decision created considerable solidarity among new social movements that crossed European national borders. Large demonstrations were held, along with illegal occupations of the proposed missile sites. Most of these activities were initiated by nationally organized peace movements (see Rochon, chapter 6 and Klandermans, chapter 7, in this volume).

In the early 1980s, most citizen-initiative groups and new political movements looked for closer contact with the Social Democrats or other large, established, left-wing parties in their nation. They expected those parties to act as an effective force against unlimited economic growth, the destruction of the environment, and the stationing of nuclear weapons. Major efforts were made to influence the nuclear policy of the Social Democrats and other labor and socialist parties, albeit without much success.

For several reasons the larger leftist parties could not (or would not) respond adequately to the demands of the new political movements. First of all, in many European countries the socialist parties held governing power during the 1970s, precisely around the time that these nations underwent a crucial economic crisis with a subsequent increase in unemployment rates. Leftist governments were forced to work more closely with the trade unions and other conventional interest groups in order to manage the economic crisis. However, the trade unions in Western Europe are strong advocates of economic growth as a mechanism for improving the status of the working class. Since the socialists and other established leftist parties are, to varying degress, dependent upon the electoral support of the trade union leaders and rank-and-file membership, an environmental or anti-nuclear power position by these parties directed against the trade unions' economic policy would be damaging for them. Therefore, the issues brought up by new political movements in the 1970s have not figured prominently in most socialist parties' platforms or policy stands, although there were anti-nuclear and environmentalist factions in these parties. Now that many of these parties have moved into opposition, these new political factions should have more influence on party policy.

Second, the hierarchical, bureaucratic organizational structure and "catch-all" character of most socialist parties made it almost impossible for new political movements to implement any major policy change in a short period of time. By and large, Michels's classic analysis of the "iron law of oligarchy" still contains a lot of truth about the internal life of socialist parties in contemporary Europe.

In sum, the negative experiences of the followers of new political movements with the established left-wing parties, as well as the perceived lack of responsiveness of other political institutions in coming to grips with a fundamentally different policy approach, became the major reasons for both the growth of green parties and the electoral success of all small parties that represent the demands of new political movements. We refer to these parties as "New Politics parties". Our use of this term is largely synonymous with the Left-libertarian parties that Kitschelt discusses (chapter 10 in this volume).

The main interest of this chapter is to determine in a differentiated manner what New Politics parties are in order to understand why new political movement followers sympathize with these parties. In addition, we examine the party preference of new political movement adherents in order to marshal empirical evidence on whether New Politics parties are the "parliamentary arm" of the new political movements. We are also interested in how the preference for New Politics parties is linked with the electorate's ideological orientations and value system. Underlying this question is the assumption that the substantive policies of New Politics parties reflect the value system and the ideological orientation of new political movement followers.

The New Politics: Theoretical Considerations

Comparative research into parties and electoral behavior almost unquestioningly assumes that support for the established parties is characterized by specific historically rooted social milieu whereby the structure of social conflict within a nation produces long-term and relatively stable political cleavages within the party system. Furthermore, the determinants of electoral choice could be traced back to basic social allegiances such as class, religion, or regional traits (Lipset and Rokkan, 1967). The end result was largely a Left-Right pattern of partisan alignment.

Yet the emergence of new value orientations in Western industrial nations together with the founding of new parties based on ecological issues and protest, directing their activities against Left and Right targets alike, has produced a new dimension of conflict. The stability of the established political constellations has consequently been challenged.

Ronald Inglehart (1979; 1984) argues that the traditional Left/Right dimension no longer adequately describes modern patterns of political conflict, because "new" issues can no longer be regarded as expressions of historical Left/Right conflicts alone. The need to combat environmental pollution and to develop a peace policy is not, at least overtly, questioned by either conservative or left-wing parties.

Inglehart shows that the "valence issues" of the New Politics are better placed on an establishment/anti-establishment scale than on a Left/Right one. Some sections of the population sympathize with the peace movement, squatters, and social fringe groups; other individuals are favorable toward the police, the administrative bureaucracy, that is, with established institutions of the state defending the existing social order. In this context, Inglehart holds that this new political dimension is partly an expression of the emergence of a sizeable and active minority giving priority to postmaterial values.

In examining the electoral behavior of this postmaterialist minority, Inglehart found that these individuals heavily preferred left-wing parties. Initially this finding appears to confirm the thesis that voters are still in the habit of attaching their political ideas and demands to certain parties via terms like "Left" and "Right". In a comparative study, however, Inglehart and Klingemann (1976) showed that the designations "Left" and "Right" have largely become stereotypes for specifying political parties, and that the decision to vote for one of these parties is still closely connected with a party identification shaped by class and religion. According to Inglehart, it is precisely this inertia of established party loyalties and group formations that prevents the postmaterialist value structure from having full effect on electoral choice.

Following the logic of this argument, the new dimension of conflict should become more pronounced when New Politics parties enter the competition for votes. It is widely accepted that the supporters of new political movements are both highly interested and very active in politics. They anticipate governmental policy more sensibly and critically than the average voter. Since these characteristics are also symbolic for the New Politics parties, one might assume that the adherents of new political movements would prefer the New Politics parties to most of the other parties. The new dimension of conflict should, therefore, intensify when these parties are in fact chosen principally by supporters of new political movements who are mostly without historically formed party identifications to one of the established parties.

New Politics Parties: a Definition and Classification

In contemporary Western Europe there are two different streams of New Politics parties: small, left-wing parties, and newly founded green parties. Currently, ten significant small, left-wing parties and fifteen green parties are organized on a national level (see table 11.1). Most of the small left-wing parties were founded in the mid-1960s (Baumgarten, 1982) and were strongly supported by followers of the student movement. These parties initially defined themselves as alternatives to the established socialist and communist parties. Their programs call for public control of political and economic decisions, industrial co-determination, and progress in the field of social policy. These small, left-wing parties reject utilization of nuclear energy, because – in their view – it provides a further connection between military and economic interests. Most of these left-wing parties are committed to international disarmament and demand the withdrawal of European countries from NATO and the creation of a nuclear-free zone in Europe. Some of these parties introduced these programmatic demands into parliamentary debate (for example, in the Netherlands, Denmark, and Italy); others were taken over by various new political movements. In sum, several small European left-wing parties developed programs that strongly emphasize new political issues. The supporters of the environmental, the antinuclear, and the peace movements are able to identify with these parties. Moreover, the parties' activities in parliament are geared toward transforming organized pressure from extraparliamentary movements into political results.

A second stream of New Politics parties consists of newly founded green parties. These parties are formed on a national level in twelve European nations (Müller-Rommel, 1989). Newly founded parties have deputies in eight national parliaments (in Austria, Belgium, Finland, Germany, Iceland, Italy, Luxembourg, Sweden, and Switzerland). The activists and the electoral success of newly founded green parties vary considerably between countries as well as between the local, regional, and national levels within any one country. Generally, the greens are relatively small at the level of voting support and parliamentary representation (see table 11.2).

Interestingly enough, the participation rate of both party streams in European parliaments is higher than for other small parties. Out of twenty-seven small left-wing and newly founded green parties under consideration, seventeen (63 percent) are represented in national parliaments, ten of them being newly founded green parties. Of the thirty-five other significant small parties in Western European party systems listed in table 11.1, only twenty (57 percent) are represented in their respective national parliaments. This is of particular interest because most of the latter small parties have competed with larger established parties for decades, some without a major electoral success. Most small left-wing and green parties, however, have increased their national vote totals in a short period of time. This finding leads us to conclude that a party's electoral support apparently increases when it introduces new political issues that are not represented by another parliamentary party. In addition, this finding rejects the often repeated statement that newly founded parties have only a limited chance of electoral success in Western Europe (Lipset and Rokkan, 1967).

The political impact of the small left-wing and green parties is predominantly related to the new issues and the new style of political participation and communication which these parties have promoted in public debate. Although there is some diversity in the programmatic demands of both party streams (especially when it comes to traditional ideological questions),

Table 11.1 A classification of New Politics parties in Western Europe, 1980–1988

Country (number of parties)	*"Family" of New Politics parties*		
	Small left-wing parties	*Newly founded green parties*	*Other small parties*[a]
Austria (3)		United Greens (VGÖ) Green Alternative (ALÖ)	Communists (KPÖ)
Belgium (5)		Ecologists (ECOLO), Agalev	Communist (KPB), UDRT/RAD, Wallon Party (WP)
Britain (3)		Greens	SDP, National Front
Denmark (6)	Socialist People's Party (SF), Venstre Socialists (VS)	Greens	Radical Venstre, Democratic-Center, Christian People's Party
Finland (4)		Green Alliance	Swedish People's Party (SFP), Democratic Alternative (Deva), Finnish Rural Party (SMP)
France (3)	Socialist Union (PSU)	Greens	Radical Left (MRG)
Germany (West) (3)		Greens	Communist (DKP), Nationalists (NPD)

Ireland (3)		Green Alliance	Sinn Fein (SF), Workers Party (WP)
Iceland (1)		Women's Alliance	
Italy (5)	Radicals (PR)	Green List	Liberals (PLI), Republicans (PRI), Social Democrats (PSDI)
Luxembourg (3)		Green Alternative	Communist (PCL), Independent Socialists (PSI)
Netherlands (8)	Radical (PPR), Pacifist (PSP), Communist (CPN)	Greens	Democrats 66 (D66), Reformists (RPF), Reformed Political Association (GPV), Centrum (CP)
Norway (4)	Venstre Socialists		Communist (NKP), Liberals (Venstre), Progress (Fremsteritt)
Sweden (3)	Communist (SKD)	Green Ecology Party	Christian Democratic Community Party
Switzerland (8)	Progressive Organizations of Switzerland (POCH)	Green Party (GPS) Green Alliance (GAS)	Swiss Car Party, Liberals (LPS), National Action (NA), Communists (SPL), Evangelical People's Party (EVP)
Total	(10)	(17)	(35)

[a] Parties with more than 6 percent of the vote are excluded.

Table 11.2 Results of green parties/lists in national elections in Western Europe (percentages of votes cast and numbers of seats obtained), 1978–1988

Country	*1978*	*1979*	*1979*[a]	*1980*	*1981*	*1982*	*1983*	*1984*	*1984*[a]	*1985*	*1986*	*1987*	*1988*
Belgium[b]	0.8(0)	–	3.4(0)	–	4.5(4)	–	–	–	8.2(2)	6.2(9)	–	7.1 (9)	–
West Germany	–	–	3.2(0)	1.5(0)	–	–	5.6(27)	–	8.2(7)	–	–	8.3(44)	–
Denmark	–	–	–	–	–	–	–	–	–	–	–	1.3 (0)	1.3(0)
France	2.1(0)	–	4.4(0)	–	1.1(0)	–	–	–	3.4(0)/3.3(0)[c]	–	1.2(0)	–	0.4(0)
Finland	–	0.1(0)	–	–	–	–	1.5 (2)	–	–	–	–	4.0 (4)	–
Great Britain	–	0.1(0)	0.1(0)	–	–	–	0.2 (0)	–	0.5(0)	–	–	0.2 (0)	–
Ireland	–	–	–	–	–	–	–	–	0.1(0)	–	–	0.4 (0)	–
Italy	–	–	–	–	–	–	–	–	–	–	–	2.5(13)	–
Luxembourg	–	1.0(0)	9.0(0)	–	–	–	–	5.8(2)	6.1(0)	–	–	–	–
Austria[d]	–	–	–	–	–	–	3.2 (0)		–	–	4.8(8)	–	–
Portugal	–	–	–	–	–	–	–	–	–	–	–	–	–
Sweden	–	–	–	–	–	1.6(0)	–	–	–	1.5(0)	–	–	5.6(20)
Switzerland	–	0.8(1)	–	–	–	–	6.4 (6)[e]	–	–	–	–	8.3 (7)[e]	–
Spain	–	–	–	–	–	–	–	–	–	–	1.0(0)	–	–

[a]Results in European elections
[b]Results of Agalev and Ecolo together.
[c]3.4 = Les Verts; 3.3 = Entente Radical Ecologiste.
[d]Results of VGÖ and ALÖ together.
[e]Results of GPS and GAS together.

this new party type has consolidated in Western European party systems alongside the traditional "party families" of socialists, conservatives, and liberals (Smith, 1984a, pp. 104–5).

Those small left-wing parties listed in table 11.1 together with the newly founded green parties possess several characteristics in common that enable us to label them the "family" of New Politics parties. This term is related to the concept of New Politics which is basically concerned with the rise and cluster of new political issues and related changes in participatory dispositions and behavior. Analysts like Inglehart (1984) and Barnes and Kaase (1979) have tried to explain these changes by referring to the surge of postmaterialist value orientations. Other authors give more weight to changes in the social structure of modern societies and the related cognitive mobilization (Kitschelt, 1988a; Dalton et al, 1984). According to Chandler and Scaroff (1986, p. 303), however, there is no fundamental contradiction between the two approaches. New Politics issue orientations and related behavioral dispositions can be explained by socialization experiences which translate period effects into aggregate changes in opinions as well as by changes in the social structure of advanced industrial democracies.

There are at least three characteristics of New Politics parties: first, most of these parties follow an ideology that consists of strong concerns for equal rights (especially for minorities), strong ecological and anti-nuclear power thinking, solidarity with the Third World, demands for unilateral disarmament, and a general left-wing egalitarian disposition. Among others, most New Political parties stand for peace through unilateral disarmament and a nuclear-free Europe; also protection of the natural environment through the introduction of transnational pollution controls, and more generally an effective environmental policy directed against an unquestioned commitment to economic growth. These parties advocate an alternative life-style through less emphasis on material goods, more individualism, self-realization, and self-determination. They display a more sympathetic orientation towards the Third World, a concern for the genuine sharing of wealth between rich and poor nations, and helping poorer countries to create their own self-sufficient economies free of financial domination by the industrialized nations. In sum, New Politics parties introduce a programmatic and ideological thinking which is less consistent with the traditional ideological framework of the Left/Right dimension; they advocate a set of alternative values that differ significantly from those of the established larger parties. In addition, the New Politics issues are widely perceived as challenging the conventional economic and security policies. Since sympathy with such policies is likely to rise with growing distance from the production process, the members of the new middle classes should be disposed to be more favorable to New Politics demands – regardless of their actual value orientation (Baker, Dalton, and Hildebrandt, 1981, pp. 152ff).

Second, all New Politics parties display a strong preference for participatory party organization. The organizational structure of most New Politics parties gives local party branches more autonomy in decision-making. It is designed to give the grassroots a maximum chance of interest articulation and, as such, an impact on policy formation within the party (Poguntke, 1987a). This process of decentralization in decision-making is seen to be the essential precondition of meaningful participatory opportunity at all levels of the party organization, because it distributes power to more units and makes politics more transparent and hence intelligible (Müller-Rommel and Poguntke, 1989, pp. 1–15).

Third, and most important, New Politics parties have a similar electorate with characteristics that differ significantly from those of the established parties. Several studies of the voters for green parties and small left-wing parties in single countries show that New Politics voters are mainly younger, new middle class, urban, highly educated, with new value orientations, and a general left-wing orientation (Deschouwer and Stouthuysen, 1984; Boy, 1981; Pastella, 1989). Comparative data on the electorate of green parties in twelve nations confirm these findings (Müller-Rommel, 1989): the voters for green parties are mostly younger, highly educated, and occupy white-collar and government jobs where the traditional class conflict is virtually non-existent. Furthermore, our analyses indicate that the voters of most New Politics parties display a "left-wing postmaterialist" profile. We refer to this group of voters as the "New Left".

Inherently, the three typical characteristics of New Politics parties – as described above – involve many continuous variables. For example, the degree of participatory party organization varies somewhat between small left-wing and green parties. The former sometimes have a more hierarchically organized party structure than the greens. Despite some diversity in organizational structure and in program demands, however, the New Politics parties have a very similar network to citizen-initiative groups at local, regional, and national levels of the political system. Furthermore, in many cases we can detect alliances between New Politics parties and new political movements. In this respect, the New Politics parties differ substantially from the established parties.

New Political Movement Supporters: A New Type of Voter

Although the relationship between new social movements and traditional organizations of interest intermediation, especially political parties, is often questioned in the literature (see, e.g., Nedelmann, 1984; von Beyme, 1986; Rucht, 1987a), little systematic cross-national research examines the voting behavior of new political movement followers. Such research is essential for predicting the implications of New Politics parties on electoral alignments and on the future of West European party systems. We assume that new political movement followers are "new voters" who are in some respects different from the majority of the electorate. Their voting behavior thus can affect the electoral alignment and the traditional coalition structures of

European party systems without systematically affecting the overall stability of these systems.

The voting preferences of new political movement supporters deserve cross-national study for several reasons. According to a recent survey, about two-thirds of the population in the countries of the European Community support the activities of the peace, environmental, and nuclear power movements (Inglehart, chapter 3 in this volume). Moreover, around 20 percent of the public are members of or might join one of the new movements. In other words, new political movement followers constitute a sizeable proportion of the European population. Since these individuals are also voters, their voting behavior becomes increasingly important for the strength of the established parties and the electoral success of the New Politics parties.

Several theoretically oriented studies have also proclaimed that new political movement followers are alienated from the established political parties and have an exclusively negative conception of traditional politics (see most prominently Offe 1983, p. 234). Consequently, one might expect that new political movement followers either do not vote at all, or that they give their vote to anti-establishment New Politics parties. This in turn would have long-range consequences for the stability of European party systems. However, some single-nation empirical studies have falsified this assumption. Reuband (1985), Rüdiger Schmitt (1987), and Müller-Rommel (1985b), for instance, document the fact that some peace movement followers in West Germany display a close affinity not only with the Social Democratic party, but also with the Christian Democrats. Similar results were found in Great Britain, where the majority of peace movement followers vote for the Labour party (Byrne, 1981). In a study of new social movement activists in Switzerland, Kriesi (1982) shows that the ties between these activists and the established Left parties are considerable; yet other activists were highly critical of the party system and refused to name a party preference. Kriesi interprets this finding as an indicator of the complex relationship between social movements and traditional organizations of interest intermediation. In his chapter in this book (chapter 7), Klandermans shows that peace movement activists in the Netherlands have extensive relations with political parties, including the PvdA, and that these activists are generally not alienated from the political system or fundamentally opposed to the traditional parties.

In another notable case-study, Kriesi and van Praag (1987) formulate some hypotheses which are most insightful for comparative research. They argue that the relationship between new social movements and traditional organizations of interest intermediation is dependent upon a mutual degree of openness in the two collective actors. According to these authors, country-specific differences in the political system determine the degree of openness. They argue that "the more intense the class conflict in a society and the more the system of interest intermediation is dominated by this conflict, the more the political system has difficulties in assimilating the demands of new social movements and the more restricted are the perspectives for their influence on the decision making process" (1987, p. 320). Furthermore, Kriesi and van Praag argue that the readiness of new movements to compromise with traditional organizations depends on the movements' assessment of the state and their specific experience with the state and political organizations: "In countries with a broad range of institutionalized channels of participation for challengers to the political system, the probability that new social movements will engage in negotiations of this kind is expected to be greater than in countries where no or only very limited such possibilities exist" (1987, p. 321). This assumption was empirically proven by Rüdig and Lowe (1986) in a study of the British Ecology party. They pointed out that the weakness of the British greens is among other factors a reflection of the institutional structure of British politics. The political system suppresses issues and integrates middle-class protest movements into established decision-making structures.

The findings presented so far have shown that widespread support for new political movements exists among Western European publics. Furthermore, these active citizens are not necessarily alienated from traditional political channels. Rather, they participate in elections, but without a clear-cut party identification. The question yet to be answered is, what are the characteristics of these voters?

THE NEW SOCIAL MOVEMENT ADHERENT

Figure 11.1. displays the age and educational characteristics of individuals who strongly support one of three new political movements (ecology, antinuclear power, or the peace movement).[1] Our analyses are limited to those six nations with an electorally significant New Politics party.

One characteristic feature of new political movement followers is their relatively young age and high educational level. The horizontal axis in figure 11.1 displays the percentage of strong supporters of the three new politics movements who have advanced education (completed schooling at age 22 or later); the vertical axis portrays the percentage of movement supporters who are between ages 18 and 34. For instance, among those West Germans who strongly approve of the ecology movement, 53 percent received advanced education and 57 percent are under the age of 35.

Because movement supporters are fairly young and have not participated in many elections, they often have not developed enduring ties with a political party (Müller-Rommel, 1985a). Expressed in terms of electoral sociology, these individuals lack a firm party identification, hitherto a clear indicator of electoral behavior. In the absence of these attachments, what are the explanatory variables for their voting behavior? There is much to be said in favor of the argument that they belong to the type of "rational voter" whose electoral choice is determined by a subjective evaluation of the parties on policy matters (see, e.g., Fiorina, 1981). We argue that most new political movement followers are not only young and highly educated;

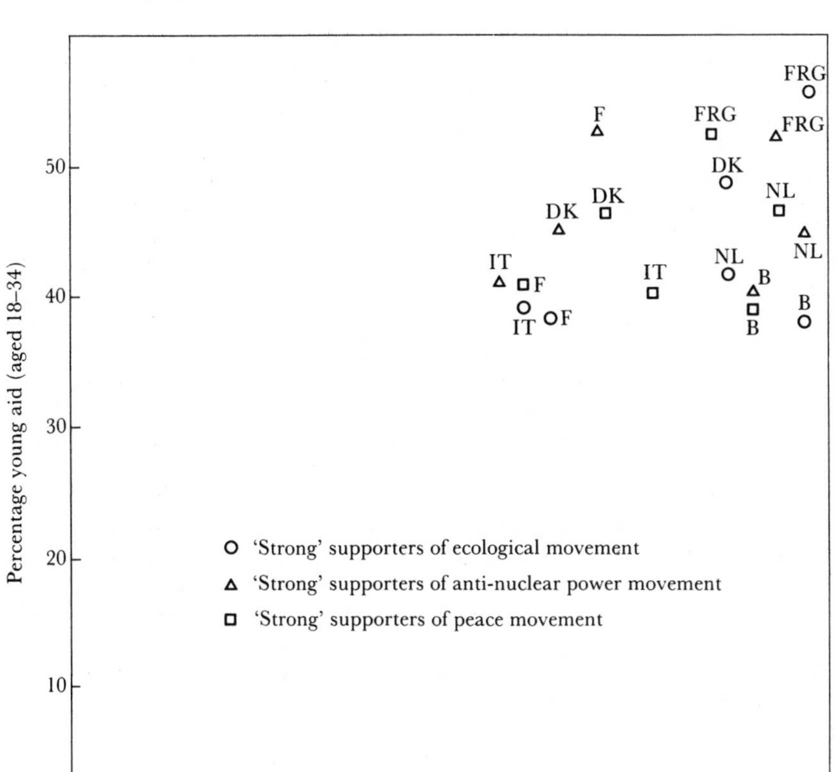

Figure 11.1 The age and educational background of new political movement supporters in six Western European countries, 1986 F: France; IT: Italy; B: Belgium; DK: Denmark; FRG: W. Germany; NL: Netherlands.
Source: Eurobarometer 25 (fieldwork in April 1986).

they are also highly interested and very active in politics. In addition, they are more critical about governmental and party politics than is the average voter. The voting behavior of this group is rationally guided; their electoral choice is influenced by the retrospective evaluation of competing parties' policies. It can be supposed that an accumulated evaluation of the policy stands of the parties and individual political leaders has a substantial effect on the voting behavior of new political movement followers. And yet, the so-called rational voters are guided in their electoral behavior not merely "retrospectively" by the substantive policies of the competing parties, but also "prospectively" in relation to the parties' future political direction as foreshadowed in party programs and public statements.[2] Therefore, we assume that the new political movement supporters vote for those parties that come closest to their own ideological and value orientation in substantive policies and public programs.

Opinion surveys in the European Community from 1982 onwards document a strong relationship between active support for new political movements and postmaterial value orientation as well as a center-Left ideological position (Müller-Rommel, 1985a; Inglehart, chapter 3 in this volume). Recent data on new political movement supporters in the European Community modify these empirical findings. In 1986, the postmaterial and left-wing ideological orientation of new movement supporters was most distinct among the followers of the Dutch anti-nuclear power and peace movements (figure 11.2).[3] In all other countries under consideration (France, Italy, Belgium, Denmark and West Germany), the combination of both leftist and postmaterial orientations of new political movement followers was not as strong as expected. In the Federal Republic of Germany and in Denmark, we find a relatively high proportion of postmaterialists among the movement followers, but a smaller proportion of ideologically left-wing people. In France, Belgium, and particularly in Italy it seems to be the other way around. Here we find more supporters with a leftist ideology and fewer with postmaterial value orientations.

It is not possible to give analytical proof that ideological composition and value orientation are the direct cause of voting behavior because there are surely additional factors that determine the party preference of new political movement supporters. We assume, however, that ideological and value orientations together with an accumulated evaluation of the parties' policy stands affect the voting behavior of new political movement supporters. If this assumption proves to be correct, then the ideological and value orientations of New Politics parties' voters should be similar to those of the new political movement followers in the six countries under investigation.

Figure 11.3 demonstrates the ideological and value characteristics of New Politics party supporters.[4] The horizontal dimension displays the percentage of postmaterialists among supporters of these parties; the vertical axis displays the percentage of leftists among this voter group. Voters for the Green party in Germany, and the small left-wing parties in the Netherlands and in Denmark contain a higher proportion of postmaterialists than in the Belgian and French green parties. The Italian case also matches our expectations. In Italy the New Politics party supporters are more leftist (46 percent) than postmaterialist (34 percent).

The data so far clearly indicate that the ideological orientation and the value priorities of new political movement followers vary by country. In fact, there are two sets of countries: one in which the percentages of ideological orientation and value priorities differ very much from one another (Italy, France, Belgium) and one where the percentages of both variables are very close (Germany, Netherlands, Denmark). In the latter set of countries, the New Politics parties and the new political movement adherents clearly belong to the "New Left," while in Italy they belong more to the

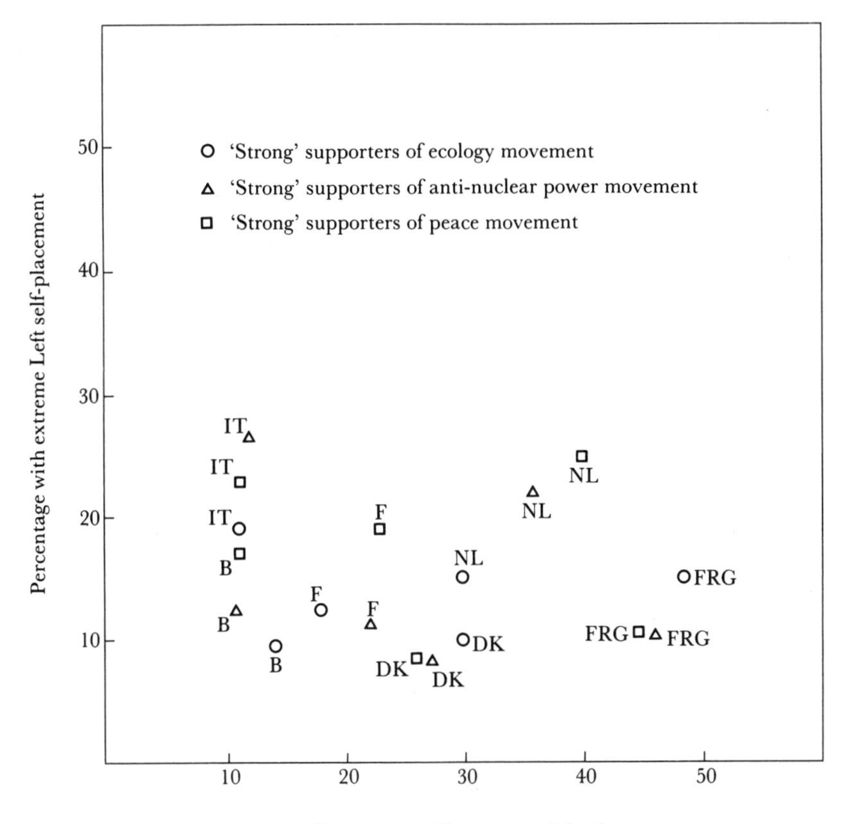

Figure 11.2 The ideological orientation and value priorities of new political movement supporters in six Western European countries, 1986 F: France; IT: Italy; B: Belgium; DK: Denmark; FRG: W. Germany; NL: Netherlands.
Source: Eurobarometer 25 (fieldwork in April 1986).

traditional Left. In Belgium and France, the movement and New Politics parties adherents are not as leftist and postmaterialist as in the other countries under consideration.

The Party Preference of New Political Movement Supporters

So far we have shown that the ideological orientation and the value priorities of new political movement supporters and New Politics party followers are relatively similar in the countries under investigation. We now have to answer the question of whether or not the New Politics parties are in fact

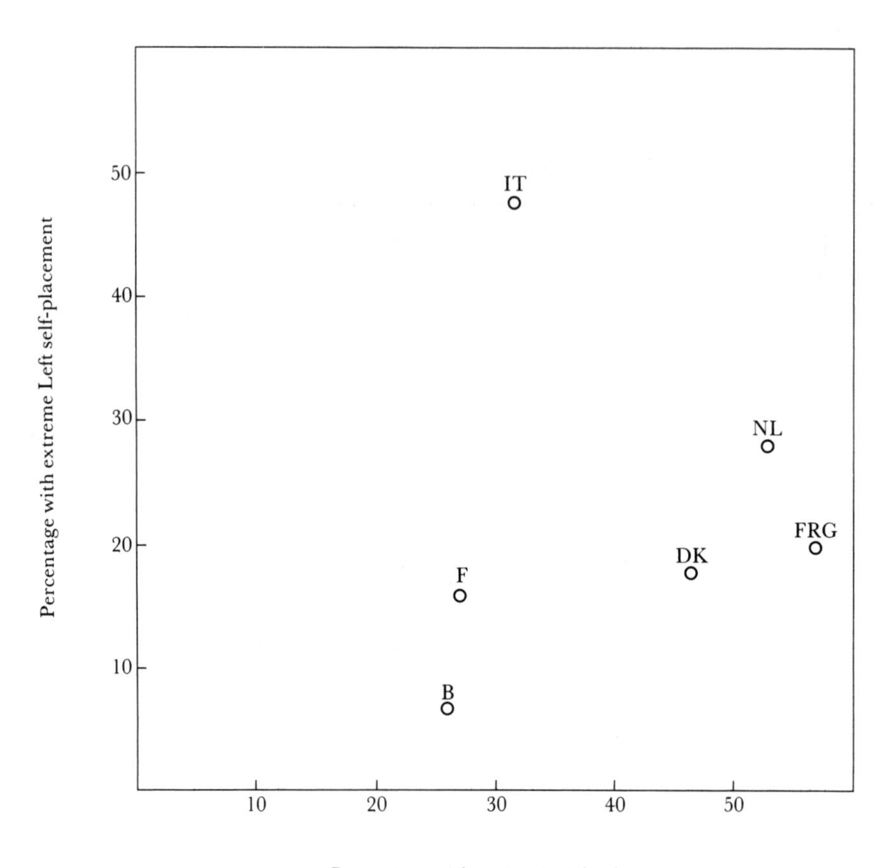

Figure 11.3 The ideological orientation and value priorities of New Politics party voters in six Western European countries, 1986
Source: Eurobarometers 16–21. See note 4 below.

the parliamentary arm of the new political movement followers or of the prototype of the "new voter".

What is the voting intention of new political movement supporters in Western Europe? Do they predominantly vote for New Politics parties or are there differences in the voting behavior of the adherents of peace, antinuclear, and ecology movements? The data on party preference of the new political movement supporters are given in table 11.3. The empirical findings can be summarized as follows: first, although the adherents of the new political movements generally prefer New Politics parties, the strength of this tendency varies considerably across nations. In Italy we find the lowest proportion of "new voters," since Italian voting behavior is more determined by traditional ideological factors is that of any other country in our sample. The Italian Radical party garners only 2–3 percent of the vote from new movement supporters. In Belgium and France the majority of new political movement followers do not consider themselves as postmaterialist (see figure 11.2); consequently they are open toward voting for one of the established political parties. Holland is a somewhat deviant case. Compared to other European socialist parties, the Dutch Labor party has been fairly successful in opening itself to the demands of the new political movement supporters. This has occurred because the Dutch trade unions are not as committed to economic growth as are the other European socialists. Roughly half the votes of the new movement supporters go to the PvdA. The highest proportion of support for New Politics parties is found among movement supporters in Denmark (the SF/VS vote averages 36 percent across the three movements) and Germany (the Greens' vote averages 32 percent).

Second, the data in table 11.3 indicate that the preference for socialist parties is stronger among peace movement followers than the adherents of the other political movements. This occurs because many traditional socialist party voters are themselves actively involved in the peace movements. On the one hand, one might argue that this led the socialist parties that went into opposition in the early 1980s to change major parts of their foreign policy. On the other hand, this phenomenon might also explain why the socialists in all six of these countries gained more support from peace movements supporters than from the supporters of the ecology movements.

Third, except in the Belgian case, New Politics parties are mainly the parliamentary arm of environmental and anti-nuclear power movements. The more political movements are organized on the national level (for example, as with the peace movement) and do not direct their protest primarily against local and/or state policy-making, the better the chances for the larger established left-wing parties to become the parliamentary spokesmen of these movements. With their recent move into opposition in most European countries, the social democrats are able to represent the issues of the peace movement in parliament with more political credibility. Shifting from government to opposition, most larger social democratic parties in Europe have modified the basic character of some programmatic stands from the 1970s and have become more flexible towards some issues of the peace movement.

Fourth, although New Politics parties are supported by many followers of new political movements, their electoral success is not so much dependent upon the strength of new movements in a given country, as upon the type of party system in which they operate. In polarized multi-party systems with a proportional representation electoral law, the movement followers' vote is distributed among several small left-wing parties, but also among larger parties (e.g., in France, Italy, Netherlands, Denmark – see table 11.3). In several multi-party systems, traditional parties demonstrate more programmatic flexibility towards New Politics issues. By making themselves

Table 11.3 Party preference of new political movement supporters, 1986 (percentages)

Countries with small left-wing parties	*'New Politics' parties*			*Established parties*						*Others*	*Total (%)*	*supporters (N)*
Italy		PR		PCI	PSI	PRI	DC	MSI	DP			
Ecology		3		27	17	5	35	6	4	3	100	(305)
Anti-nuclear		2		37	18	3	27	5	4	4	100	(248)
Peace		2		32	19	5	31	5	4	2	100	(386)
Denmark	Greens	VS	SF	SD	RAD	CONS	VEN					
Ecology	4	5	33	30	5	14	4			5	100	(183)
Anti-nuclear	2	3	32	36	5	13	5			4	100	(305)
Peace	2	3	33	35	3	13	7			4	100	(284)
Netherlands	PPR	CPN	PSP	PvdA	CDA	VVD	D'66					
Ecology		8		47	28	7	8			2	100	(439)
Anti-nuclear		13		61	16	5	5			2	100	(212)
Peace		12		64	14	2	8			0	100	(264)

Countries with Newly Founded Green Parties

France	Greens	PSU	PS	MRG	UDG	RPR	PCF			
Ecology	12	3	41	3	11	16	10	4	100	(224)
Anti-nuclear	12	3	51	1	6	14	8	5	100	(128)
Peace	8	3	47	3	11	16	10	2	100	(231)
Belgium	ECOLO/AGALEV		PRL/PVV	KPB	PS/SP	PSC/CVP	VU			
Ecology	11		15	3	37	29	4	1	100	(150)
Anti-nuclear	13		9	3	45	25	4	1	100	(149)
Peace	16		13	3	43	21	4	0	100	(186)
Germany (FRG)	Greens		CDU/CSU	SPD	FDP					
Ecology	38		14	47	1			0	100	(112)
Anti-nuclear	34		8	55	1			2	100	(112)
Peace	23		17	56	3			1	100	(209)

Source: Eurobarometer 25 (fieldwork in April 1986).

the spokesmen of New Political issues, they keep the electoral support for New Politics parties low. In a multi-party system, where more than two parties exist, with, however, only two party blocs (e.g. Conservatives/Social Democrats), which concentrate on maximizing votes among the moderate electorate, the new movement followers strongly support newly founded green parties (e.g. in Germany and Belgium). In those party systems the established (mostly catch-all) parties failed to respond to the New Politics demands of the "new voter" over a long period. This basically explains why New Politics parties in Germany and Belgium are electorally fairly strong although the total amount of new political movement followers is much lower than in countries like Holland and Italy. In sum, we argue that New Politics parties are the "parliamentary arm" of new political movement followers. However, in two-party-bloc systems New Politics parties play a stronger part as interest representatives of the "new voter" than in polarized multi-party systems where New Politics issues are also taken up by some established parties.

The Impact of New Political Movements and 'New Politics' Parties: A 'New Politics' Realignment?

Our data show that the ideological composition and the value orientation of New Politics party voters are not homogeneous in the countries that we surveyed. This finding indicates that political change has reached different stages in different European countries. In some countries, a close link exists between leftist postmaterialism, new political movements, and New Politics parties. In other countries, however, the majority of new political movement followers hold material and rightist values (as in Belgium). In general terms, this underlines the importance of the "Old" and the "New" politics dimensions when it comes to analyzing the electoral potential and the conflicting competition structures of the New Politics parties in the different party systems. We are able to demonstrate a close link between support for new political movement voting preference for New Politics parties and ideological self-classification as belonging to the "Left" and the "center-Left" as well as links with the postmaterial value orientation. This allows the conclusion that the New Politics dimension by no means cuts across the Old Politics dimension. In more general theoretical terms it means that the voting preference of new political movement followers in favor of New Politics parties can be traced back to traditional class conflicts as well as to the structurally "new" social conflicts – conflicts which no longer run along the lines of "haves" and "have-nots" alone but also separate the representatives of a new subculture of protest movements from those wishing to maintain a traditional political culture, namely, the controlling organizations of the state.

The findings of our study also indicate that the overall impact of the relationship between new political movements and New Politics parties on party system alignment is still controversial in some countries. On the one hand, New Politics parties mobilized many movement adherents by making it possible for them to find rational expression for their views at the ballot box. New Politics parties thus serve as a political vehicle for those movement supporters whose grievances have been ignored by the larger established parties. New Politics parties give assurance to their voters that they are doing something on a parliamentary level about the causes of their discontent. By making themselves the spokesmen of the discontented, New Politics parties, however, additionally promote the process of change of party loyalties and prepare the way for increasing volatility and for a dealignment within the party system.

On the other hand, New Politics parties and new political movement followers also affect political issues and the tone of political life by bringing controversial matters before the public. If the issues prove popular, they may well be adopted by one or more of the larger established parties, as larger parties in Europe currently seek to adopt some environmental issues first raised by New Politics parties. This, however, could lead to determining the process of realignment within European party systems.

It seems to be evident that New Politics parties compete in the first instance with larger socialist parties. Both party types are committed to changing the political system. However, while socialist parties seek system change through reform policies addressing the traditional conflict between capital and labor, the New Politics parties ask for a fundamental rethinking of the economic growth theory. New Politics parties generally direct their political activities against both Left and Right targets. In this role the New Politics parties are "repatterning" the traditional party system by adding a "new" conflict dimension to the "old" party system cleavage structure without breaking down the primary cleavage structure.

This process, initiated by New Politics parties, particularly affects the larger socialist parties. In most European countries, the socialist party's rank-and-file members as well as party elites split into two groups: those with a traditional left-wing outlook who are concerned with the security of the working class and economic stability (the Old Left), and those with a New Politics orientation who rather emphasize the quality of life, the nature of economy, and the extent of democracy (New Left). The "New Left" in socialist parties stands in competition with New Politics parties regarding the "new voter", while the "Old Left" is still fighting along the old cleavage dimensions. The socialists are, therefore, trapped between two cultures, although only a minority of the electorate is on the New Politics side. The majority in most West European democracies stands in the center of the political spectrum. Whatever the socialist parties might be able to gain from the New Left, they risk losing from among the Old Left voters. Consequently, the only viable strategy for the socialists is to attempt some reconciliation of Old Politics (in order to integrate the majority of the Socialist party's voters) and a moderate version of New Politics (in order to attract New Politics parties' voters). A radical realization of "New Politics issues" is beyond the reach of the socialist parties.

In functioning as promoters of New Politics issues, however, New Politics parties offer radical answers to radical questions concerning ecological problems, military concerns, and the questions of democratic and civil rights. The success of New Politics parties is nourished by radical issue positions that larger socialist and conservative parties are not able to take fully into consideration. It thus seems theoretically cogent and empirically substantiated to predict that New Politics parties are here to stay as long as the political issues of new political movement followers remain on the political agenda and are not adopted by a major party.

NOTES

I am grateful to the Department of Sociology at the University of New England, Australia, for awarding me a visiting fellowship during which I was able to write this chapter. I would also like to thank Thomas Krickhahn (University of Lüneburg) for his help in the analyses presented here.

1 These data were taken from the *Eurobarometer* 25 (1986), a survey that was kindly provided to the author by the Zentralarchiv für empirische Sozialforschung at the University of Cologne. Representative national samples in each nation were asked the following question:

There are a number of groups and movements seeking the support of the public. For each of the following, can you tell me . . .
(a) whether you approve (strongly or somewhat) or you disapprove (somewhat or strongly)?
(b) whether you are a member or might probably join or would certainly not join?
1. The nature protection movement,
2. The ecology movement,
3. Movements concerned with stopping the construction or use of nuclear power plants,
4. Anti-war and anti-nuclear weapon movements.

2 See Fiorina (1981) for the concept of retrospective voting in the United States. Unfortunately, because of data limitations we cannot test our assumption about the "rational new political movement voter" in rigorous methodological procedures.

3 The "leftist" percentage is based on those individuals scoring 1 or 2 at the left end of a 10-point Left/Right scale.

4 These data were drawn from *Eurobarometers* 16–21; the individual surveys were combined to yield a sufficiently large number of New Politics party voters for analysis. The number of party voters in the combined data are:

Green parties		*Small left-wing parties*	
West Germany (Greens)	461	Netherlands (PSP/PPR)	297
Belgium (Ecolo/Agalev)	285	Denmark (SF/VS)	611
France (Ecologists)	599	Italy (PR)	133

12

Reflections on the Institutional Self-transformation of Movement Politics: A Tentative Stage Model

CLAUS OFFE

New Movements: For "Negative" Features

It has become quite commonplace in the 1980s to refer to movements such as the peace, civil rights, environmental, and women's movements as "new" social movements. This terminology is used by activists, political commentators, and social scientists alike. But its justification is far from obvious. It can hardly be taken to signal just the recent or unexpected nature of the socio-political phenomena which are summarily categorized as "new" social movements, since, at least in retrospect, there seems to be a virtually uninterrupted history of significant movement politics in most Western democracies for at least twenty years. In the United States, these not-so-new social movements date back to the civil rights, anti-war, and student movements of the 1960s; and non-institutional movement politics, most importantly those focusing on military integration and rearmament, were found in various European polities in the mid-1950s.

To use concepts in a reasonably rigorous way is to rely on implicit hypotheses. A number of hypotheses are implied by the term "new social movements." First, they are new compared to earlier post-war movements in that they are neither created by nor dependent upon the resources of the established political parties (which was the case with most peace movements of the 1950s), nor eventually absorbed by these parties (which was, at least in West Germany, the case with the student movement of the 1960s, whose political activists, energies, and motivations came ultimately to be absorbed and coopted to a large extent by social-democratic and liberal political parties). Hence, in contrast to these older waves of intense political activity and mass enthusiasm, the designation of the new movements as "new" is justified to the extent that they persist outside the universe of "old" political parties and their electoral politics.

Second, these are "new" movements to the extent that they persist as political movements, that is, they do not retreat into literary, artistic, religious, or other cultural forms of collective expression and the folklore of life-styles, but continue to claim a role in the generation and utilization of political power. Movements are "new" in that their very existence and persistence testifies to the limited and perhaps shrinking absorption and political processing capacity of established political actors and the procedures of "normal politics," as well as of institutions within civil society (e.g., art and religion).

Third, these movements are "new"' in that they are clearly different from "reactionary" forms of social protest which have regularly emerged and disappeared in the history of socio-political modernization of Western societies and which also remain outside the universe of party-dominated "normal politics" (e.g., nationalist, protectionist, xenophobic, racist, and tax-revolt movements). They represent a non-reactionary, universalist critique of modernity and modernization by challenging institutionalized patterns of technical, economic, political, and cultural rationality without falling back upon idealized traditional institutions and arrangements such as the family, religious values, property, state authority, or the nation.

If a specific progressive (as opposed to reactionary) orientation can be claimed for new social movements, as I think it can (Offe, 1987a, 1987b), and if this orientation still cannot be captured and absorbed by the established political forces of either the conservative, liberal, or socialist/social-democratic varieties, a question must be raised about the relationship between these movements and the older liberal-bourgeois and social-democratic labor movements from which these established political forces themselves have emerged. I think that the distinctiveness of the "new" movements can accurately, if schematically and overly briefly, be further conceptualized in the following terms. The axis of socio-political conflict that was proclaimed by the liberal-bourgeois movements of the late eighteenth and nineteenth centuries was freedom vs. privilege. The associated vision or utopian design of a just order was that of a civil society relying on the economic dynamics of the market within a framework of egalitarian legal guarantees and liberties. In contrast, the dominant axis of the social-democratic labor movement was social justice and economic security vs. private property and economic power, and the associated socio-political design of an interventionist and redistributive state that would provide citizens not with liberties, but with rights to resources. Thus socialist and labor movements demand what the liberal-bourgeois movement leaves to be desired once its goals are implemented and its morally and politically less appealing features become apparent. In this sense, the socialist movements may be interpreted as the collective and historically consequential articulation of disappointment with the concrete results to which the liberal-bourgeois design has led.

Now an analogous continuity exists between the demands of new social movements and the joint accomplishments of the liberal-bourgeois and the social-democratic movements. The disappointment of new social movements concerns the perhaps unanticipated, but now apparent and evident failures and negative impacts of the modernization process that was carried out, be it cooperatively or be it antagonistically, by these two great antecedent movements. The axis of conflict on which the new movements concentrate can, in my view, be best and most comprehensively described as fear, pain, and (physical or symbolic) destruction vs. integrity, recognition, and respect. This set of claims and demands clearly radicalizes the emancipatory thrust of the earlier movements and their historical designs, which distinguish them from anti-modernist and reactionary movements. But it is also, apart from this continuity, radically different from them in that the politics of new movements do not crystallize into anything like a historical design, a positive utopia, or a new mode of production that would be introduced by revolutionary or reformist tactics.[1] The confirmation of this fourth hypothesis would also justify the description of these movements as "new" because of their lack of a comprehensive vision or institutional design for a new society. They are incapable of using the grammar of political change that was common to the liberal and the socialist traditions. This grammar basically consists of two dichotomies: the dark past vs. the bright future, and the progressive "we" against the selfish and reactionary "them." Instead of such grandiose ideological constructs, we find a scattered set of issues and the incoherent expression of complaints, frustrations, and demands which do not add up – either ideologically or, for that reason, organizationally – to a unified force or vision. The "enemy" which is to be overcome is no social class or category of people, but some more abstract kind of dominant rationality in which, at least to some extent, "all of us" do actually partake or upon which we depend. As a suggestive description of this situation, Wiesenthal has used drug-dependency as a metaphor: he compares the condition of modern man in a capitalist economy to the situation of an addict who would fatally suffer both from the sudden withdrawal of the drug as well as from its continued use (Wiesenthal, 1988). Under such conditions, the absence of a basic and global "alternative" is not just a matter of the failure of intellectual imagination and political vision, but is a result of substantive difficulties inherent in the situation itself which do not easily lead to feasible and attractive transformative strategies. Equally obsolete is the positive notion of a universal class which, by striving for power and by establishing its own institutions, would simultaneously perform a civilizing and liberating mission for all mankind.

The post-ideological and perhaps even post-historical nature of their protest and critique is, it appears to me, the most significant reason why these movements deserve to be described as "new." It is significant that most of them, in spite of occasional alliances with socialist and radical political forces, seem to find the very idea of 'revolutionary' transformation, as well as the use of the Left/Right code of the political universe, rather useless. They can best be described as the rediscovery and eclectic application of certain demands and values from the liberal and socialist traditions which are now used as a critical standard against the outcomes of the socio-political, economic, technological, and military processes of modernization in an organizationally and ideologically unintegrated way, without a genuine vision or design for a new society.

So far, I have summarized four features of the new movements which both justify their designation as "new" and share the logical (though, of course, not evaluative) characteristic of being "negative" – that is, negations of some attributes of older political and movement phenomena. The new movements are not aligned to or absorbed by traditional political parties; they do not retreat into distinctive artistic, religious, or life-style practices; they are not "reactionary" in that they invoke and advocate forms of particularistic collectivities that are threatened by or have already fallen victim to processes of modernization; and they are also not ideologically "progressive" in the sense that a comprehensive design of a just and stable order is thought of as the necessary and desirable outcome of revolutionary or reformist change. They rather represent causes which seem to defy the accommodation within conventional forms such as the mass party, the sect, the community, or the avant-garde.

The Transitory Nature of Movement Politics

The above characteristics make it extremely difficult for new social movements to develop institutional forms in which their particular mix of radical (but not in any serious sense revolutionary) motives and demands can be accommodated. With the following remarks and observations, in which I take the institutional dilemmas of the Green party in West Germany as my implicit and only occasionally explicit reference point, I wish to propose a stage model of the institutional dilemmas, ambiguities, and crises that are typically encountered by these new movements. While the empirical observations on which the model is built are entirely taken from the German case in general and the development of "green" politics in particular, the generalizability of the model may be broader than this case. A comparative exploration of this question, however, cannot be undertaken here (see the essays by Kitschelt and Tarrow in this volume, chapters 10 and 13).

This model should not be mistaken for a descriptive, predictive, or prescriptive account of the organizational problems of new movements, that is, for their predicament in having no readily available institutional shelter which could be used for the accommodation of their particular mix of issues, demands, and motivations. This essay is, rather, limited to a heuristic exercise and the construction of a stage model which may – or may not – be implemented in the actual development of new movements. For in the spring of 1988 it appeared far from certain that an institutional solution to these problems was at all likely to be accomplished, or even feasible. All I can do, instead, is to propose an intelligible and I hope generalizable pattern of the search processes and its stages which can be observed when new movements strive not only to achieve their goals, but also to design and implement institutional forms that might enable them to pursue their goals in a continuous and cumulative way in the future. That is to say, the four characteristics specified above are not only objectively there, but they are also self-reflectively integrated into the strategy and political practices of new movements as a set of problems that they must consciously and actively cope with.

THE TAKE-OFF PHASE OF MOVEMENT POLITICS

Socio-political movements as forms of collective action usually start in an institutional vacuum, with no other institutional resources available to them than the usually partly contested legal and constitutional rights of citizens to assemble, communicate, protest, petition, and demonstrate. The other initial ingredient of the situation from which a movement emerges is a widely publicized and highly visible event (or anticipation of an event) that triggers expressions of opinion and protest and helps to define the collectivity of those who are actually or potentially affected by it. In the initial phase of social movements, the absence of organizational form and institutionalized resources is typically not perceived as a liability. On the contrary, according to what may be sour grapes logic, some movement activists consider the established forms of political conflict as either unnecessary, given the evident urgency of the movement's causes or demands, or even manifestly harmful, through their suspected tendency to divert and coopt the political energies mobilized by an emergent socio-political movement. The emphasis is overwhelmingly on content, not form. The style of discourse and action is characterized by militant rhetoric, spontaneity, decentralized experimentation, and often vehement confrontation. Action is not triggered according to plans, strategies, or leadership decisions, but by perceived provocations which are responded to by radical demands.[2]

The radicalism of these demands is indicated by the fact that they are immediate in two senses. First, in their substance, they are phrased in an absolute language, that is, a language using phrases such as "no," "never," "stop," "ban," and "end," which does not leave room for processes, gradual accomplishments, or compromises, but insists upon immediate fulfillment of demands.

Second, these demands are not processed by intermediaries or through a machinery of deliberation, representation, and tactical calculation, but are the often plebiscitary expression of the moral values and protest sentiments of the movements' constituents. To the extent that there is some discernible group of core activists its members do not consider themselves, nor are they considered by others, as formally appointed, elected, or otherwise procedurally legitimated leaders; there are only "spokespersons" or "organizers" who voluntarily perform certain services, such as producing information materials, which are essential for internal communication and/or communication with the outside world. Characteristically, in this initial phase of the life-cycle of sociopolitical movements there is no formal separation between: (a) leadership and rank-and-file followers; (b) rank-and-file followers and "the people in general" or those affected by the event or development that arouses the protest; finally, there is no explicit and recognized separation between (c) contending groups, factions, or divisions within the movement itself. The last situation occurs since the prevailing rhetoric of the movement stresses the value of unity and consensus; and because of the lack of any elaborate mechanism for collective decision-making, the only method by which emerging controversies can be decided is through the unanimity rule and by pushing aside those issues upon which unanimity cannot easily be reached.

All three of these rather crucial distinctions – between leaders and followers, between members and non-members, and between adherents of different policy preferences – are ignored and at times actively repressed and denied recognition and legitimacy in the name of "grassroots democracy" or *Basisdemokratie*. Apart from its well-known theoretical as well as normative problems, this concept of *Basisdemokratie*, and the practices of collective action that is employed to legitimize, suffer from a serious operational problem: it is simply not clear who the "base" is unless this question is first settled by a constitutive or constitutional decision made by a "non-base" group. This can be, for example, a constitutional assembly that defines and incorporates, by the use of categories such as territory, place of birth, language, or age, the universe of citizens enjoying the rights of citizenship. Thus the base that is claimed to be the ultimate source of any collective decision is clearly also the outcome of some prior decision, and hence by no means an ultimate source. Conversely, unless such a prior constitutive definitional act has taken place, the concept of the "base" or the "people" remains operationally fuzzy and the constant object of disputes, which is exactly what happens when quasi-empirical collectivities such as "all those affected," all mankind, all members of a specific ethnic, racial, age, or gender category are used as the referents in the names of which political action is staged. Such non-constituted collectivities invite rather than settle the question of whom exactly we mean when referring to them: all those who share the intentions, interests, and other subjective attributes and value characteristics typically assigned to the members of the category; all those who happen to be present as participants in some collective protest; or all those who share the objective characteristics associated with the category? Given the impossibility of deciding this question, the authority of the "base" remains contingent upon the definitional power of those who manage to constitute groups by using the force of semantics.

STAGNATION

Socio-political movements are extremely ill equipped to deal with the problems of time. In their action and protest, they respond to present dangers and injustices or to events that are anticipated to be part of the immediate future and thus the source of intense present fear. In their demands, new movements do not anticipate a lengthy process of transition, gradual reform, or slow improvement, but an immediate and sudden change. A common and widely used rhetorical tactic of both peace movements and ecological movements consists in depicting long or at least uncertain periods of time as being in fact very short; in this way, possibly dangerous developments in the unknown and perhaps distant future are dramatized as being imminent catastrophes. Analogously, the time-span that is needed for basic changes is represented as being minimal. This representation of the political time structure, on the one hand, helps to increase the perceived marginal productivity of present protests; on the other hand, any delay in achieving a solution to the problem in question will be attributed, within the framework of such distorted imaginations, to the bad intentions of the opponents.[3] In both of these respects, it appears as if an attempt is being made to turn the structural short-sightedness of movement politics, which is conditioned upon their low organizational complexity, from a liability into an asset. The mode of decision-making within the movement is insufficiently complex to permit anything but rapid responses on the spot, as any type of prognostic theorizing, long-term planning, or a political investment calculus would presuppose some clear-cut internal division of labor between leaders, followers, and permanent staff, including administrators and analysts.

Apart from the aforementioned thorny problem of defining the "base" in ways that are not essentially contested and divisive, the other pressing problem experienced by new movements stems from their extremely constrained time horizon. The only time-span that is likely to be quite short is that of the movement's survival. As we have seen before, movements thrive upon three resources: rights to protest, dramatic and highly visible events that offer themselves as reasons for protest, and the spontaneous motivation of relevant segments of the population to engage in protest in response to these events.

All three of these resources may well turn out to be of a highly perishable nature or may be easily withdrawn from the movement, thus making its continuity precarious. To be specific, strategic responses of political, juridical, and economic elites will limit and redefine the citizens' rights to engage in protest in ways which make the use of these rights less easily available or more costly. A case in point is the recent decison of a West German federal court that declares all cases of sit-down demonstrations (e.g., in front of military installations) a criminal offense. In addition, a highly effective elite response may make the triggering events less visible or less frequent, or actually absorb some of the concerns and demands raised by the movements. Substantial – and not just symbolic – reorientations on the part of political elites have occurred in response to feminist, anti-nuclear energy, and peace movements; this development has, at least temporarily, put each of these movements into the position of becoming a "victim of its own partial success," thus weakening the forces striving for more ambitious and far-reaching goals. Lastly, spontaneous protest motivations display a strong tendency to decline both in the case of success and in the case of failure. After substantive success, that is, a visible and effective redress of the situation that initially occasioned the protest movement, activists may soon come to feel that protest activities are no longer necessary. The less substantive, more formal variety of success that consists in high levels of mobilization, turnout, and participation will confront the problems of collective action that emerge as individuals reason that as "everyone else" seems to be actively concerned about the movement's cause, "my own" participation becomes dispensable because of its negligible marginal productivity. Various "frustrations of participation" may add to the growing inclination to leave the burden of active involvement to others (Hirschman, 1982). This inclination may be even stronger in the case of perceived failure, if the distance that needs to be travelled in order to achieve success turns out to be longer than anticipated, or the support for the movement weakens and its opponents gain strength.

The problems that are posed by the receding tide of movement enthusiasm are particularly hard to cope with under conditions where all three of the essential characteristics of formal organization are lacking: leadership roles, membership roles, and established procedures to deal with conflict and divisions.[4] The experience of precarious continuity resulting from abruptly shifting levels of support and activity leads movement activists to respond in order to overcome these deficiencies. In contrast to the initial phase of communication with the outside world in substantive terms of protest and mostly "negative"' demands, the second phase in the life-cycle of movement politics will therefore focus upon internal communication and on organizational formalization.

As a movement's continuity is perceived to depend upon its effective self-transformation into an organization, there will be a strong tendency to adopt at least rudimentary features of formal organization (see Roth, 1988). Among these are the following: First, there is the acquisition of funds and legal expertise that is needed for the purpose of the legal representation of activists who either are being prosecuted for alleged violations of the rights of others or are trying to challenge existing practices and arrangements through the court system. Legal resources are appropriated in the service of the movement's causes, be it by defending and augmenting the space of protest action or by supplementing the means of protest by way of court procedures; also, in both cases the time horizon of action is extended and an investive type of rationality is employed.

Second, there is the incipient formalization of membership roles and the concurrent differentiation between, on the one hand, members and non-members and, on the other hand, members and leaders. A significant transition in this process of formalization is that from occasional donations in the context of face-to-face interaction to dues paid by people who thereby become "members"' on the basis of a more or less permanent commitment, and similarly the transformation of the form of communication from leaflets and posters to newsletters and periodical publications to which supporters can subscribe; again, the continuity-securing aspect of these moves is evident.

Third, conferences, regular meetings, and similar types of horizontal internal communication are introduced at the same time for the purpose of debating and reconciling internal divisions on ideological and tactical issues among those who become (by virtue of such events and due to their participation in them) "leaders" of the movement. The contribution of this aspect of the formalization process to the solution of the problem of continuity is twofold: on the one hand, it gives rise to some rational interest of leaders to perform as leaders and assert themselves in that role; on the other hand, such formalized opportunities for debating conflicts of opinion and ideology will typically converge upon some commonly shared interpretation of the movement's present situation in which two rival points of view are combined – namely that "much has been achieved already" and "much remains to be done (and can be done)" – thus avoiding the two complementary demobilizing interpretations of premature despair about the chances of success and of premature euphoria over what actually has been accomplished.

After these three transformations have been achieved, the movement has come a long way from its initial phase of spontaneity and informality and has reached a certain degree of organizational maturity.[5] The focus of activity has then shifted from substantive demands and protest activity to the formal and reflective concern with the conditions under which the movement can secure some measure of permanence and an extended time horizon. To the extent that the corresponding efforts are successful, the movement is likely to be perceived as a somewhat durable collective actor, whose continued existence and activity can and must be counted upon both by its members and leaders and by its opponents and the general public. This general perception, however, does not mean that a state of equilibrium has actually been reached, as the accomplishment of organizational formalization will presumably cause as many problems as it helps to solve.

THE ATTRACTIONS AND TEMPTATIONS OF INSTITUTIONALIZATION

With these features of organizational formalization in place, movements find themsevles caught in the following dilemma. There is the opposition of those who fear – in the spirit of a vulgarized version of Michels' "iron law of oligarchy" – that any step towards formalization might involve the danger of bureaucratization, centralization, alienation, and deradicalization. From this point of view, the spontaneous, local, quasi-syndicalist, and *ad hoc* form of protest activity is the most effective and promising one for the movement's causes; any organizational formalization that goes beyond a loose network of independent initiatives is suspected as counter-productive. Conversely, the opposition advocates the gradual transformation of movement politics into the institutional modes of "normal politics," involving party competition, participation in elections, parliamentary representation, the formation of alliances and coalitions with rival political forces, and eventually even the occupation of government positions.

The transformation of the West German new social movements that emerged in the 1970s into the Green party that was established in 1980 testifies to the overwhelming attractiveness of the institutional mode of normal politics. Entering the official institutionalized channels of political participation and representation seemed to offer opportunities that no other conceivable form of political activity could possibly match. The use of the political institutions of liberal representative democracy appeared as a rational strategy to permit the fullest and most effective utilization of the ecological, pacifist, feminist, and "alternative" movements' resources. Relative to the adoption of this institutionalizing strategy, the adherence to any alternative organizational form would have appeared as irresponsibly wasteful from the rational point of view of the optimal extraction and utilization of political resources. More specifically, the relative advantages of proceeding from organizational formalization all the way to political institutionalization were seen to consist in the following points.

Gains resulting from alliance-formation

According to the logic of the "rainbow-coalition," it was tempting to expect that many individual issue movements would reinforce each other by pooling their respective electoral support, thereby neutralizing the ups and downs of individual protest cycles.

Gains resulting from the fuller extraction of support

Movements in their initial stage are primitive and precarious forms of collective action precisely because the only resource they are capable of absorbing is their participants' willingness to join the movement's activities. In contrast, a movement that has become a formal organization makes available for itself an important additional category of resources, namely the members' commitment to pay dues. A significant further step in the direction of increased resource absorption capacity is the transformation of a movement organization into a political party, which permits the tapping of resources of those who are willing neither to act nor to pay, but just to vote. In that sense, the evolutionary advantage of a political party can be compared to that of a car engine that would run, if need be, on vegetable oil instead of ordinary gasoline.

Gains resulting from the special status of political parties

The West German political system is commonly referred to as "party democracy." Section 21 of the Basic Law (*Grundgesetz*) as well as other laws, traditions, and practices grant special privileges to political parties that are unknown in most other liberal democracies. Most important among these are considerable financial subsidies granted to all political parties that win more than a tiny fraction of the vote, so as to make them relatively independent both from revenues out of membership dues as well as private donations and campaign contributions. Another privilege gives political parties free time on public radio and television programs to advertise their programs and candidates. Because of these and related peculiarities of German political institutions, the decision not to adopt the party form would amount to the decision to forgo significant resources.

Gains resulting from the logic of party competition

One important difference between political markets and common commodity markets is that in the former I can win even if some competitor imitates and succeeds in marketing "my" product, while a business firm would hardly ever find comfort in having launched a new product on which some imitator is now making a profit. In fact, profiting from the profits of others by forcing them (i.e., persuading them) through the mechanisms of electoral and parliamentary competition to redesign their own "product" (i.e., platform or program) is one of the major, if often less spectacular or even visible, avenues of successful political change. A necessary though obviously not sufficient condition for accomplishing this indirect change of political discourse and conflict is the preparedness to join and confront the opponent on the same institutional terrain of party competition, as failure to do so would imply the impossibility of accomplishing this indirect kind of antagonistic accommodation.

Given the above four apparent advantages that result from the transition to political institutionalization, the pressure to actually make this transition is likely to mount with the perceived opportunity costs of not making it. The advantages to be gained from the transition must be, however, counterbalanced against the reverse opportunity costs, namely those of giving up movement politics in favor of institutional politics. The anti-institutional (or "fundamentalist") argument maintains that the apparent advantage of political cartelization of the causes of various movements will be paid for by significant losses in the identity, autonomy, and distinctiveness of each individual movement. The pooling of various sorts of support ("acting," "paying dues," "voting") will lead to a relative deprivation and loss of influence of the activist core of the movement. The resources and privileges associated with the form of the political party will corrupt representatives and compromise the movement's demands. Finally, the logic of assimilation inherent in the institutional modes of electoral and parliamentary competition is likely to work both ways, thus penetrating the movement's parliamentary representation to at least the same extent as the movement is able to persuade its competitors.

Underlying this debate is an important controversy in democratic theory that partly relates back to – and is sometimes framed as a revival of – the theoretical argument that evolved in the first two decades of the century between Bernstein, Luxemburg, Kautsky, and Lenin.[6] I cannot enter this debate within the present context, but must limit myself to a brief consideration not of which side is "right" in this debate, but of which side has prevailed in the specific case of the transformation of West German new social movements into the Green political party. Concerning this empirical question, there cannot be any doubt that the institutional strategy has become the dominant one, and that its attractions or, as its opponents would see it, its temptations have been sufficiently strong as to exert a continuous learning pressure in the direction of institutional accommodation. This relatively rapid process of self-transformation seems to support the theoretical argument that the political power a movement gains through its successful mobilizing efforts can be maintained, exploited, and expanded only if it undergoes an often demanding and sometimes painful strategic self-transformation that eventually enables it to cash in that power within the channels of dominant political institutions. A growing understanding and appreciation of this condition – and of the growing opportunity costs that are associated with any protracted failure to comply with it – will promote a process of collective learning aiming at and converging upon organizational formalization and subsequently political institutionalization.

To be sure, the transition process from "movement" to "political party" tends to be – and has actually been in the case of the Green party – full of inconsistencies and uneasy as well as unstable compromises. In the early stages of electoral and parliamentary participation, the Greens liked to think of themselves as a "new type of party," an "anti-party party," or a form of collective action that comprised both extra-institutional and institutional practises. This attempted synthesis has turned out to be characteristically fragile. In the early years of their existence as a political party, the Greens introduced a number of regulations and special organizational features that were unknown within the "established" political parties and that were meant to preserve – in accordance with some of the doctrines and historical examples of "direct," "participatory," or "council" democracy – some of the spirit of movement politics. Among these regulations and features were the following:

- Members of the Bundestag and other legislative bodies within the Länder should not be free in their parliamentary work, but committed to following decisions of party conventions and other bodies of the party.
- Green members of parliaments should serve less than the full term for which

they have been elected (two instead of four years), and they should be barred from seeking re-election.

- No cumulation of party office and parliamentary function should be permitted.
- The selection of candidates for seats in parliament should not primarily be made according to criteria of professional qualification or political experience, but according to gender and other quota as well as to the symbolic significance of their minority status (e.g., handicapped persons, persons convicted of political crimes, etc.); sometimes candidates were nominated from outside the membership of the party.
- Members of parliaments should be remunerated according to some moderate fixed income, which would commit them to making major deductions in favor of party funds for the income they are entitled to as parliamentarians.

Interestingly and significantly, virtually all of these stipulations, with the important exception of gender quota, have been compromised, questioned, or silently dropped from the practice of Green parliamentary politics. The populist, direct-democratic, and anti-professional emphasis of these regulations has given way to a much more conventional pattern of candidate selection and political professionalization. As of 1988, it could be argued that the party's political control over its own members of parliament in the Bundestag is less strict and direct than is the case with any other party. The simple if paradoxical reason is that the ban on re-election (still adhered to in a watered-down version) means that members of parliament have nothing to lose from non-compliance, just as they do not have anything to win (e.g., nomination for re-election) from compliance in the first place. The rather absurd spectacle of the party leadership publicly denouncing its own parliamentary group in an advertisement in a daily newspaper for no longer being "representative" or the party's "base" epitomizes this dramatic loss of anti-institutional control over institutional actors.[7]

In the course of its short parliamentary history, the Green party has not only, as these examples show, abandoned most of its partly naïve experiments in mingling the forms of movement politics and parliamentary politics, but it has also adopted much of the conventional tactical repertoire of (oppositional) parliamentary politics and party competition. This becomes most clearly evident if we look at the type of demands and proposals that are typically made within the context of movement politics, and compare it to the routines and the logic of parliamentary politics. The typical political discourse of movement politics consists of negative demands on isolated and disjointed issues; these demands are voiced in response to events and are framed in a short-term and confrontational ("yes or no", "them or us") logic. The discourse of parliamentary politics, in contrast, tends to be agenda-generated rather than event-generated; it consists of competing proposals rather than the expression of protest and rejection, and these proposals are formulated with the more or less implicit intention of winning over the members of parliament belonging to other parties or at least parts of their electoral base, which therefore may not be antagonized. Moreover, parliamentary political debate will often focus on long-term consequences, budgetary burdens, and side effects of proposed legislation and programs. Parliamentary parties, even if their core concerns are limited to some particular policy areas, will always try to demonstrate some competence and distinctive political preferences in even the least appealing policy areas, which presupposes a certain degree of expertise, professionalism, preparedness, and hard work. Finally, the possibility of forming issue-specific alliances or even more general coalitions, even if it is only meant as a tactic to threaten opponents with divisive initiatives, is always part of the game of parliamentary politics.

On all of these dimensions, casual observations as well as numerous journalistic commentators have demonstrated that, apart from some stylistic ingredients of "alternative" culture, the Green members of parliament have quickly and effectively adopted all the essential elements of the parliamentary discourse, and simultaneously abandoned much of the discourse of anti-institutional movement politics.

This remarkably smooth and rapid transition can be accounted for in terms of at least three different and cumulative factors. First, there is the pragmatic advantage of facilitating the survival of the political causes and activities of the movement by making use of the protection and recognition of established political institutions (as well as non-political ones, such as churches, universities, and the institutions of art and the media).

Second, there is the striking absence of models and designs for alternative political institutions, such as those which eventually emerged out of the revolutionary struggles of all the "old" social movements (Alberoni, 1984). This lack of alternative designs and projects for institutional reform is probably best explained in terms of the pervasive preoccupation of the new movements with specific issues, aspects, and sectoral irrationalities and injustices, which, however, does not give rise to a global revolutionary critique, and hence to the vision of entirely new relations of production, or relations of political authority. From this point of view, accommodation within existing institutions is not only pragmatically attractive, but there also seems to be hardly anything else available and feasible.

Third, seen from the point of view of a radical, New Left, libertarian, or a similar progressive perspective, there are compelling reasons to embark on this (only available) road in good political conscience. There is a long-standing and intellectually powerful tradition on the political Left on the European continent dating back, at least, to the work of Rosa Luxemburg according to which the Left must consider itself neither as the heir nor as the opponent, but as the protector of those modern and liberating political institutions, such as parliamentary government, that the ruling classes are always on the verge of abandonning, betraying, or corrupting in authoritarian ways.

A Resurgence of Movement Politics?

So effective seems to be the logic of institutional politics, and so pervasive its impact upon individual actors who learn and practice the rules of the institutional game, that this rapid evolutionary self-transformation does not need to be explained, while the hard-nosed refusal of anti-institutional and anti-reformist minorities within the party (who are commonly referred to – and refer to themselves – as "fundamentalists") to follow the same pattern of organizational and institutional learning does need explanation. In other words, we need to make sense not of the evolving configuration of perceived threats to the movement's survival, emerging opportunities, incentives, perceived irreversibilities, and institutional logics that propel the transformation of movement politics into institutional politics, but rather the halting, ambiguous roundabout, and highly conflictual process by which this is apparently taking place. In still other words: is there a rationality of resisting the learning pressure towards self-rationalization?

Two popular explanations for the phenomenon of persistent "fundamentalist" minorities can be dismissed as uncompelling. One is the biographical explanation that points to the fact that some of the most outspoken leaders of anti-institutional politics within the Green party have a background, some 15 years ago, in Marxist-Leninist and other "revolutionary" splinter groups of at best regional importance. While conservative opponents of the Green party like to emphasize such discoveries in intellectual and political biographies,[8] the striking absence of any trace of revolutionary rhetoric or theorizing and the clear discontinuity between "fundamentalist" and traditional leftist radicalism leaves little if any plausibility to this explanation. Rather, the rapid disappearance of any claim to "revolutionary" politics may arguably be one of the important political and cultural changes that took place in Western Europe somewhere between the late 1960s and the early 1980s. One other explanation, which is sometimes invoked or implied by the intra-party "realist" opponents of the fundamentalists, focuses on psychological variables such as the alleged weak ego of the latter, their overwhelming need for asserting collective identities by purely expressive modes of action, their inability to tolerate cognitive dissonance, and their urge to display an ultra-radical "ethic of conviction" (*Gesinnungsethik*) for the sake of the specific psychic rewards that presumably are associated with this syndrome. Relevant though this explanation may appear in individual cases, the question remains why these personality types seem to be attracted to, rather than gradually repelled from, a party that predominantly embarks upon an institutional strategy.

Let me conclude by briefly discussing two alternative, less reductionist, and I believe more plausible, explanations for the stubbornness of the "fundamentalist" syndrome. The first of these two explanations is based on what may be termed a "poverty of public policy" argument. It refers to the often observed exhaustion – or perhaps even categorical inadequacy – of the means of public policy for satisfactorily solving some of modern society's most pressing problems. I will develop this argument in some detail.

Governments (including what are sometimes referred to as corporatist "private governments" such as encompassing associations and "quangos"), whatever their power may be, are restricted to the use of three categories of resources: (a) legal regulation, bureaucratic surveillance, and the use of state-organized violence; (b) the manipulation of fiscal resources through spending on collective consumption and investment, taxation, and subsidies; and (c) the use of information and persuasion. There are clear *absolute* limits to the use of all three of these media of governmental intervention. Bureaucratic commands, surveillance, and sanctioning fail as reliable steering mechanisms wherever the context of action becomes turbulent and unpredictable, so as to make the "rationality" of this (relatively rigid and inflexible) mode of intervention questionable; they also fail where the social actors that are the target of such intervention become powerful enough to resist, subvert, or obstruct the effectiveness of this mode. Next, economic modes of intervention such as incentives for desired behavior and punishment for undesired action fail either where the targets of control refuse to operate according to some utility-maximizing economic calculus, or where they in fact do operate according to this calculus but find it feasible and profitable to either "pass on" the costs imposed upon them or to "push up" the incentives which are presumably designed to steer their action to fiscally unaffordable levels. Finally, the effectiveness of information and symbolic politics is contingent upon a number of conditions, among them a viable sense of moral obligation by the public (which may or may not be depleted in the course of cultural processes of modernization) and the absence of the suspicion that the appeal to "facts" or "moral values" is employed solely for the purpose of altering the behavioral disposition of actors, rather than automatically and for their own sake.

In addition, there are limits to the effectiveness of these media – *relative* to the nature of the social and economic problems that call for political solution. In policy areas where the passions, identities, collectively shared meanings, and moral predispositions within the "life-world" of social actors (rather than their economic interests) are the essential parameters that need to be changed in order to achieve a solution, the three conventional modes of intervention are virtually ineffective or even counterproductive. Moreover, there is good reason to believe that this type of collective problem is proliferating in modern Western society, while those problems (e.g., incomes policy, economic growth, etc.) for which the conventional tools of government intervention are most adequate tend to decline in their relative significance, while certainly not being "solved" in any definitive sense. No amount of legal regulation, taxation, subsidies, or even state organized promulgation of information and education will succeed in controlling or altering problematic and pathological behavioral patterns in areas such as health (including sexual) practices, nutrition, gender and family relations, socialization and educaton practices, environmentally relevant styles of consumption, drug use, various forms of crime and violence, or the treatment of ethnic and other minorities; the same seems to apply to the norms and codes of technical, scientific, and professional groups.

The "poverty of public policy" consists precisely in the fact that these spheres of social action appear to generate the most significant, highly costly, and conflictual aggregate effects, thus considerably raising public concern, while at the same time being located outside the reach of public policy and almost entirely immune from its conventional forms of intervention. To the extent the objective limits of public policy become apparent as a result of frustrating experiences with programs that failed or turned out to be counterproductive, the conclusion becomes plausible that these types of social problems can no longer be approached through the means of public policy, but only through remedial initiative that originate within civil society itself, such as consciousness-raising campaigns, moral crusades, demands for a change of the dominant way of life (rather than public policy), and communitarian forms of action.

To the extent that this line of reasoning on the nature and conditions of social change makes sense, in the light of political perceptions and the experience of social problems, it will provide a strong argument against the otherwise powerful drift towards more formally organized and eventually institutionalized modes of action. It will thus raise basic doubts concerning the potential usefulness of institutional participation and of any eventual influence upon government action, and hence strengthen the case of those who advocate a return to "fundamentalist" political practices, the persistence of which could consequently be explained and, up to a point, perhaps even justified on the basis of this analysis.

While this explanation of the phenomenon of anti-institutional "fundamentalism" is based on some assumptions about the reach of public policy that are as general as they are pessimistic, the second explanation refers to a set of conditions that would appear to be rather specific to the German condition and experience. Such an additional explanation, however, may be called for in view of the fact that the schism between the "fundamentalist" advocates of anti-institutional strategies and the "realists" who are actively promoting institutional modes of action is nowhere as pervasive and hostile as in the West German Green party, where it has recently reached virtually suicidal proportions. The stubborn resistance of "fundamentalist" forces within the new social movements in general and the Green party in particular may largely be due, or so I wish to submit, to a negative reflection of the "fetishization of the state" and state institutions that impregnates much of German constitutional theory and practice.[9]

Built into this theory and into the constitution of the Federal Republic itself are two cumulative traditions, the common denominator of which is a strong distrust of the forces and capacities of civil society. The more classical thread of this tradition refers back to the political theories of Hobbes, Hegel, Max Weber, and Carl Schmitt. On top of this tradition, the post-fascists and explicitly anti-totalitarian context out of which the

West German Basic Law (*Grundgesetz*) originated after World War II gave rise to an extremely state-centered version of democratic constitutionalism. Conservatives, and to some extent social democrats with their own version of *etatist* traditions, saw the constitution's major task as taming, controlling, and containing disruptive and potentially "totalitarian" forces that might arise out of the conflicts of interest within civil society. This conception of the new constitutional order implied a strong emphasis upon the "combative" (*wehrhafte*) quality of a democratic form of government that would be able to deal with its "enemies" in reliable ways, most importantly through the strength of the state's monopoly of force. The consistent distrust that the constitution displays against disorderly and potentially dangerous movements, demands, and conflicts that might emerge from the sphere of civil society is illustrated by the strong constitutional position of political parties (which are upgraded into the status of virtual organs of the state, rather than of civil society), by the absence of any significant plebiscitarian or direct-democratic modes of democratic participation, and by the strong position of the Federal Republic's constitutional court with its power to review and challenge the conformity of parliamentary legislation and political parties to what are believed to be substantive value commitments of the constitutional order.

As a consequence of this decidedly state-centered conception of "democracy" and the political order, constitutional theorists and segments of the political elite have developed political doctrines that border on the equation of the democratic citizen with unqualified loyalty and faithfulness to existing arrangements (as indicated by the ominous criterion of *Verfassungstreue*, or "faithfulness to the constitution", that even a state-employed postal worker is required to endorse). Democracy in this sense is often held to be a way of life, if not a state-defined and state-enforced *Weltanschauung* or "constitutional culture" that must be promulgated through heavy doses of elite-supervised civic education. The implication of these philosophical, cultural, and legal interpretations is the almost methodical elite distrust in a citizenry that is held to be in need of a constant paternalistic supervision, control, and education; the democratic process itself must contain a strong element of institutional distance from the citizenry, in order for the citizens to become "safe for democracy".[10]

It seems obvious that the prevailing, though certainly not uncontested, interpretation of the meaning of democratic government in these terms will provide strong counter-arguments to those who refuse to consider an institutional strategy for the new social movements as viable and potentially productive. If acting within the existing institutions, or so their argument could be summarized, would automatically imply wholesale compliance with the standards of democratic worthiness and respectability as they are defined by existing political elites, the use of institutional forms of action would amount to virtual corruption of the causes of any movement. Such an implication is in fact claimed by "fundamentalists" strategists who are often able to derive some plausibility from this argument for their own refusal to join in what I have described as an institutional learning process along the stages of our model. It thus appears that such an institutional learning process is itself contingent upon a favorable institutional environment, and that, conversely, the excessive practice and persistence of fundamentalist *Gesinnungsethik* ("ethic of conviction"), and the failure to overcome its deficiencies in a gradual process of institutional learning, is partly the by-product of an excessive *Verantwortungsethik* ("ethic of responsibility"), as it is built into the prevailing practices of political institutions.

NOTES

1 Giddens (1986, pp. 13–16) emphasizes the point that "peace movements lack a deep structure of objectives" and that they, as well as ecological movements, "find it easier to state what they are against than what they are for." They also lack the rudimentary outlines of a distinctive political theory, as they "tend to set themselves against the institutions of parliamentary democracy, [while] their future probably lies substantially in the transformation of those very institutions;" see also Kitschelt (1984).

2 See Wiesendahl (1987, pp. 364–84) for a lucid description of the anti-institutional features of the new social movements and a review of the literature that deals with the syndrome of anti-elitist enthusiasm for direct forms of mass action and mass expression.

3 On the "production functions" of protest politics see Oliver and Marwell (1985).

4 For an analysis of the nature and impact of these problems (such as declining issue-attention cycles, issue-displacement, and issue cooptation) see Zeuner (1984).

5 For an analysis of the transformation of environmental protest groups into formal organizations and networks of organizations see Rucht (1987b).

6 For modern elaborations of these debates in democratic theory see Przeworski (1985) and Bobbio (1986).

7 This advertisement appeared in the *Frankfurter Rundschau*, the Left-liberal of the four major national daily newspapers in Germany, on May 27, 1988. It was not placed in the *Tageszeitung*, the national daily closely associated with the new movements.

8 For a collection of pamphlets written along these lines see Langner (1987).

9 For a recent account of the implicit political theories of German constitutional and legal thought see Hammans (1987).

10 For a critical view on these assumptions and implications, see Dreier (1987).

13

The Phantom at the Opera: Political Parties and Social Movements of the 1960s and 1970s in Italy

SIDNEY TARROW

The research on the "new" social movements of the 1970s and 1980s reveals three major gaps: first, though many observers trace these movements to the student movements of the late 1960s, little attention has been given to how they related to the party politics and policy issues of that decade; second, although the new movements were thought to reflect the structural conditions of advanced capitalism in general, most of the research centers on Germany and France, as if southern Europe and Scandinavia were undergoing different phases of development;[1] third, although the new movements were often seen as reactions to the ossification of the mass parties, little attention has been given to the adaptive strategies of the party system in responding to them.

By focusing on the beginning and the end of the cycle of mass protest in Italy from 1965 to 1975, and on the relations between parties and social movements during these phases, I hope to address these problems for a country which is too often ignored in studies of the new social movements. In Italy, a new generation of movements dated from the realignments of the early and mid-1960s – and not from 1968, as is sometimes supposed. Rather than arising directly as a "new political paradigm" out of the structural development of advanced capitalism, they first appeared as an insurgency within the party system – a heritage that they never fully transcended. Finally, rather than responding helplessly or defensively to the movements, the Italian parties adapted quickly to their challenge by cooptation and preemption.[2] In the overture and the denouement of the drama of the new movements, the party system was not the villain of the piece; it was the phantom of the opera.

Two New Paradigms

In the wake of the student and worker movements of the late 1960s, two broadly diffused models of interpretation arose to explain the period that had just ended. Each had roots in both the collective actors of the period and their academic observers (in reality, the line between the two was never so sharply drawn). Let us sketch these two models to help us in analyzing the role of political parties in the wave of social protest that traversed Italy in the late 1960s and 1970s.

The movementist model is based on the precepts of autonomy from political parties and authorities, on the rejection of hierarchy, and on aggressive confrontational politics. The movementist paradigm in Italy resembled the new social movement school that was simultaneously developing in France and West Germany.[3] What linked the approach in all three countries was its deductive cast and the underlying presumption that the changes of advanced capitalism were creating a new paradigm of interpretive frames, movement organizations, and collective action (Touraine, 1971, Melucci, 1980; Offe, 1985).

In the German Federal Republic the paradigm leaned heavily towards macrostructural interpretations, while in France, Touraine and his followers were more deeply impressed by cultural change. While the Italian paradigm had no single prophet, its two foremost interpreters both conceived of social movements in more cultural than structural terms. Francesco Alberoni spoke of a "state of birth" that gives rise to new movements (1968; 1979), while Alberto Melucci wrote about how movements are created through the formation of new collective identities (1980; 1985; 1988). Though they were deeply aware of the central role of politics in their country, like their French colleagues, Alberoni and especially Melucci perceived of new movements as action systems which coexisted with the "old" politics. For all of these observers, social movements were representative of a creative *statu nascente*. The political parties were, at best, the harbingers of a static middle age; at worst, they emitted the stale breath of old age.

The *political exchange model* developed, at the same time, from studies in industrial sociology. Alessandro Pizzorno (1978) and a group of his students and associates (Regalia, Regini and Reyneri, 1978; Beccalli, 1971) developed a model in which mass mobilization was seen as the contingent result of enduring conflicts in industrial society. For these scholars, social groups engage in movement activity when the structural balance tilts against them and advantage can be gained by insurgency. Though expressive in form and often in content, the collective action of these groups is part of the ebb and flow of conflict in capitalist society.[4]

Though they were aware of the theories of "new" social movements in Germany and France, Pizzorno's group focused less on how new collective actors were created and more on political exchange: on the interchange among institutional actors (parties and unions), semi-institutional ones (like "worker vanguards") and non-institutional ones (like the unskilled immigrant workers). Unfortunately, their work was largely limited to conflict; hence its relevance to the general cycle of protest that was occurring in Europe remained largely unexplored.[5]

This was unfortunate, for the proponents of the political exchange model had something to teach new social movement scholars. They were focusing on a "new" movement among the very social actor – the industrial working class – that many new social movement theorists were relegating to the past! Their research seemed to show that the "new"' characteristics that the NSM theorists were finding among the new middle class, and that they linked to post-industrialism, were also prominent in the 1960s and early 1970s in Italy in the labor movement. If this was the case, perhaps the new movements were not "new" in a historical sense, but were reflecting the genetic "newness" that is found during any wave of mass mobilization.

Political exchange writers also differed more broadly from the new social movement school: they were unconvinced that a fundamentally new political paradigm had overwhelmed the class-borne conflict patterns of the past. As Pizzorno warned those who would detach their theories from the conflict patterns of the past, if we do not pay attention to the lessons of history, then "at the upstart of a wave of conflict we shall be induced to think that we are at the verge of a revolution; and when the downswing appears, we shall predict the end of class conflict" (1978, p. 291).

NEW MOVEMENTS, OLD PARTIES

Two decades after 1968, we see neither the end of class conflict nor the revolution. Many of the new movements that were spawned a generation ago are already forgotten; others have given way to interest groups; still others have been institutionalized as new parties. Although such underlying features of advanced industrial society as "postmaterial values" are by now deeply ingrained in Western consciousness (Inglehart, 1977; 1981; chapter 3 in this volume), empirical studies have not yet demonstrated their translation into new political realignments and new paradigms (Dalton et al., 1984).

Why is this? No one doubts that new values and new political attitudes, new forms of collective action, and challenges to the hegemony of traditional party dominance have marked the mass politics of the past two decades. But there is a broad gap between these values, behaviors, and challenges and the actual degree of realignment that they have provoked in the party system – much broader than what the ringing metaphors of new social movement theory would predict. Perhaps the breadth of this gap is the result of the theorists' overestimation of the original distance between the new movements and the old party system? Or perhaps they underestimated the capacity of the parties to adapt to change and took too literally the new movements' rejection of political exchange?

In this chapter, I wish to investigate the role of political parties in the cycle of mass protest in Italy which lasted from 1965 to 1975. In Italy, as in France and the Federal Republic, the party system was often cast in the role of villain in the scenario of the movementist model. But a look at several aspects of the relationship between the parties of the Left in Italy and the new movements that appeared in the late 1960s suggests that the parties are more usefully seen as offstage but creative prompters in the origins, the dynamics, and the ultimate institutionalization of the new movements.

Before Mobilization

In an important and unfairly neglected article, Zald and Berger (1987) pointed out that new movements frequently arise – not directly out of the structural changes of advanced capitalism – but from insurgencies within existing organizations. In Italy, at least, even those Italian movements that rejected their origins deeply reflected the alignments, the interpretive themes, and the repertoire of actions of the existing party system.

To understand how this occurred we must first briefly summarize the role that the party system played in the post-war settlement and in the years preceding 1968. The New Left did not arise in the mid-1960s because the curtain had gone up on a new act of capitalist society, but because a political realignment in progress provided it with a new political space in which to act and with the script on which to construct a new scaenario.

THE POST-WAR SETTLEMENT

The three major characteristics of the post-war settlement in Italy are usually seen as: the reproduction within its domestic politics of the international Cold War; a domestic cleavage structure that pitted a Communist-led working-class bloc against a Christian Democratic-led conservative one; and weak and inefficient political institutions. Each of these is true enough as a starting point, but each was modified in the course of the 1950s and early 1960s.

The bipolar alignment

The post-war Italian settlement was similar to West Germany's in every respect but the most important one: the US and its domestic allies never succeeded in delegitimizing the Communist-led opposition. This not only broadened the political arena to include a historic bloc which – in West Germany – could be artificially identified with the hated East; it also produced a competitive dialectic *within* the Left that helped keep alive ideological themes that were rapidly sublimated in West German political culture.

Left and right subcultures

The Italian settlement did resemble the West German one in one important respect: Rather than being governed by an intransigent pro-business Right, the governmental coalition was built around a *centrist* position in which religion had a far deeper foundation than pro-capitalism. Of course, just like the CDU/CSU, the Democrazia Cristiana (DC) had close ties to business. But it was never the party *of* business, both because Italian businessmen had learned to have a healthy disrespect for the profession of politics and because, from the first de Gaspari government on, the DC understood that it needed to govern from the center to remain in power.

Nor was the Left ever as proletarian or as militant as it was often portrayed. The Italian Communist party (PCI) organized thousands of farmers and agricultrual workers, influenced the *petit bourgeoisie* in many regions, and had a strong basis among intellectuals. Despite the polarized pluralism of the country's political party space (Sartori, 1966), Italy developed a species of the open and occasional partisanship characteristic of liberal democracy (Tarrow, 1974; Parisi and Pasquino, 1980).

In fact, there was a symbiosis between the two ideological monoliths that, beneath the polemics of the cold war, took some time to become apparent. Although neither the DC nor the PCI ever became a "catch-all" party, in the jargon of the 1960s (Kirchheimer, 1966), both were trying to "catch more" support than in the past, to some extent from among the same electorate. The long-run effect of this competition for the political center was to moderate their competition and to create a need in the governing party to extend its coalition towards the Left. The opening was made in 1960–3 when the DC succeeded in wooing the Socialist party (PSI) away from the left-wing opposition.

Weak institutions, strong parties

The weakness of Italian governmental institutions is famous. But Italy's frail executive was not a sign of weakness, but a desired outcome of the 1948 constitutional settlement. Its counterpart was the strengthening of the parties in parliament. In fact, the real institutional success story of the post-war settlement was the tremendous increase – *vis-à-vis* its pre-fascist past – in the strength of the party system, both in parliament and in the country.

Although elections soon lost the "Christ versus Communism" militancy of the immediate post-war years, the two main parties did retain a kind of monopoly – over mobilization. Outside the party system representation was stifled, and it was increasingly difficult for new issues to be placed on the agenda or for new groups to gain a hearing except by granting a political party a mandate for representation. Even much of the press was party controlled or party influenced, and many associations – for example, farmers and Catholic Action – gained their benefits through a family relationship (*parentela*) with their party sponsors (LaPalombara, 1964).

The parties' monopoly of representation began to collapse in the early 1960s, as interest associations began to assert their independence and new social groups began to clamor for representation. The most important source of this change was the new salaried middle class of the cities, which doubled in size as a proportion of the active population between 1951 and 1971 (Sylos-Labini, 1975). Another less articulate source was the immigrant workers of the northern factories. Lacking a "culture of work" and the party loyalty of the traditional northern working class, from 1962 on they showed that they were available for "wild" forms of industrial action (Sabel, 1982).[6]

The same period revealed a profound secularization of Italian society, especially in the metropolitan areas of the North, where church attendance was declining, family ties were being loosened, and citizens were forming non-party-linked associations. The partisan subcultures that had structured people's lives since the Liberation – as well as their voting behavior – were actually in profound decline as scholars crystallized them into formulas like "polarized pluralism" (Sartori, 1966) or "imperfect bipartyism" (Galli and Prandi, 1970). Secularization also produced new demands for civic modernization, for increased access to services and education, and for a government that could attack the problems of an advanced industrial society.

THE REALIGNMENT OF THE 1960S

These changes in the post-war settlement produced a change in the quality and intensity of political debate, which prepared the way for a new generation of political movements by identifying overarching issues, establishing a mutual awareness among diverse social and political actors, and creating new political space. Across the political spectrum, there was a widespread debate over the transition to mature capitalism and its costs and promises. In governmental circles, the debate focused on economic planning, on the technical needs of a modern society, and on the defects of the existing industrial relations system. On the Left there was an open debate on the tendencies in modern capitalism.

The academic tone of these debates at first disguised the novel fact that they bridged political subcultures and had the potential for broadening the arena of controversy to include new actors. In governmental circles, debates on the economy were stimulated by a small but influential group of liberal thinkers in and around the small Republican party, led by Ugo la Malfa, who would be budget minister in the first center-Left government. But the ferment also had expressions in Catholic circles, particularly in a group of younger economists who had come out of Catholic Action and were active in the Interministerial Commission for the South and in the semi-public SVIMEZ (Association for the Development of the South).

On the Left, the debate took on more theoretical tones, but was no less politically portentous. The Communists, through their Gramsci Institute in Rome, held an important conference on "Tendencies in Italian Capitalism," which put them ten years ahead of their French comrades in recognizing the effects of economic change (Gramsci Institute, 1962). A strategic debate on the role of the "productive" middle class began in the party at the same time. Sensing the decline of its traditional rural supporters, the party began to look for new electoral allies in the expanding parts of the new middle sectors, long before the "new" social movements were ever imagined.

As for the Socialists, they were more concerned with readying themselves for a future role in government than in understanding the future of Italian capitalism. But political opportunism has strange effects: the prospect of joining the government led the PSI to put reforms on the government's agenda – education, planning, pension reform – which would later become rallying points for mass protest. Ironically, but not for the first time in history, the themes of future protest movements were popularized by the very people who would eventually be the object of attack for the positions they took on them.

THE CENTER-LEFT GOVERNMENT

Such was the atmosphere in which the DC brought the Socialists into government in 1963–4. The "Opening to the Left" was neither the first nor the last stratagem the Christian Democrats used to hold on to power. The DC had always been willing to share resources with other political groups, first with the conservative Liberals and then with the moderate Social Democrats and Republicans. But when coalition politics was extended to the PSI, it became a much more risky undertaking. First, because the PSI was too large to relegate to subaltern status like the Republicans or Social Democrats; and second, because its shift from opposition to government provided political opportunities to others that stimulated mobilization. The inclusion of the Socialists in the coalition placed new issues on the agenda and exposed them to public criticism in a way that made plain how divided the government was and how strong were the forces on either side of the debate.

Some of these issues revealed the conflicts in the center-Left without touching the mass public or stimulating a broad debate: for example, the compensation of the stockholders of the electric companies that were nationalized in 1962; the shape and extent of the economic planning apparatus; the implementation of the ordinary regions. But other issues – such as pension reform and the divorce issue – triggered a wider circle of conflict. Most important in this respect was educational reform, for it gave the PCI a popular parliamentary platform from which to appeal to middle-class concerns and provided the university Left with a ready-made theme around which to organize (Tarrow 1989, ch. 6).

The center-Left also encouraged mobilization because of its more tolerant attitude to dissent, for the Socialists could ill afford to be identified with repression. Neither the old recessionary solution to wage increases, nor unleasing the forces of order against demonstrators, nor the use of political anti-communism were possible with the PSI attempting to preserve its claim to a share of working-class votes and trying to gain support from the new middle class. For example, when workers were killed by police in Avola and Battipaglia in 1968, the PSI called for an investigation and joined the Communists in a call to disarm the police.

The Widening Circle of Conflict

The debates that followed the center-Left experience in Italy were similar in many ways to debates about the New Deal in the United States: both revolved around the academic issue of whether they were progressive or conservative. But like the New Deal and the French Popular Front, the center-Left government was important because it placed issues on the agenda that it could not resolve and – by its own internal divergences – gave groups outside the coalition the chance to intervene in debates that had begun within it. This was true both for the Communists and for the extreme Left.

THE COMMUNIST PARTY

Although the center-Left government challenged the PCI, driving it out of many municipalities that it had governed jointly with the PSI, it also left the party with a predominant position in the unions and induced it to adopt more aggressive policy postures to expose the PSI's betrayal and protect its own remaining bastions of working-class support.

That the Communists did not lose electorally from the cooptation of the Socialists into the government assured them in their conviction that the country might still have a left-wing future. When the protest cycle broke out in the late 1960s, they were therefore encouraged to adopt more radical positions at their Congress of 1969. But the Communists had their ear to the ground even earlier than the "Hot Autumn" of 1969; for example, in the debate over pension reform in early 1968, they quickly sensed a mass pressure for change that had not been evident a few years before. Despite the fulminations of its moderates about the danger of subversion from the Left, the party was far more open to worker and student militancy than the French Communists, and more sensitive to the new currents than the unions.

THE EXTREME LEFT

As for the *non*-Communist Left, its appearance dates, not from 1968, but from the early 1960s, when the debate on the center-Left government began.

For as one traditional party of the Old Left was preparing to turn to social democracy and the other was seeking support from the middle class, new actors were preparing to challenge both for political space on the Left. They did so both *within* the structures of the Old Left and outside of them, through a combination of old and new themes of mobilization and by using both conventional and confrontational forms of action.

The New Left within the Old

Soon after the effective start of the center-Left experiment, conflict began to stir within the party system. Splits in the Socialist and Republican parties in 1963 and 1964 were followed by an attempted merger between the Socialists and Social Democrats in 1966 and by talk of forming a unified party of labor by the moderate wing of the PCI. The most important split created a new left-wing socialist party, the PSIUP, from the Left wing of the PSI, after its majority decided to join the government in 1963. Drawing its membership from among the Socialist party's most radical cadres, the *socialproletari* of course sought to occupy the political space to the PSI's left; but the hegemony of the Communists led them to take a position to the left of the PCI as well. Well before a new "extraparliamentary" Left was dreamt of, the PSIUP had extended the boundaries of the parliamentary Left to new themes and forms of action designed to outflank the Communists.

The new party soon became active in the cities and universities in which the PCI was strongest, challenging it for supporters among new groups of workers and students. It was not uncommon for demonstrations to be jointly organized by *socialproletari* and Maoists or for PSIUP members to militate within outright extraparliamentary groups. The PSIUP goaded young Communists to adopt more radical positions than their party would tolerate and to participate in joint demonstrations that older PCI leaders frowned upon.[7] For many young intellectuals, the PSIUP would be a bridge from the institutional political arena to the new movements forming outside the polity.

Insurgent currents also developed within the PCI. For example an engaged communist intellectual wrote in an anonymous pamphlet that the Communist party "attracts a generic adherence among the working class, but fails to seek a more engaged participation . . . Why should anyone want to militate in a party that doesn't get you any further than Parliament? It is enough to vote for it" (Accornero, 1967, p. 10). Another insurgent group was the one around Lucio Magri, Luigi Pintor, and Rossana Rossanda, the leaders of the future Manifesto group, who would be 'radiated' from the party.

Even in the Catholic camp there were defections and insurgencies against orthodoxy. The ACLI, the association of Catholic workers, split between left-wing and a right-wing factions in this period. Catholic Action, a traditional bastion of anti-communism, lost thousands of members (Cattaneo Institute, 1968). Many future student radicals received their first political socialization in Catholic student associations like *Intesa Universitaria* and *Gioventu Studentesca* (GS). When a revolt broke out among students at the Catholic University of the Sacred Heart in Milan in 1967, the leaders were all former Catholic Action or GS militants.

Even before Vatican II, new Catholic journals began to give a sympathetic hearing to radical doctrines. In the religious Veneto region, the journal *Questitalia* sympathetically followed the development of the student and workers' movements, while in Tuscany, *Testimonianze* was in touch with Florentine marxists and giving a sympathetic hearing to the "theology of terrestrial realities." Here and there, religious "base communities" were preaching the gospel of the poor, using the parish structures of the most conservative institution in the country to mount an insurgency against it (Sciubba and Pace, 1976; Tarrow 1988b).

It is no accident that these new currents – in the universities, the Church and the institutional Left – began to stir just as the party system was undergoing its most delicate realignment since 1947. More surprising in a country that complacent observers still saw as divided into air-tight political subcultures, people were beginning to cross the once-unassailable dikes between the Marxist and Catholic subcultures. Insurgents were not only distancing themselves from the moderates who controlled their respective political organizations; they were beginning to find one other across the traditional political divide between the nation's subcultures.

Challengers from outside

A deeper ferment was stirring outside the party system, where new political groups were appearing and mobilization was beginning to escape traditional channels. This was first apparent in a generation of "little" reviews of both secular and Catholic Left that developed in the early 1960s (Lumley, 1983; Becchelloni, 1973). Each of these represented less a separate intellectual current than a cadre of militants who hoped to use the press to create an alternative to the existing party system. These reviews circulated even to small towns and distant regions through the mails and by way of students returning home on vacation, and were passed around from hand to hand.

The best known of the "little" reviews was *Quaderni Rossi*, published in Turin. Although it was founded by a dissident socialist, Raniero Panzieri, it also attracted support from among communists, unionists and independent leftists. It built on a traditional leftist theme – worker centrality – and attacked both the PCI and the PSI for deflecting the workers from their revolutionary tradition (Magna, 1978, p. 315). Unionists who would later play an important role in consolidating the factory councils were attracted to Panzieri's teaching no less than future extraparliamentary leftists like Adriano Sofri, a founder of *Lotta Continua*, and Michele Salvati, who would become an influential political economist. Splits within the *Quaderni Rossi* group soon created two new workerist reviews, *Quaderni Piacentini* and *Classe Operaia*.

Some of these young leftist groups limited themselves to making propaganda, but QR began a tradition of "intervention" in the factory that ranged from involving workers in surveys of factory conditions to fomenting strikes and insurgency against the unions. Particularly aggressive were various "workers' power" groups that formed around the country in the mid-1960s, one of which played a major role in radicalizing industrial conflict in the Mestre-Marghera industrial pole near Venice (Tarrow, 1989, ch. 5).

But worker insurgency could only be stimulated during peak periods of contract negotiation. This put the new radical groups in the untenable position of having to wait for union initiatives against management to attack the hegemony of the unions over the workers. Although they had scattered successes, each one quickly evaporated between contract negotiation periods.

It was in the university that the mass base was available with which to outflank the party system. But even here mobilization was not generated spontaneously: the student movement arose out of revolts against the party-affiliated student associations, and was generated by the latters' own agitation against the government's plan for educational reform. The student movement illustrates better than any other how issues placed on the agenda by the "old" party system provided the tools for insurgents to mobilize a new mass base.[8]

The campaign against the government's educational reform plans began with a coalition between the PCI- and PSI-led student association, the *Unione Goliardica*, and the DC-affiliated *Intesa Universitaria*. Many Catholic and leftist student militants came to know each other, and moved together to more radical positions, in this intra-institutional debate. When both party-led associations resisted attempts to radicalize their positions, the militants joined with external activists to form a new mass movement. The culmination came at an occupation of the administration building of the University of Pisa in 1967 against the government's reform plans, when the insurgents outflanked the old student movement by adopting a radical workerist set of theses (Lumley, 1983).[9]

The old student associations were soon scored with conflicts that turned the debates of the national student assembly, the UNURI, into heated ideological conflicts (Pero, 1967). Underneath their workerist rhetoric, the radicals were proposing that all the traditional student associations merge into a single national student union in which the parties would lose their tutelary role. At a congress in Rimini in the spring of 1967, they purposely presented a motion that the majority could not accept (Cazzaniga, 1967). It was soon after this that the traditional associations collapsed.

Utopia and Confrontation

The proposal for a unified student union with a workerist ideology illustrates how New Left advocates constructed new themes out of a traditional leftist matrix to use against the Old Left. The PCI had long put forward a broad strategy of alliances (Hellman, 1975; Tarrow, 1967). The idea of a single national organization, which would express the students' solidarity and autonomy, was perfectly compatible with its proposals and in fact had first appeared within the Communist-led *Unione Goliardica*. The PCI could not easily oppose a single national student union. Yet such an innovation had the potential to give the radical students a forum in which to challenge party control. The university student movement grew like an acorn from an oak tree of the Old Left. Utopian, expressive, and demonstrative elements were more prominent in the university student protests than in any other part of the social movement sector. However, they first developed around political, instrumental, and policy-oriented demands like those generated by the debate on educational reform.

INSTRUMENTAL AND EXPRESSIVE DEMANDS

The policy-oriented demands of the student movement can be seen in a statistical breakdown of its demand structure calculated from daily Italian newspaper data.[10] Table 13.1 compares the various types of demands made in the university protests to those found in all the protest events analyzed from a national newspaper-based archive collected between 1966 and 1973. The table shows that, in over 60 percent of the university disputes, the students put forward claims for new rights, for substantive benefits, or for or against policies of the government. In 17 per cent of the disputes, the protests were simply expressive, affirming the students' identity, showing their sympathy for others, opposing other groups, or demanding the overthrow of the system. But the newspaper data also underscore the uniqueness of the university students' movement and its links to the new social movements of the 1970s. This can be seen in three ways.

First, the students demanded new rights of participation far more often than other social actors. The theme that underlay most of these protests was that of autonomy, sometimes expressed as a struggle against academic authoritarianism, sometimes against the party-led student associations, and sometimes extended to a demand for a student-run university (Grazioli, 1979, pp. 30–5, 186–91). The students sought the right to assemble, to influence curriculum and teaching procedures, and in some cases to reorganize the university. They wanted to be able to publish freely, to criticize professors within and outside the classroom, and to take exams when they pleased.

Second, table 13.1 shows that the university students were more interested in *general* policy issues than were most other social actors. They made their own policy demands or tried to stop government actions of which they disapproved. This concern with general policy was most evident in the area of university reform, but students also demonstrated against the war in Vietnam, in favor of pension reform, and against fascism and police violence. Their policy proposals were radical and were sometimes put forward in a

Table 13.1 Italian university students' protests compared to all protest events: Types of Demands, 1966–1973 (percentages)

Demand type	*Students*	*All events*
Substantive protests		
Getting rights	17.3	3.5
Getting more	28.1	42.0
Policy demands	18.0	13.3
Expressive protests		
Identity or solidarity	11.5	10.8
Opposing others	5.0	25.8
Getting out	0.7	0.9
Mixed types		
Substantive and expressive	19.5	3.7
Total	100.1	100.1
(N)	(139)	(4,296)

Source: Data collected by the author; see note 10 below and Tarrow (1989).

spirit of play, but they were far more likely than other social groups to make demands that went beyond their "hard"' material interests.[11]

Third, far more than other social actors, the university students' demands collapsed "the wall between the instrumental and the expressive" (Zolberg, 1972, p. 183). Some new social movement theorists have argued that movements have an underlying "logic" that is either instrumental or expressive (Rucht, 1988). But the Italian students used expressive elements to advance instrumental goals – as in their use of workerist symbolism to demand a unified national student union. Rather than following either a straightforward instrumental or an expressive strategy, like many new social movements, they were "radically pragmatic" (Lumly, 1983, p. 434).

CONTENTION AND CONVENTION

New social movement theorists have emphasized the radical forms of action used by these movements. From the very beginning, a great majority of the Italian students' protests were built around confrontations with authorities. Table 13.2 compares the use of conventional, confrontational, and violent actions by students in the universities to the incidence of these types of action in the protest events as a whole.[12] The students used confrontational forms twice as often as did other social actors. In the first half of 1968 alone,

Table 13.2 Conventional, confrontational and violent forms of action as a proportion of total protest forms: university protests and all protest events, 1966–1973 (percentage)

	University students	*All events*
Conventional	40.7	56.1
Confrontational	37.4	18.9
Violent forms	18.5	23.1
Other forms	3.4	2.1
Total	100.0	100.2
(N)	(356)	(9,006)

Source: As for table 13.1.

of the university-based protests for which we have detailed information, confrontational forms of action were used in 90 percent of the cases.[13]

The main form of action that the students developed was the occupation. It was seldom a single isolated event. Occupations quickly became a basis for other actions in which solidarity reigned, social constraints were released, and group activities were organized. Teams of enthusiastic students printed newsletters and produced wall posters; courses were organized and debates were carried on; activists formed new networks crossing previously established political lines. Occupations were crucibles for the release of new energy.[14]

But the occupation was no new creation. It had been a traditional action form of the Italian Left since 1919–21, when the plants of the sprawling Fiat empire in Turin were occupied by the workers (Spriano, 1964). It reappeared in 1943–5 as workers took control of their factories from managers who had worked hand-in-glove with the corporate state. The traditional Left continued to use the occupation during the post-war period as a means of expressing its policy demands and gaining publicity. In 1965, for example, the Communists had symbolically occupied the municipalities of Pisa and Genoa to embarrass the PSI, which was negotiating for the formation of new municipal governments with the DC. In an ironic incident in Pisa, they "instructed" the student militants of the *Unione Goliardica* in how to use the occupation (cited in Tarrow, 1989, ch. 10).

But the leaders of the student movement added much to the use of the occupation. They used it not only as a dramatic and disruptive form of action but also as a "practice of the objective" of direct democracy to create social and educational space for themselves. The occupation both linked the student movement to Italy's revolutionary past and provided it with the political and physical space in which to develop its own identity and network of activists.

The main outcome of the occupations was to form new social networks – for example between secular leftists and progressive Catholics – which would not have been possible before. The carefully crafted documents produced by these occupations were designed both to express the movement's goals and, more pragmatically, to bridge the differences between groups of varying ideological derivation. The occupation became less an aggressive form of action against authorities than an institution for the formation of the collective identity of a new social movement.

The same was true for the other forms of direct action invented by movement activists in these years; they were less a spontaneous expression of the "life-space concerns" of the activists than instruments developed out of the process of confrontation to form new collective actors. For example, until they were turned out into the street by the authorities, left-wing Catholic activists routinely occupied churches to express their demands. Turning defeat into victory, they "invented" the protest form of the mass in the piazza, which had the advantage of attracting both non-practicing Catholics and the irreligious into their movement (Tarrow, 1988b).

In summary, in the student movement, activists experienced from within the old party system radicalized debates that had begun within it. They developed a language of utopia and confrontation that was designed to outflank the Old Left, hold the movement together, and appeal to new supporters. If the old parties were often a target of the movement, it was less because they opposed its aims than because they were a necessary target against which to build new identities. But the Old Left was also the foundation upon which the new movement was built. For, as two observers of the French scene in 1968 (Schnapp and Vidal-Naquet, 1971, p. 17) wrote:

> 'In France and Italy the very fact that there exist large workers' parties which are revolutionary in speech, if not in practice (allows the student movement) to present itself as the University detachment of a revolutionary workers' party that does not exist.'

After Mobilization

If the party system was a participant in the overture to Italy's protest cycle, it also took part in its denouement. In the wake of the protest cycle, many observers emphasized either the violence to which it gave rise or the institutionalization of the movements that it produced. Italianist Joseph LaPalombara (1987, pp. 169–70) stresses the former theme:

> In its earliest phases, terrorism appeared as little more than an extention of the protest movements and the hundreds of more or less revolutionary groups that mushroomed in the universities in the late 1960s. Initially, these groups talked and talked. Some then turned to kidnapping and kneecapping. Murder came later.

But another school of interpretation stresses the institutionalization of social movements. As Theodore Lowi (1971; p. 54) writes:

> When movements act on the government or any of its parts, there tends to be action with very little interaction – that is, very little bargaining. But this is not violence . . . The effect of the movement is of another sort altogether: *the demands and activities of a movement tend to activate the mechanisms of formal decision-making.*

How does such institutionalization occur? Through the decline of mass mobilization and the attempt of movement leaders to keep their organizations alive by providing members with selective incentives? Through the cooptation of leaders into established institutions and the transformation of militancy into compliance? Or through the efforts of political parties and interest groups to adapt to insurgency and to absorb the energies of the movements?

All three processes could be observed in Italy in the mid-1970s. But it is only through a detailed analysis of the political process that we can understand how a phase of mobilization gives way to a phase of institutionalization. If we emphasize here the adaptive efforts of the party system, it is to counterbalance the pessimism and disillusionment that mark most research on the end of mobilization. As Zolberg writes, for people emerging from a period of mass mobilization, "post coitum omnia animal triste" (1972, pp. 205–6).

COMPETITION BETWEEN PARTIES AND MOVEMENTS

The movementist model left us with an image of the political parties as "old gardeners" who attempted to prune the fertile wild flowers of the new movements that arose in the 1960s. This is certainly part of the picture – especially in cases like that of the French Communists, who were particularly resistant to the message of liberation that the new movements bore. But the PCF is a poor example of party response, for its anti-movement tactics were a recipe for failure. Indeed, it was in part the Communists' hostility to the new movements that provided a social base on which the resurgent PS could build.

The British Labour party is also frequently cited as a party that had no sympathy for the new movements that arose in the 1960s and 1970s. But Labour's anti-movement reputation is largely based on its "cloth cap" image and on Gaitskell's campaign to reverse the unilateral disarmament plank adopted by the party conference in 1960. Research in the 1970s showed that a high proportion of CND members in the 1960s were Labour party members (Taylor and Prichard, 1980). The same appears to be true of CND membership in the 1980s (Maguire, 1988).

The Italian Communists' adaptation to the new movements represented a situation between that of the PCF and British Labour. Unwilling to encourage movement organizations that it did not control the party competed with them for support among new social actors. Unlike the French Communist leaders, the PCI was questioning inherited dogmas and seeking support from the new middle class from 1956 on (Blackmer and Tarrow, 1975), well before the new movements of the 1960s and 1970s appeared. But since the party assumed that it could count on the support of the working class, its efforts aimed at the new middle class, which left a strategic opening for its competitors.

This competition had its brutal side. For example, the PCI would sometimes send tough factory militants as a *servizio d'ordine* to "protect" its youth demonstrations from being broken up by the radicals. But it also had a more political aspect, as the party, the unions, and the movements bid for support of new social actors like urban peripheral dwellers, ecologists, and women, with competing programs and broad mass forms of action. The only area the party refused to follow the new movements into was the realm of "vanguard violence," which it roundly condemned, both among its own militants and in the extreme Left.

The tenants' movement is a good example of the inter-organizational competition that developed among movements, parties, and trade unions. From the beginning, it was not a homogeneous radical movement, but a cross-section of the social movement sector, with institutional, semi-institutional, and anti-institutional elements competing for support (Daniele, 1978). It would thus not be fair to say that the movement was institutionalized only as the cycle wound down; like the student movement, at least a part of it was linked to the Old Left from the start.[15] For example, Perlmutter (1988) cites figures that show that PCI militants were active in a large number of "spontaneous" neighborhood groups in Turin from the start.

The early stages of the tenants' movement were dominated by confrontational protests based on rent strikes and the *autoriduzione* (self-reduction) of rents and rates. Organized occupations followed those of the student movement, as the urban groups expanded their constituency from public housing tenants to the homeless. Occupations and refusals to pay increased public transportation and utility costs continued into 1975. But as mobilization declined, such initiatives were more and more difficult to mount, and the movements organized themselves and moved into local politics (Perlmutter, 1988).

ACROSS THE FRONTIERS OF THE POLICY

Social movement theorists often conceive of the institutionalization of movements taking place only through the cooptation of their leaders or the preemption of their policy goals by elites (Gamson, 1975, ch. 1). However, rather than suffering either debilitating preemption or corrupting cooptation, movements often act in uneasy coalition with elements in the old party system. The outcome of their struggle is often reform – partial, disappointing, but incremental reform. If we only visualize new social movements growing up outside of, and against, politics, it is difficult to see these mottled and composite effects. But seen in combination with changes in ordinary politics, we find objective – and sometimes self-conscious – coalitions between political parties and interest groups that sometimes, under some conditions, achieve some of the goals of the movements. The main function of the social movements for reform in Italy was to radicalize and publicize issues that had already appeared on the public agenda and to make alliances – either objetive or subjective – with groups within the frontiers of the polity, leading to partial, and often reversible, reforms.

Aborting Dissent by Reform

An example of this can be found in the case of the women's movement. An independent Italian women's movement was virtually nonexistent before 1968, with most women's organizing dominated by the Catholic CID (*Centro Italiano della Donna*) and the Communist-Socialist UDI (*Unione Donne Italiane*). In the reflux against the New Left of the early 1970s, women organized in the unions and extraparliamentary parties and in a plethora of collectives, consciousness-raising groups, and in *ad hoc* mobilization campaigns (Hellman, 1987). This led, among other things, to pressure for a new abortion law outside the parliamentary parties.

The years 1975–7 are full of reports of demonstrations in favor of abortion by non-parliamentary feminist groups, institutional interest groups like the UDI, and radical extraparliamentary groups attempting to overcome their well-deserved macho image. These were not concerted efforts, and many of the groups had conflicting goals. Sometimes there was a common front attempted between challengers and members of the policy elite.[16] But, more commonly such groups operated independently of one another, each in complete awareness of what the others were doing. The net effect was to force abortion on to the policy agenda and to force the institutional parties to deal with it.

The demonstrations of the new feminist groups had vivid expressive elements (e.g., organizers might dress up as witches, sport pink armbands, and refuse admission to men marchers). But we cannot understand their impact if we look only at their "expressive logic." For they forced the question of abortion on to the policy agenda and combined with developments within the party system to revoke the old Fascist-era family legislation and produce a moderate abortion law. The bill that resulted, and passed in 1978, did not grant women the right to "open, assisted, and free abortion" that the feminists had demanded. But its moderate language made it possible for the DC to abstain from the parliamentary vote, for the PCI to combine with the Socialists and the lay center parties in support of a compromise, and for abortion to become legal in a country in which the Vatican was still a powerful political force (Ergas, 1982, p. 271).

We cannot understand the success of the abortion campaign unless we relate the politics of the piazza of the new feminist groups to the politics of electoral advantage of the major parties. As in the passage of civil rights legislation in the US in 1964, reform was achieved both through the efforts of an external movement and through the desire of party leaders to preserve their political positions (Piven and Cloward, 1977, ch. 4). But the parties also successfully coopted some of the participants in the new movements of the 1960s and 1970s, as we shall see below.

MOVEMENTS INTO PARTIES: THE CASE OF PCI RECRUITMENT

In his provocative book about cycles, Albert Hirschman observed in the early 1980s that political involvement is characterized by "oscillation between periods of intense preoccupation with public issues and [periods] of almost total concentration on individual improvement and private welfare goals" (1982, p. 3). For Hirschman, "Western societies appear to be condemned to long periods of privatization . . . followed by spasmodic outbursts of 'publicness' that are hardly likely to be constructive" (p. 132).

But however we judge its impact, the wave of mobilizations of the 1960s cannot be said to have led only to a trough of privatization. On the contrary, it had major impacts on public life, not only through the triggering of a series of new movements – like the ecological, anti-nuclear, peace, and women's movements – but also upon cultural life, interpersonal relations, and attitudes towards authority. Most important in the long run, the new movements and new value orientations have significantly affected the institutional settings against which, but also *within* which, the new movements arose.

For example, the Italian Communist party, which was the source of much of the insurgency of student and worker militants in the 1960s, was also the recipient of much of the residual activism at the end of the period.[17] For as mobilization waned in the early 1970s and some activists entered new movements, while others accepted positions in the unions and a few took up the desperate cause of armed struggle, a significant number of those socialized into politics in the 1960s became militants in the Communist party. Between 1969 and 1978, in fact, party membership rose from just over 1,500,000 to just under 1,900,000 (Lange, Tarrow, and Irwin, 1989).

The meaning of this membership increase is subject to varying interpretations. Barbagli and Corbetta (1978), basing their conclusions largely on regional correlations between the strike rate and party membership, conclude that it was from the *periphery* of the mass movements of the 1960s in central and southern Italy that the party recruited the largest numbers of its new members. Hellman (1980), using more disaggregated data from Turin, finds a positive correlation between strikes and increases in membership.

Neither of these analyses, however, had much to say about the *character* of the new members who entered the PCI in the wake of the movements of the 1960s. For example, do they remain on the periphery of the party's

Table 13.3 PCI delegates to 1979 provincial party congresses: generation of party entry by pathways of recruitment (percentages)

Background	*1921–46*	*1947–56*	*1957–66*	*1967–9*	*1970–3*	*1974–6*	*1977–9*	*Total*
Traditional								
Young Communists	24.1	56.8	50.1	37.9	21.8	17.5	27.5	29.5
PCI family	35.3	9.4	12.4	10.8	7.6	0.5	6.2	10.8
Union	1.7	5.9	10.1	14.8	20.5	21.3	15.0	15.3
Other party	3.4	2.3	1.8	1.9	2.5	1.1	0.7	1.9
New Movement	2.1	1.0	1.9	0.9	10.0	11.3	8.0	7.3
Mix of traditional and new	4.9	12.6	12.4	16.0	25.0	25.8	27.3	20.0
Multiple backgrounds	0.6	1.7	1.3	1.8	4.9	4.8	5.6	3.5
None	28.1	10.3	10.1	9.9	8.3	11.6	9.6	11.7
Total	100.2	100.0	100.1	94.0	100.6	93.9	99.9	100.0
Row percentage	9.4	7.9	13.1	8.7	23.6	25.7	11.4	99.8

Source: Peter Lange, Sidney Tarrow, and Cynthia Irwin (1989).

membership, or have they begun to occupy roles within its organization? A survey of the delegates to a party congress allows us to at least see how rapidly they moved into its grassroots decision-making level. Table 13.3 summarizes data from the CESPe survey of 1979 provincial congress delegates, showing both the generational breakdown of the congress delegates and their paths of recruitment. Reading along the bottom row of the table, we can see how the generational composition of the congress delegates bulges with those who were recruited into the party after 1969 and before 1977. Although one would normally expect a traditional mass party to be staffed at the grassroots level by militants who had been in the party for many years, the PCI delegates in 1979 were overwhelmingly drawn from among those who were socialized into politics within the decade after the student-worker movements of 1967–9.

Reading along the top four rows of the table shows how people from the traditional sources of PCI recruitment (the FGCI, the Young Communist Federation, and Communist family backgrounds) were giving way to those with union or other political backgrounds. And reading along the lower rows in the table shows that recruitment from movement backgrounds or a mix of traditional and movement background increased dramatically after 1966. If one disaggregates the data still further, the curious amalgam of those with a background in both the Communist Youth Federation and in a social movement mushrooms after 1966: while tiny proportions of the militants recruited into the party before 1967 had such a mixed background, they rose to 7.5 percent of those recruited in 1967–9, all the way to 13 percent of those recruited between 1977 and 1979 (data not shown in table). Socialization into politics through new movement activity became an important source of old party recruitment.

Conclusions

Beneath the surface appearance of a rigid opposition between the wild, spontaneous, and anti-partisan new movements of the 1960s and 1970s, and the staid, tired, conservative institutions of the party system, there was, in Italy at least, a much more complex and interpenetrating set of relations.

First, the mobilization of the 1960s was triggered by the realignment of the party system, which produced the policy issues, the political space, and some of the militants who later formed a new generation of new social movements.

Second, these movements were formed partially against, but also partially within the representative associations provided by the party system, the unions, and even the Catholic Church. The new movements have rightly been pictured as having an anti-institutional vocation; it has far less often been observed that they had sources and communication channels within the traditional institutions of Italian democracy and used these as resources with which to build a following.

Third, political change occurred, not through the dramatic clash between institutional and anti-institutional armies, old and new political paradigms, but through the competition between parties, unions, interest groups, and movements, through reform coalitions, and through the absorption of at least a part of the "people of 1968" within the party system after mobilization had ended.

New social movement theorists have cast the parties as critics of the creative dramas of the 1960s and 1970s, but the parties, rather than the villain of the piece, were the phantom of the opera. They sketched the themes that the new movements would vary and orchestrate; they trained some of the actors who then walked alone on to the stage of insurgency; they developed a repertoire of actions which the movements — like a jazz ensemble — would improvise upon and radicalize; and after the show was over, they welcomed many of them back into the party fold.

But this was no *opera buffa* — it had many aspects of tragedy and drama. As one veteran of these movements (Sofri, 1985), reflecting on their history many years later, observed:

> The tragedy is that the words that we inherited [from the Old Left] were never contested, but were simply accompanied by other words. We took the nouns and instead of proving them false . . . we added adjectives to them, an incredible quantity of adjectives.

NOTES

I wish to thank Russell Dalton, Manfred Kuechler, and Dieter Rucht for comments on an earlier draft of this paper.

1 For the major sources on new social movements in Scandinavia, see Olofsson, 1988.
2 These are two of the four outcome categories proposed by William Gamson (1975). The parties also responded by repression in some cases, but this came much later – during the phase of organized terrorism that followed.
3 Readers of this book will be aware of the sources on which I base these generalizations and will be equally aware that they do not apply in equal measure to any single theorist. For the review and analysis on which this section is based, see Klandermans, Kriesi and Tarrow (1988, introduction).
4 Though coming out of a European structuralist tradition, Pizzorno's model had striking, but unrecognized parallels with the way American students of politics were studying protest movements (for reviews, see Marx and Wood, 1975; Tarrow, 1988a). The Americans emphasized not only the use of protest as a political resource (Lipsky, 1968), but the relatedness between institutional and non-institutional conflict (Piven and Cloward, 1977). And just as Pizzorno pointed to conditions in which new social actors used political exchange, the Americans emphasized the role of political opportunities in stimulating the rise of social movements (Eisinger, 1973; McAdam, 1982; Katzenstein and Mueller, 1987; Tarrow, 1983). The only Europeans to relate political opportunity structures to social movements were Della Porta (1988) and Kitschelt (1986).
5 However, see Bob Lumley's excellent doctoral thesis, (1983, ch. 1), which extends the reach of Pizzorno's theory beyond industrial conflict.
6 Note, however, that Sabel underestimates the importance of the "old" skilled working class in triggering the industrial conflicts of the late 1960s and in demonstrating the efficacy of the "wild" forms of action that characterized it. For the role of skilled workers, see Reyneri (1978). For the planned nature of the "wild" forms of action, see Dubois (1978) and Tarrow (1989, ch. 6).
7 For examples, see Tarrow (1989, chs 3, 7).
8 This was also true of the anti-Vietnam movement on which the New Left cut its political teeth. It was stimulated by the PCI's campaign in 1966–7 against American intervention. The party placed the issue on the political agenda; but it was outflanked by more radical groups who were willing to engage in more radical protests (Tarrow 1989, chs 7, 9).
9 For an analysis of the events in Pisa and their role in destroying the power of the traditional student associations, see Tarrow (1989, ch. 7).
10 The data were gathered from a daily reading and enumeration of all strikes, civil protests, and violent collective action from Italy's major newspaper of record, *Corriere della Sera*. For a description of how the data were collected and analyzed, see Tarrow (1989, appendix B).
11 For additional evidence that the students were more likely to "mind other people's business" than other social actors, see Tarrow (1989, ch. 5).
12 For a discussion of these three main protest repertoires, see Tarrow (1989, ch. 3). The concept of the repertoire of contention is developed by Tilly (1978; 1986).

13 Excluding protest events outside the universities which appeared to include students, we counted 29 protests about educational issues in the first half of 1968. This is the operational indicator of "university-based protests."

14 As Lumley (1983, p. 164) writes of the influential occupations of the Architecture faculties in 1967: 'An environment was created which was functional to collective living, debate and shared work; all major decisions were taken by the general meetings . . . . commissions were set up to examine political and educational issues.'

15 As early as 1969, for example, the unions were demanding housing reform – in fact this was one of the tenets of their "strategy of reforms." Tenants were organized by party-led tenants' associations from the early 1960s. The rise of mass mobilization in the late 1960s expanded the number of new groups trying to represent tenants. But the new movements never destroyed the old tenants' groups; on the contrary, their competition gave new life to them and forced them to expand into new issue areas and to adopt new forms of action. For information on the mobilization campaigns in housing in Turin, see Perlmutter (1988). On Milan, see Daniele (1978).

16 See Tarrow (1989, conclusion) for an important case that occurred in 1975.

17 The data summarized below come from a jointly written article (Lange, Tarrow, and Irvin, 1989) based on data provided to us by the CESPe of Rome. I am grateful to my collaborators for their willingness to allow me to refer here to our joint work and to the CESPe for the use of their data. For a more extensive analysis of the CESPe survey, see Accornero, Mannheimer, and Saraceno (1983).

Part V

New Social Movements in Perspective

14

New Social Movements and the Political Order: Inducing Change for Long-term Stability?

MANFRED KUECHLER and RUSSELL J. DALTON

The chapters in this volume provide a critical assessment of the impact of new social movements on the politics of advanced industrial democracies. Deliberately, scholars representing different schools of thought, drawing on intimate familiarity with different national contexts, were asked to contribute. It is not surprising, then, that the chapters offer somewhat conflicting views. In particular, there is no general consensus on whether the social movements discussed here – the environmental movement, the peace movement, and the women's movement – are indeed "new" in a theoretically meaningful sense.

This concluding chapter attempts to develop a coherent view of the contemporary movements, piecing together ideas and insights presented in the previous chapters. However, rather than just sorting out points of agreement and issues of dissent as impartial referees, we will offer one particular synthesis, putting the findings into our own frame of reference. As we will demonstrate, much of the apparent disparity in findings is due to a difference in vantage point. The chapter authors employ different approaches and focus on different traits of these movements, and they use different methodologies to assess the role and function of these movements. We will not take issue with any particular position or argue over the "right" vantage point. Rather, we will explain our frame of reference first and then put the chapter findings into this perspective.

The introductory chapter noted that there is little agreement over the precise meaning of the term "social movement." We will not offer yet another attempt to define this term, but some clarification is essential if we are to synthesize the findings. Following others (e.g., Foss and Larkin, 1986, p. 2) we conceive a social movement as a significant portion of the population developing and defining interests that are incompatible with the existing social and political order and pursuing these interests in uninstitutionalized ways, potentially invoking the use of physical force and/or coercion. In this sense of the term, a social movement is a collectivity of *people* united by common belief (ideology) and a determination to challenge the existing order in the pursuit of these beliefs outside institutionalized channels of interest intermediation.

In addition, social movements may (and most of them will) develop organizational correlates. Institutional forms (such as lobby organizations and movement parties), may emerge from within a movement; and movements may seek to cooperate with existing organizations and institutions. Furthermore, social movements typically generate a support base beyond their core membership: the larger segment of the population attracted to some of the movement's goals, but not willing or ready yet to join the struggle against the existing order.

With respect to social movements, then, we analytically distinguish between four different facets: (a) the core members and their shared beliefs – their *ideological bond*; (b) the larger segment of sympathizers; (c) the movement-produced organizations; and (d) the organizations and institutions externally supporting the movement and/or pursuing related goals. Obviously, all four aspects are important in order to determine the overall impact of a movement. These four aspects are not unrelated, but there is a considerable degree of independence as well. For instance, the strategic choices of building coalitions and obtaining institutional support vary considerably (see e.g. Rucht's chapter 9); they are, at best, loosely associated with a movement's goals and its underlying ideological bond. We will consider the various facets of the contemporary movements in turn.

The Ideological Bond

We contend that the ideological bond between the core members determines the nature of the movement. It provides the prime criterion in determining whether the qualifier "new" is theoretically meaningful. In this section, we will explain our emphasis on the ideological bond in more detail, contrasting our view with other approaches in the study of social movements. Drawing upon various findings, we will characterize this bond and then address the question of whether the contemporary movements are indeed "new".

Our emphasis on the individual members and their ideological bond in assessing social movements is in marked contrast with other approaches, in particular with resource mobilization theory which has shaped the scholarly debate to a large extent. Resource mobilization theory is most prominently exemplified by the work of McCarthy and Zald (McCarthy and Zald, 1977; Zald and McCarthy, 1987; see also Klandermans, chapter 7 in this volume). They have adopted a view, which Turner and Killian (1972, p. 251; 1987, pp. 234–6) label as "extreme," that any society contains enough discontent to generate grassroots support for a movement, provided that entrepreneurial elites are ready to define, create, and manipulate these grievances (McCarthy and Zald, 1987, p. 18). Hence, the emergence of a social movement is an almost arbitrary event, and the ideological bond of a movement is manufactured and manipulated by elites. Consequently, resource mobilization theorists focus on the organizational aspects of a movement: *social movement organizations* (SMOs) and *social movement industries* (SMIs). Moreover, Zald and McCarthy readily admit (1987, p. 12) that their notion of social movement organizations is fairly broad, "sometimes coming perilously close to groups many would call 'pressure groups'." In contrast, we feel that the term 'social movement' should be restricted to collective actors trying to change the given social order in *un*institutionalized ways, referring back to Smelser's (1962) long dominant conceptualization. A more restrictive use of the movement concept clearly increases its value as an analytic characterization.

Neglecting the difference between a movement and its organizational representations will almost inevitably lead to the conclusion that there is nothing "new" about the contemporary movements. Typically, organizations work within the system. They take an incremental approach; they aim at redistribution (of power, of wealth) rather than at a redefinition of basic priorities and procedures. In addition, a continuity or a similarity of issue concerns in the activities of organizations tends to obscure any changes in the underlying motives of individuals involved and in the macro political and social context.

A closer look at European organizations concerned with environmental protection may illustrate this point. Organized representatives of environmental issues have been in existence for quite some time, some of them going back to the nineteenth century. The French Society for the Protection of Nature traces its origins back to 1854; the British Fauna and Flora Preservation Society was founded in 1903. National organizations concerned with the protection of birds were founded in the late 1800s in Germany, Great Brtain, Belgium, and Holland. Other large environmental groups, such as the influential Council for the Protection of Rural England, were created in the first half of this century. In post-war Germany the 3.3-million-member *Deutscher Naturschutzring* was founded in the early 1950s. Although these organizations are concerned with environmental issues, they do not represent the particular ideological beliefs of the contemporary environmental movement. Other organizations — such as Friends of the Earth, the *Bundesverband Bürgerinitiativen und Umweltschutz (BBU)*, or Greenpeace – are closer to the core members of today's environmental movement, but even these organizations only partially reflect the movement's goals. Looking at organizations, their policies, strategies, and tactics is an important research topic, but it leaves a glaring hole. In the end, the impact of movement organizations depends on the activities of their members and adherents, on their motivation to invest time and effort. Unless one accepts elite manipulation as a sufficient explanation, it is necessary to focus on the people making up the movement and their shared beliefs.

Focusing on the substance of these shared beliefs, we are faced with a serious research problem. Conceptually, the notion of a core membership and its ideological bond is easily explained. On the empirical level, operationally, *core membership* is much harder to define. There are no formal membership criteria (e.g., paying dues) and no unique behavioral indicators. Being a core member is not an exclusive societal role. Core members may also hold positions in established organizations; they may use institutionalized channels of interest intermediation as well. Similarly, the *ideological bond* cannot be sufficiently determined by some statistical classification of responses to a battery of standardized questions in a survey. Our description of this ideological bond, then, will be preliminary.

Several chapters, most notably the ones by Offe and by Brand (12 and 2), offer a holistic[1] description of this ideological bond, drawing on a multitude of sources. In addition, a few empirical case studies (e.g. Cotgrove, 1982; Milbrath, 1984; Kriesi, 1985; Kriesi and van Praag, 1987; Klandermans, Kriesi, and Tarrow, 1988) have provided some insight on the core members of these movements and their ideological beliefs.

In essence, there is *one* ideological bond uniting the core members of the peace, the environmental, and the feminist movements, notwithstanding differences in more immediate goals, strategies, and tactics between the single movements as identified by their primary issue concern. This ideological bond has two major traits: a humanistic critique of the prevailing system and the dominant culture, in particular a deep concern about the threats to the future of the human race, and a resolve to fight for a better world here and now with little, if any, inclination to escape into some spiritual refuge. Our current social order is perceived as *inhumane* in various ways: as fostering a "survival of the fittest" mentality on the level of individual interaction, as pursuing a course of mindless waste and exploitation with respect to the use of natural resources, and as relying on domination backed by military strength on the level of international relations. Offe describes the main axis of conflict on which the new movements concentrate as "fear, pain, and (physical or symbolic) destruction vs. integrity, recognition, and respect."

The humanistic component of this ideological bond is nothing particularly new. We agree with Brand that such a "critique of modernization" has emerged before. However, today the threats to human survival are universal: a major nuclear fallout, as a result of war or accident, or other environmental disasters will affect everyone. In essence, then, the movements' basic concerns about the future coincide with the interest of the population at large. Also, given that the threats are universal, only a religious or spiritual world could offer refuge. Living a simple life in the country ("Zurück aufs Land!") is no longer a viable option. The contemporary movements are not just critical of modernization; they have not settled into a gloomy anticipation of a world falling apart, into a *fin de siècle* mood. These movements are determined to face and confront the existing order in predominantly pragmatic and pracical ways – objectively representing the interests of the population at large. This sets them apart from historical predecessors of similar idealistic orientation.

However, this ideological bond is nothing like a strict doctrine. The new movements do not follow some grandiose plan for a better society; they do not adhere to some Marxist (or any other) vision. Their concept of the future society is largely negatively defined. They know what they do not want, but they are unsure and inconsistent with respect to what they want in operational detail. The absence of a strict doctrine may suggest that we could call these movements "post-ideological." This characterization, however, is meaningful only in contrast to a classic old movement – the labor movement – and to some extent to the student movement of the 1960s and 1970s with its emphasis on Marxist doctrine and – in part – its attempts to replicate the hierarchical organizational principles of communist parties. At any rate, "post-ideological" indicates the absence of strict doctrine and hierarchical organization rather than a lack of shared beliefs, the absence of an ideological bond.

The new social movements go beyond revolutions in the classical sense. More generally, the focus on political violence central to Gurr's (1970) approach and – following Gurr – the stage model which posits that alienation leads to frustration which in turn leads to aggression (see, e.g., Muller, 1979) are of little help in understanding the new movements. For one, violent confrontations with the authorities are *not* at the core of their action repertoire (see the chapters by Klandermans, Rochon, Gelb, and Rucht (7, 6, 8, 9)). Thus, political violence is not a central dependent variable. Secondly, most members of the contemporary social movements are indeed likely to be alienated from the dominant system. However, this alienation manifests itself as defiance of dominant norms and values. It does not necessarily lead to frustration and on to aggression, because the ideological bond of these movements provides an anchor and a meaning for being.

An essential feature of any social movement is its reinterpretation of social reality. As Offe puts it, typically the progressive "we" is contrasted with the selfish and reactionary "them." And he sees a break from this established pattern in today's movements. Obviously, movement activists are not immune to episodes of siege mentality, to dichotomous thinking in terms of "we" and "them." Yet, the familiar scheme of conflicting *group* interests is no longer adequate. The contemporary movements do not reinterpret social reality from a position of deprivation (measured against prevailing societal norms); they do not picture a better future just for themselves, for their group. They envision a better society for all. This positive identification with alternate norms and values (irrespective of their vagueness and the degree of their realization) achieves a "disalienation" (Foss and Larkin, 1986). The coexistence of radical critique of the existing order, on the one hand, and *de facto* integration into the existing society and into the political arena, on the other hand, is a genuine characteristic of the new movements.

Obviously, additional in-depth field studies of new movements are needed to arrive at a more detailed description of the ideological bond between their members and of their everyday practices.[2] In particular, we need to clarify the difference between just sharing a generally critical view of the present political order and the associated cultural and societal arrangements (which probably holds for most people who consider themselves as politically "Left") and the ideological bond between the core members of the new movements.

The survey data presented in several chapters in this volume (including those by Inglehart, Wilson, and Müller-Rommel, chapters 3, 4, 11) are valuable in their own right, but they provide only very general clues with respect to the exact nature of the movements' ideological bond. In these surveys, membership is idiosyncratically defined by the individual respondent; the shared beliefs remain largely unexplored. Also, given the inherent limitations of nationwide general population surveys, the sheer number of (self-identified) members is too small for any reliable detailed quantitative analysis. However, these data allow us to assess the size and the characteristics of the much larger group of individuals displaying a favorable attitude towards the new movements. We need to keep in mind, however, that there are important analytical differences between the core members of a movement (its activists), the potential members, the supporters, and the sympathizers.[3]

To summarize our view: we argue that the core members and their shared beliefs – their ideolgoical bond – are the essence of a social movement, that the organizational manifestations are an epiphenomenon. As a consequence, the "newness" of social movements is largely dependent upon the nature of this ideological bond. We contend that the radical idealistic critique of prevailing norms and values on the individual, the societal, and the international level and the focus on problems of concern to the population at large coupled with the determination to confront the existing order here and now in mostly very pragmatic ways sets the contemporary movements apart from historical predecessors of similar critical conviction. Thus, we claim that the specific characterization of the contemporary movements as "new" is warranted.

Obviously, this character of the contemporary movements is more apparent in some nations than in others. In particular, the notion of a unified core membership of the environmental, peace, and feminist movement may seem at odds with the empirical reality in some countries – if one uses the term "movement" in a broad sense. The feminist movement of the United States is a case in point. Within this movement several strands are discernible (Ferree and Hess, 1985), only one of which – the radical strand – easily fits our description of a new movement. The other strands make up the generally supportive context of individuals and organizations pursuing related goals.

Longevity of New Movements: A New Age or Another Cycle?

The chapters by Karl-Werner Brand and Ronald Inglehart address the origins of new social movements and thus the values and beliefs shared by the individuals constituting these movements. Using different theoretical approaches and different data, they reach different conclusions regarding the longevity of these movements. They agree, however, that these movements reflect a fundamental change of values, that movement followers advocate basic changes in the fabric of industrialized societies.

Brand argues that a general set of ideological beliefs critical of the prevailing social norms and values, a "critique of modernization," is the common denominator of new social movements. He posits that the environmental movement, the peace movement, the women's movement, and other "alternative" movements are basically different manifestations of the same fundamental ideological theme.

Inglehart maintains that the recent emergence of what he labels as "postmaterialist" values is one of the major driving forces behind new social movements. He presents data that show an impressive and relatively stable trend towards postmaterialist values over the past two decades. He links this process of value change to a fundamental transformation in the socio-economic structure of advanced industrial democracies. Having grown up in a society where security and stability seem relatively assured, the priorities of many young people are changing to reflect alternative values of individual freedom, self-expression, and an emphasis on the quality of life.

Inglehart's definition of postmaterialist values is broadly compatible with the belief syndrome Brand labels "critique of modernization." Both identify the upper middle class as the population segment most susceptible to these ideas. However, Inglehart and Brand differ in their descriptions of the processes producing these changes in the public's political orientations. Inglehart sees the development of postmaterialist values as the beginning of a permanent change in the value priorities of the publics in Western democracies. Brand maintains that similar waves of modernization critique cyclically reoccur in the modern history of these countries. In his view, the current wave is already receding.

Brand's findings imply, in contrast to Inglehart's, that the "new" label may not be appropriate for the contemporary social movements. His analysis of social movements in Germany, Great Britain, and the United States establishes that the environmental, peace, and women's movements have historical predecessors. Looking at the temporal pattern, one might discern some regularity, some specific cycle length. However, as Brand frankly acknowledges, it is difficult to precisely time the periods of high movement activity. Also, given the very few points in time, an exact cycle length cannot be reliably determined.

More generally, the very emphasis on cycles appears to be problematic.[4] For one, over-emphasizing the cyclical aspect of social movements tends to obscure the fact that the success of a movement is not determined by the length of its life span. Movements are transient phenomena by their very nature; all movements ebb and flow. It is more important to determine whether, and how much of, the impact of a movement on the pre-existing sociopolitical system persists even after it passes its apex. Secondly, and this is the more important objection, an emphasis on cyclical occurrence also tends to veil differences in content, relying too much on immediate (more formal) similarities, disregarding the specific historical setting of a movement.

The environmental movement is a case in point. The fundamental goal of this movement today is different from that of the environmental movement at the turn of the century — reflecting a drastic change in the historical context. In the past, local solutions were sought. It was (or at least it seemed) possible to geographically restrict the impact of environmental hazards. As an example, many European cities display a particular pattern of city zones: upper–class neighborhoods in the west; plants, factories, and working-class neighborhoods in the east — in line with the prevailing weather pattern of westerly winds in Central Europe. The establishment of parks, preserves, and sanctuaries seemed to be a viable tool to protect nature and wildlife and to provide recreation areas for the citizenry in the past. Obviously, such local stategies are insufficient today. Acid rain, dying forests, a decimated ozone layer, or – much worse – a nuclear fallout are threats to everyone irrespective of class, status, and geographical location. The fundamental goal of the movement today is not particularistic. It serves the very survival of the human species. The nevertheless limited – and sometimes lacking – support for today's environmental movement results from differences in risk assessment rather than from conflicting goals. Clearly, there is also continuity between the various environmental movements over time, which is accentuated in Brand's analysis. However, we feel that it is more important to emphasize the very distinct differences in context when assessing the character of today's environmental movement.

On the other hand, a close look at historical predecessors is valuable. As we pointed out above, movements ebb and flow over time. During periods of high activity a movement is typically able to reach some of its goals before subsiding into a latent, submerged state. The achievements of previous cycles of high movement activity may serve as a starting base for the ensuing one. This way, movement cycles can be seen as a steadily progressing process with phases of rest and activity. We argue that this view is *not* adequate for the environmental movement. However, it may be (more) fitting for the feminist movement (Taylor, 1988). The fundamental goal of this movement has remained the same: to obtain equal rights for women in society. Each wave of the feminist movement has produced some progress towards this general goal. Subsequent waves took off at a level of more equality than the ones before. For example, women's right to vote was a goal achieved by the feminists movement of the early twentieth century (some deviant cases like Switzerland notwithstanding), constituting a (taken for granted) resource for the more current women's movement. Similarly, there has been steady, though mostly rather slow, progress in terms of equality of economic opportunities. Still, a focus on cycles and cumulative progress may be of relatively minor importance for the feminist movement as well. Gelb's chapter details the diversity of the feminist movement both within and across nations. A more radical perspective, most prevalent in the British model, deliberately defies a cumulative, incremental approach.

Paradoxically, the consideration of historical cycles as in Brand's chapter tends to neglect the *specific* historical context of today's movements. We do agree with Brand's assessment that there is one unifying ideological bond behind the seemingly different movements. However, we feel that this unifying ideological bond is not sufficiently characterized as just another wave of "anti-modernism."

In contrast, Inglehart's theory of value change is a deliberate attempt to capture the effect of macro-level changes in the political and social setting, the impact of *today's* specific historical setting on the values and beliefs of individuals. With respect to the new movements, Inglehart's work allows him to quantitatively assess the size of the population sympathetic to the movement — one of the four different facets in the study of social movements we introduced above — and to understand its dynamics in theoretical terms. By now, the shift in public opinion towards "postmaterialist" values is well documented for most industrial democracies – notwithstanding significant national variations. Value orientations are strongly correlated with attitudes towards the new movements and associated parties; "postmaterialists" are generally sympathetic towards the new movements and are more inclined to vote for movement parties.

Given the lasting and irreversible changes in the social structure, it follows that the new movements can rely on a stable and ever-growing support base. Consequently, the movements' goals are likely to stay on the political agenda for some time to come. Even if the movements as such will lose momentum, we expect that they will have a longer-lasting impact: key goals will be absorbed by the larger segment of sympathizers who will provide permanent stimulus for society to change in the direction of the new movements' visions. In this sense, the recent emergence of the new movements could market the beginning of a new age.

However, we see some problems with Inglehart's theory of a "silent revolution," of a permanent shift towards postmaterialist values as well. First, on the conceptual level, the notion of a *one*-dimensional shift from materialist to postmaterialist values is controversial. Other empirical evidence in value research suggests that value change occurs on more than one dimension and that "postmaterialist" values complement, rather than replace, traditional material values (e.g. Flanagan, 1987; Westle, 1989). The current debate on value change is rather complex, but there is one simple implication of these opposing views: if the value space of the individual must be conceptualized in more complex ways, then the overall effect of changes in (some) values is more difficult to assess, and the impact on political behavior is less distinct and less predictable.

Secondly, on the level of operationalization, it is doubtful that Inglehart's basic four-item survey question measures postmaterialism in its full conceptual meaning. A more cautious assessment of the (face) validity of these items suggests that they primarily reflect different views on the relationship between citizens and government. "Postmaterialists" in Inglehart's terminology favor a more active role for the citizen; they see citizens as participants in rather than as subjects of the governmental process. This interpretation is fully consistent with Inglehart's broader assertions, but it redefines the scope of what is (already) backed by hard empirical facts. The associations between the postmaterialist typology and movement-related attitudes (presented in the chapters by Inglehart and by Müller-Rommel) are compelling, but given the wording of the survey question they are less surprising. Movements by their very nature challenge a strictly representational form of democracy in which citizen participation is largely restricted to voting. It is fairly obvious, then, that an attitude favoring a participatory form of democracy tends to coexist with positive attitudes towards movements and movement parties.

Inglehart's data provide overwhelming evidence that the new movements have developed in a favorable context, with a fairly large segment of the population principally endorsing their goals. Yet, they do not sufficiently account for the emergence of these movements. In our view, this emergence is not just a reflection of widespread affluence and the ensuing rise of participatory ideas of democracy. Serious problems about the further course of advanced industrialized nations in general (economic growth, use of natural resources) and about the role of Western democracies in world politics (maintaining peace, distribution of wealth) also characterize today's specific historical situation. These aspects must be seen as key elements in the emergence of the "new" movements and the possible beginning of a new era.

In a restricted sense, with respect to mobilization and the use of uninstitutional means of interest intermediation, the new movements may be past their apex, as Brand and others in this volume assert. In terms of Offe's tentative stage model (chapter 12) they are well into the stage of institutionalization, thereby receding as "movements" (in the analytically restricted sense of the term). However, the challenge of the political order persists. The growing public preference for more participatory forms of democracy (for postmaterialist values in Inglehart's terms) as well as a more observant attitude towards the established parties provide fertile soil for this challenge to grow in.

New Movements between Fundamentalism and Pragmatism

The "new" movements discussed here pursue goals that put them in competition with the established powers in society. The envisioned better order of tomorrow will not be achieved in some uncharted territory; all the movements strive to transform the present society. They aim to utilize – though maybe in different ways – the given resources: human, natural, and economic. This section addresses the predicament arising from pursuing fundamentally different goals within an established order, the related strategic and tactical choices new movements face.

The early literature on new social movements, and the initial reactions of political elites, often focused on the political isolation and unconventional political tactics of these movements. The proliferation of citizen action groups, environmental lobbies, and women's groups challenged the political order by their use of protests, spectacular events, and civil disobedience, as well as by their new issue demands. Presumably the populist ideological bond of these movements and their position as challenging organizations isolated them from established political interest groups and institutions, and led these movements to adopt unconventional methods of political action (Brand, 1982: Brand et al., 1986; Raschke, 1980). Alberto Melucci (1980) and Claus Offe (1985, p. 830) even argued that new social movements do not participate in conventional pluralist politics because they command no negotiable resources – thus these movements must resort to unconventional methods of political persuasion.

Despite this early stress on the political isolation and unconventional style of new social movements, the bulk of the evidence presented here substantially tempers these early assumptions. Rochon and Klandermans (chapters 6, 7) both find that the European peace movement involved a number of established interest groups, ranging from the unions to the churches to other social movements. Klandermans, in fact, maintains that the Dutch peace movement was virtually a fully integrated component of the overall structure of interest intermediation in the Netherlands; an assertion derived from the extensive connections between the adherents of the Dutch peace movements and established organizations (parties, unions, and religious groups). Similarly, Joyce Gelb (chapter 8) recounts the close ties between the American women's movement and the Democratic party, and the virtual incorporation of the Swedish women's movement within state institutions. New social movements might advocate a new social paradigm, but they are not totally isolated from other organized interests in advanced industrial societies.

The action repertoires of these movements also appear more conventional than initially theorized. While a dramatic event — such as the protests at Greenham Common, or the mass demonstrations against INF deployment, or the hanging of banners from polluting smokestacks — might capture our attention, new social movements devote the majority of their efforts to more mundane activities. Dieter Rucht (chapter 9) argues that contemporary movements are distinct from historical predecessors in their parallel and flexible use of both unconventional *and* conventional forms of political action. Rochon, for instance, finds that during the peak of its mobilizing phase in the early 1980s, the British peace movement devoted the bulk of its activities to educational meetings, conferences, showing films, and organizing social functions. Joyce Gelb discusses the extensive lobbying activities of the American women's movement, with some parallel in the Swedish case. In a separate work, Dalton et al. (1986) document the extensive conventional political activities (lobbying parliament and ministries) of the European environmental movement.

The existence of political ties to established interests and the use of conventional political activities do not mean that new social movements are identical to established interest groups. It does mean, however, that new social movements are responding to their environment. In her comparative analysis of the feminist movement, Gelb emphasizes the "political opportunity structure" as a major determining factor of the particular style and significance of movement action. The American movement is perhaps most conventional in its style and tactics, while the British movement has remained more unconventional in its approach. This theme reoccurs in other chapters, though it may be couched in different terms or it may be implied rather than explicitly stated. Kitschelt (chapter 10) employs the notion of a country's structure of interest intermediation in discussing the fate of movement parties (see below), and the highly cooperative style of the Dutch peace movement (Klandermans, chapter 7) results from a very favorable political opportunity structure.

It is more realistic to assume that new social movements face a choice of political tactics, but from this choice also flows a basic dilemma. In his analysis of the peace movement, Rochon (chapter 6) concludes that the aspirations of new social movements for broad social change are not compatible with the achievement of mass mobilization for a specific policy goal. The more a movement engages itself in conventional politics and the broader the attempts at (formal and informal) coalition building, the greater the likelihood that the fundamental social critique manifested by the movement is obscured. In the case of the peace movement, the fundamental critique of militarism was overshadowed by the pursuit of a more immediate (and much more restricted) goal: the prevention of deployment of a particular kind of missile in particular locations within the combined territory of the Western alliance. Similarly, the anti-nuclear power movement has been successful in generating broader support in fighting particular plants at particular sites (e.g., in the landmark cases of Wyhl in West Germany or Windscale in Great Britain), but much less so in terms of global strategy (shutting down of all nuclear power plants). As women's groups focus on activities that are necessary for the passage of legislation, the social-consciousness goals of the movement inevitably suffer (Ferree, 1987a; Gelb, chapter 8 in this volume).

To a large extent, the movements are faced with an insoluble predicament. They face a choice between being pragmatically successful (in broadening their popular base and reaching more modest policy goals) and being true to their fundamental beliefs. This dilemma is particularly apparent in regard to internal organization. By their very nature, movements value openness and immediate participation, which at the same time seriously restrict their effectiveness and efficiency in reaching policy-oriented goals. Social movements on the whole are rather amorphous; leadership is established informally, based on charisma rather than formally defined procedures. The characterization of social movements as "networks of networks" (Neidhardt, 1985) is very much to the point. In a formal sense, a social movement is a collective actor, but in contrast to other collective actors (e.g., political parties, trade unions) it typically does not display uniquely determined structures, strategies, and tactics. This makes it difficult to determine the leaders, the strategy, or the tactics of any particular movement.

To determine choices of strategy and tactics we need to move beyond the core membership of a movement and include the various organizational correlates of a movement. From our perspective, the most important organizational correlates are the movement parties. Movement-related parties provide the link between the movement and the established political system. Their policies are (not necessarily truly representative) manifestations of the latent and often rather vague general goals of the movement. The struggles within these parties over policies and procedures are thus a reflection of the movements' basic dilemma discussed above. However, the decision to form or support a party is an important strategic act in itself. In discussing their overall impact on the existing political order, therefore, we need to also address the choices new movements face in their attempt to broaden their partisan representation.

New Movements and Partisan Politics

In many Western democracies, the rise of new social movements has led to the creation of new political parties, though there is great variation in the mass appeal and the electoral success of these parties across nations. Herbert Kitschelt's chapter refers to these parties as "Left-libertarian" to reflect a fusion of two ideological traits: a commitment to egalitarian redistribution (Left in the traditional sense) and a rejection of bureaucratic regulation of individual and collective conduct (libertarian). Ferdinand Müller-Rommel prefers the term "New Politics" parties, introducing an additional distinction between small left-wing parties (founded earlier, but now advocating a New Politics agenda) and more recently founded parties, most of which label themselves as green or environmentalist.

Both desriptions are related to the ideological bond of the new movements we described earlier. However, both labels do not fully capture the essence of this bond. The term "New Politics" predominantly refers to a new issue agenda replacing the alignment of citizens' interests in accord with traditional social cleavages. Yet, there are environmental parties which except for this particular issue concern are not close to the new movements.[5] Similarly, traditional socialist concerns about egalitarian redistribution are not a major component of the new movements's ideological bond. Left-libertarian may be a fitting description for most of the parties Kitschelt discusses, but the ideological bond of the new movement is more than some form of humane socialism free of hierarchical structures.

We will refer to parties which reflect the ideological bond of the new movement to a significant degree in their program, in their internal structure, and in their actual policy preferences simply as "movement parties." This does not imply that these parties are fully controlled by the new movements nor that they fully reflect the new movements' agenda. As a matter of fact, at times there may be considerable tension between the movements and a movement party.

Looking at the relationship between new movements and partisan politics several aspects should be separated analytically, though they overlap empirically. First, we need to assess the immediate impact of movement parties on the established party system and the general political order as measured by their share of the popular vote and their strength of parliamentary representation in national elections.[6] Second, we need to consider the indirect impact of movement parties on the established political order by shaping public opinion and by setting the political agenda. Third, we need to consider to what degree movement parties reflect the political goals of new social movements with respect to both form and substance. We will discuss these three aspects in turn.

First, the electoral success of Left-libertarian parties varies considerably (see table 10.1). Among the seven major (in terms of economic output) Western nations – Canada, France, Italy, Japan, United Kingdom, United States, and West Germany – there is only one country in which a movement party has had significant success. In five of these nations movement parties have not even contested elections or gained barely 1 percent of the popular vote. The West German Greens are the sole deviant case. Apart from West Germany, movement parties have been relatively successful only in a number of smaller countries in Northern and Central Europe, such as Belgium, the Netherlands, Denmark, and Sweden.

It is difficult to conclusively identify particular traits of the national context as major factors explaining the success of movement parties. However, all countries with successful movement parties have electoral systems based on proportional representation which favors small parties.[7] In addition, there is a well-established democratic tradition in all of these smaller countries (in contrast to, e.g., Portugal and Spain, where the transition to democratic rule is much more recent). It seems plausible, then, that the electorate in well-established democratic systems, which are less central to the Western economic and military alliance, are more receptive to challenges to the established order because the potential "cost" of political experiments is low. For example, a decision by Belgium to leave the Western alliance is not as consequential as a similar decision by West Germany. In turn, the electoral system adopted in these countries (encouraging smaller parties) may be another reflection of this openness, not a separate contributing factor.

Kitschelt (1988a) has examined several quantitative economic and political indicators for a possible correlation with the existence of a significant Left-libertarian party.[8] Combining five such indicators – all in some dichotomized form and relating to the late 1970s rather than the 1980s – he determines odds for the emergence of Left-libertarian parties. High per capita GNP, high social security expenditures, low strike activity, high Left parties' participation in government, and high intensity of nuclear controversy define a political and economic setting most amenable to the success of movement parties. Denmark, the Netherlands, Sweden, and West Germany all fall into this category – and in all these countries movement parties have been successful. However, without going into methodological detail, we need to be aware that this kind of statistical (correlational) analysis is fairly problematic: each country is reduced to being just one "case" characterized by five rather crude measures; moreover, due to the small number of cases statistical computations are highly sensitive to slight changes in definitions and extreme values (outliers). Still Kitschelt's results are by and large plausible as an *ex post* explanation.

Going the opposite methodological route, we will take a closer look at one extreme case: West Germany – given its unique position among the major Western nations. The rise of the West German Green party might be understood as the organizational manifestation of a growing split within the Social Democratic party. This factional split pits the traditional, union-oriented working-class membership against a growing number of better-educated, new middle-class members (Feist et al., 1978; Hermann Schmitt, 1987) of a Left-libertarian orientation. Although the Greens are not a split-off from the Social Democrats in a formal sense (like the USPD in the Weimar Republic), both parties do seek support from the same clientele of New Politics voters (the party-in-the-electorate). The Greens and Social Democrats compete for the same social-liberal segment of the population, notwithstanding the Greens' additional appeal to various protest voters. This assertion is well supported by various public opinion data showing the SPD as an almost unanimous second choice of Greens voters.[9] From this perspective, the emergence of the Greens seems less spectacular, not indicative of a fundamental restructuring of the German party system.

The SPD – like all established political parties in West Germany – tends to allow little room for dissent from different factions within the party. The Italian and the French cases show (Tarrow and Wilson, chapters 13 and 4 in this volume) that the established leftist parties have pursued a different course, which provides more responsiveness to new ideas and thereby curtails the potential appeal of new movement parties. A similar point can be made for Great Britain and the United States. In both cases the diversity within the established (Left) parties provides ample opportunity for new movements to seek and find partial cooperation. In addition, an electoral system strictly based on a simple majority in single-member districts further discourages attempts to launch new parties in these countries. Therefore, sweeping changes in the constellation of parties in the major Western nations will be difficult to accomplish.

In regard to formal coalitions, the empirical base is rather limited at this time, due to the failure of movement parties to gain a sizeable number of parliamentary seats in the first place. Even West Germany provides only a few examples on the state level. Here, a formal coalition in the state of Hesse was formed between the Social Democrats and the Greens in December 1985, which was abruptly terminated some 14 months later by the dismissal of the Green cabinet members.[10] In the city state of Hamburg, the Greens informally supported a minority government by the Social Democrats for a few months after the 1986 elections, until new elections in 1987 produced a coalition between Social Democrats and Liberals. However, as of spring 1989, there were prospects for a more stable cooperation between the Social Democrats and the Greens. In Berlin, after devastating losses for the ruling Christian Democrats and a surprisingly strong showing for an extreme right-wing party (the Republicans), Social Democrats and *Alternative Liste* (the local affiliate of the Greens) formed a coalition and jointly elected a Social Democratic mayor by a comfortable margin in March 1989. In the state of Hesse (which includes the city of Frankfurt – a movement stronghold), municipal elections confirmed the Berlin pattern: significant losses by conservatives and Liberals, slight gains by Social Democrats and Greens and strong gains for the extreme Right.[11] As long as the Liberals side with the Christian Democrats – or fail the 5 percent threshold, as in the last three regional elections – the SPD needs the votes of the Greens to control power at the state level. There is much debate within the SPD about whether to seek cooperation with the Greens or whether to confront them outright. Until very recently it appeared unlikely that this controversy would be resolved soon. However, as the ruling Christian Democrats slipped in the polls during 1988 and 1989, a Red-Green alliance for the federal elections in December 1990 appeared as a viable option. However, any negative experience with the Berlin experiment can quickly change this scenario. After all, according to the polls, a majority of SPD voters in Berlin would have preferred a grand coalition with the Christian Democrats to the Red-Green alliance.

Turning to the second aspect, the indirect impact of movement parties, we first want to re-emphasize that the electoral success of a party heavily depends on the electoral system. Small parties (except those with a specific regional base such as the Scottish Nationalists) are at a severe disadvantage in single member, simple majority systems as well as in proportional systems with high thresholds. Thus, chances of parliamentary representation for movement parties are structurally low in some countries, which limits their appeal to voters who might not want to "waste" their votes. Even with limited electoral success, however, small parties can bring about changes in the policies and the internal structure of the established parties and provide new options for inter-party cooperation. New movements and movement parties were and still are redefining the political agenda.

Support for the issue positions of movement parties often exceeds their votes at the ballot box. Public opinion polling that documents these issue preferences provides an additional channel for the manifestation of public sentiment. At times, these polls can serve as substitute referenda, allowing the public to express support for the issue agenda of movement parties outside of the electoral process. In the age of media-induced polling, movement parties can thus exert considerable influence on the political agenda. This is fairly obvious in regard to ecological issues, while the impact on military and disarmament issues is more difficult to assess. The peace movement failed to reach its immediate goal of barring the INF deployment, and major policy changes in regard to basic NATO commitments do not seem imminent. It would also overstate the case to attribute the progress in arms limitations talks between the United States and the Soviet Union to pressure exerted by the peace movement. Yet the increased flexibility of European governments in this area (e.g., the West German decision to forgo a modernization of the Pershing missiles to facilitate the Geneva arms limitation negotiations between the United States and the Soviet Union in 1987 or, more recently, the German reluctance to make a firm commitment to the modernization of short-range missiles in the 1990s and the ensuing rift within NATO over the future course of disarmament talks with the Soviet Union) reflects growing elite awareness of changing public sentiments.

Data for West Germany indicate a very favorable perception of the Greens by the voters on a number of issues beyond environmental protection, but at the same time a reluctance to actually vote for the Greens based on this perceived issue proximity (Kuechler, 1986). Moreover, perceived party competence in handling ecological – or more generally non-economic – problems is an increasingly important factor in explaining vote choice in Germany (Kuechler, 1990). Yet this shift in salience does not translate into a major swing towards the Greens. This is inconsistent with a narrow (and normative) concept of "rational voting" (Himmelweit et al., 1985; also Müller-Rommel, chapter 11 in this volume), but appears to be quite rational given the questions about unreliability and governing ability that follow the Greens and other movement parties. It seems that much of the support for the Greens is ambivalent. A significant part of the electorate appreciates their role in providing incentives for the established parties to readjust their focus of attention, to redefine their priorities, but stops short of trusting them with a legislative mandate.

This considerable popular support for movement issues cannot be ignored by the established parties, especially leftist parties. The Dutch PvdA and the Danish Social Democrats have been most responsive to new social movements (see Klandermans, chapter 7, for a detailed account of the Dutch case). Tarrow (chapter 13) discusses the close interrelationship between Italian social movements and established Leftist parties. Other parties have been responsive to some issues, but not to others. Large parts of the British Labor party and the German Social Democratic party are receptive to the peace movement, but are more reluctant to embrace environmental issues. Strict environmetnal controls, the closing of nuclear power plants, and similar measures are detrimental to short-term economic growth and prosperity, creating a policy dilemma for parties with special ties to trade unions. In Britain the Social and Liberal Democrats, rather than the Labour party, are most responsive to environmental concerns. Although there is much variation across nations, direct forms of contact and cooperation seem to be restricted to established parties on the Left (Dalton, 1988b). However, even conservative parties cannot afford to ignore issues like environmental protection and safe energy resources. The governing Christian Democrats in West Germany appointed a highly regarded expert to a newly created ministry of environmental affairs after the 1987 elections and have taken a lead within the European Community in pushing for stricter emission standards for automobiles. In early 1989, the European Community as a whole (with the support of the conservative British government) moved towards an accelerated schedule for the ban of chemicals threatening the ozone layer.

As a general rule, mass institutions like parties change slowly, though. Therefore the extent of change is hard to assess at this time. Moreover, there are conflicting aspects in regard to conveying these changes openly. In terms of programmatic and policy continuity, party elites may be reluctant to advocate and document changes and revisions that they now consider advisable.[12] Yet we see many indications that the established parties are contemplating change, and this change was initiated by the new movements.

The third aspect we want to discuss is the nature of the relationship between the new social movements and movement parties: to what extent do movement parties reflect the fundamental goals of the new movements? Given the fairly general nature of *substantive* movement goals, these are widely represented by movement parties. These parties de-emphasize economic growth and national security by means of military investments; they emphasize a "sustainable society" (Milbrath, 1989), a society with proper protection of natural resources where the human race will be able to survive in peace. And, with some variation, they advocate equal rights and equal opportunities for women.

The more interesting aspect lies in the organizational and communicative structure of these parties. Is the form of decision-making, is the involvement of rank-and-file party members, and is the degree of women's participation and impact significantly different from the pattern found in the established parties? The new movements emphasize form as much as substance. In their view, not only should society set different goals, but the ways in which any goals are pursued must be changed as well. Dominance of the socially fittest must be replaced by mutual support and understanding. As an immediate consequence, there is no room for hierarchy, for leadership fairly separated from its base. In sum, movement parties are expected to practice direct democracy, to make "New England Town Hall meetings" work in a mass society. Or, as Kitschelt (chapter 10) phrases it, Left-libertarian parties ought to be characterized by a "decline of party organization."

However, even more than the movements proper, movement parties face the basic dilemma of reconciling idealistic beliefs with pragmatic politics.

The price for short-term success in the political arena, for being efficient and reliable in the eye of a larger public, is the alienation of significant parts of the movement. Several contributors to this volume discuss the tension between fundamentalism and pragmatism (see Kaase, Rochon, Gelb, and Offe). The West German Greens' successful consolidation of the movement into a party with parliamentary representation seems to be leading to a more pragmatic, policy-oriented party that plays by established rules. In the process the Greens seem to be slowly abandoning their original promise to be a party of a "new type" (Langguth, 1984; Papadakis, 1984; Poguntke, 1987b). In contrast, Kitschelt (chapter 10) provides a more positive view on the ability of movement parties to realize their novel organizational concept, based on his study of activists of the Belgian Ecolo/Agalev and the German Greens. Using the established parties as basis for comparison, Kitschelt's evidence of a present "decline of organization" is convincing. Also, compared to established parties, women do play a much larger role in these new parties. Still, Green party activists are seemingly the first to acknowledge that the pressures of the "iron law of oligarchy" are felt even within their party. In early summer of 1988, the (rather small) Green Party in Austria went one step further and officially adopted an organizational structure very much in the traditional mold. In late 1988 the fundamentalists within the German Greens lost a critical vote of confidence at a party convention and their majority on the party's executive committee. By spring 1989, the pragmatists gained control and led the Greens into a formal coalition with the Social Democrats in Berlin. However, the factional struggle is bound to continue for some time to come.

Realistically, one cannot expect that the movement parties will fully succeed in establishing and practicing non-hierarchical forms of communication and decision-making. Measured against a utopian ideal, the parties may fall short, but the differences from established parties are clearly discernible. Yet, in their attempt to function effectively in the long run, they may have to compromise, partially resorting to more traditional organizational patterns. With respect to women's participation and impact the prospects are much better. Women have obtained key positions within movement parties, providing a very valuable personal resource. This has pushed established parties to increase their efforts to attract and to provide opportunities for women within their own ranks,[13] thus establishing a probably lasting trend.

Overall, new movements and movement parties seem to undergo a process of partial detachment owing to the differentiation within the movements and the need for movement parties to achieve greater coherence. Given their generally small electoral base, further diversification of movement parties may not be a viable strategy, not even in Germany. In addition the radical ideological impetus of the movements proper is being superseded by a tendency towards pragmatic arrangements. The movements themselves seem to undergo a process of institutionalization (beyond the formation of political parties); they seem increasingly amenable to utilizing the offerings and benefits of the welfare state wherever and whenever possible, rather than solely relying on autonomous self help – a trend particularly visible in the women's movement (see e.g., Ferree 1987a; Knafla and Kulke, 1987; Gelb, chapter 8 in this volume.)

Taking all three aspects (electoral success, indirect impact, and form) into account, Western party systems have been challenged by the emergence of movement parties – but they have not (yet) been fundamentally changed. Some movement parties have gained parliamentary representation, many may stay in parliament, others may come along. Given the appealing issue agenda of many movement parties, their chances of continued electoral survival are good. But the success of these parties is also partially contingent upon their factual assimilation and cooptation into the existing political system.

Challenge and Response: What Lies Ahead?

The emergence of an environmental, an anti-nuclear power, a peace, and a feminist movement in recent years has challenged the established political order. By definition, any social movement does. But what is the exact nature and extent of this challenge? Has the political order changed in significant ways; has a process of maybe slow, but irreversible change been set in motion; or is the rise of the contemporary movements just an episode with few longer-ranging effects on the existing political order?

The chapters in this volume offer a host of valuable facts, analyses, and insights which lead us to a still preliminary, but empirically grounded answer. First, in different ways, the chapters by Offe, Brand, and Inglehart point to the fact that all these movements feed on a common ideological theme; that they are different metamorphoses of one – what we would call – "alternative" movement. In essence, none of these movements is a single-issue movement as their labels may suggest; none of them primarily serves special group interests – though some qualification with respect to the feminist movement may be in order. The vision of a "sustainable world" (Milbrath, 1989) and of a better and kinder society (going beyond President Bush's rhetoric) is common to all these movements. Waves of idealism, romanticism, and anti-modernism have emerged in earlier historical periods as well (Brand), but the contemporary movements go beyond this: they are a reflection of real threats to global survival. Threats of nuclear devastation and destruction of natural resources are no longer the apocalyptic fantasies of a few. They are increasingly perceived as real by the general public. The movements manifest these concerns. They symbolize basic problems of advanced industrialized nations which have long been neglected by the political forces in power – and for which there are no easy solutions. Regardless of actual support in terms of mass mobilization, favorable ratings in public opinion polls, or votes for movement parties, the new movements represent widespread, if not universal concerns. Since they, unlike historical predecessors of similarly "modernization-critical" conviction, are active in the political arena, they have become self-appointed representatives of mass interest.[14] In this sense, the contemporary movements are indeed "new." The political order is challenged to address long-neglected fundamental social and political problems. This ideological challenge will continue, even if the movements proper subside into a low activity phase.

Secondly, the new movements challenge the established political order in a more immediate sense, actively intervening in the ongoing processes of policy formation and decision-making. The challenge in terms of political action manifests itself in the strategies, tactics, and action repertoires of the different strands of the "alternative" movement. These strands, identified by specific issue concerns like the environment or peace, have extended the movements' sphere of influence beyond the core membership to a much larger segment of the population that supports and to some degree participates in movement activities. In using issues like the environment or peace as a rallying point, the new movements are also able to build cooperative links to organizations and institutions within the established order. The chapters by Rochon, Klandermans, and Gelb document the many linkages between the new movements and the existing order. In contrast to social movements of earlier historical periods, the action repertoire of the new movements is characterized by a very flexible use of both uninstitutionalized and institutionalized means. Following Rucht, this flexibility is one of the most prominent traits of the new movements.

Thirdly, in addition to organizational coalition-building (a very dominant feature in the Netherlands as Klandermans shows, a more ambiguous and controversial strategy in Britain as Rochon and Gelb demonstrate) the new movements have undergone a process of institutionalization themselves. Offe's stage model is grounded in the West German case, but it is compatible with the course of events in other nations as well. Among the institutional forms emerging from the new movements, several contributions to this volume focus on movement parties and their impact on the established party system. The form of "party government" is the dominant common trait of Western European democracies. Parties are at the heart of the political process, and any significant change in the political order will affect the composition of the party system and/or the role and function of parties in general. The most direct challenge to the political order by the new movements, then, lies in the entry of movement parties into electoral competition. This challenge was less successful. With the exception of West Germany, movement parties have failed to gain significant parliamentary representation in the major Western nations; but these parties have gained considerable ground in the smaller countries of Northern and Central Europe. Overall, the established parties have kept their ground in terms of popular votes and parliamentary seats. The general public overwhelmingly supports the existing order, although citizens are appreciative of the movements and related parties to the extent that the movements push the establishment toward innovation and reorientation. Finally, movement parties have also had more indirect effects on Western democracies. Kitschelt describes the differences in internal structure between the movement parties and the established parties. Most striking are a much greater involvement of women in movement parties, and greater participation of the membership at large in intra-party decision-making. These organizational features of movement parties correspond to a growing sentiment in the general public favoring more participatory forms of democracy, to the trend towards postmaterialism, in Ingelhart's terms.

Overall, the new social movements have left their mark on the political arena in restructuring the issue agenda, in pushing for more direct citizen participation, and in redefining the "boundaries of institutional politics" (Offe, 1987a). Their ideological challenge was clearly successful across all nations. There are important national variations with respect to the new movements' prominence as an independent collective actor and in leaving distinct institutional marks (e.g., an emergence of successful movement parties). Wilson, Kitschelt, Gelb, and others point to political culture and the political opportunity structure as important determinants of a political system's degree of responsiveness and adaptability. West Germany seems to be the system least open to innovation and to integration of dissenting forces. As a consequence, new movements have achieved a particularly high profile in this national setting. But, as of spring 1989, even France (analyzed in detail by Wilson as a case representing the other extreme) now shows indications of a significant rise of movement parties.[15]

For the time being, there still seems to be considerable diffuse system support in all nations. This support of the citizens can no longer be taken for granted, however. Changes on the individual level in perceptions of the political world, in the propensity for more direct participation, and in expectations of convincing policy responses to pressing fundamental problems extend far beyond the core members of social movements. This will affect the overall social fabric of Western societies in significant and permanent ways.

The challenge has been mounted and the established political forces have begun to respond, indicating at least a certain amount of flexibility in all Western democracies. Still, the established political parties will have to prove themselves in the years ahead. No drastic changes in the political order appear imminent in the immediate future, but a slow evolutionary process of adaption has been induced: producing stability by way of change. The reaction of the established parties to the challenge of the student movement of the 1960s and 1970s in Italy (Tarrow, chapter 13 in this volume) may provide a blueprint. The unintended consequence of securing the long-term stability of the political order may turn out to be the most important impact of today's new social movements.

NOTES

1 The characterization they offer is grounded in empirical observation, but it is not based on a standardized or codified form of empirical analysis.
2 Fortunately, several such studies are now under way, e.g. a study in West Germany focusing on the personal networks of core movement members in three select localities (co-principal investigators: Peter Grottian and Roland Roth at the Free University of Berlin).
3 Data displayed in tables 4.2 and 4.3 of the Wilson chapter illustrate this point: asking about "strong approval" of the movements currently seeking support from the public produces vastly different results from assessing "possible membership." The difference is most dramatic for West Germany, where "possible membership" levels are highest and "strong approval" levels are lowest among the eight European nations considered. This seems to indicate that – at least for Germany – there is widespread support for the general goals of the movements, but considerably less support for the movements as factual collective actors.
4 For an opposite view see the various contributions in a thematic issue of the *European Journal of Political Research*, in particular the introduction by the issue editor (Bürklin, 1987b).
5 For instance, a conservative environmental party (ÖDP) received 1.4 percent of the popular vote in the March 1988 state elections in Baden-Württemberg, West Germany, with 3 to 4 percent in the Freiburg district which includes the village of Wyhl, the site of early mass protest against a proposed nuclear power plant and now a new movement symbol.
6 Obviously, municipal and regional elections are of minor importance with respect to a challenge to the existing political order. However, viewed over time, success in lower level elections may well be an early indicator of national impact to come – as the case of the German Greens shows.
7 Of these countries, only Sweden employs a modified proportional system with a significant threshold. As in West Germany, parties not meeting a threshold of 4 percent (West Germany: 5 percent) of the popular vote nationwide do not receive any parliamentary seats. However, even including a significant threshold, proportional systems are much more amenable to small parties than majority systems.
8 Kitschelt (1988a, p. 198) used a (combined) gain of "about 4 percent or more of the vote in a natioal parliamentary or presidential election at least once in the 1980s" for Left-libertarian parties as criterion for their significant existence in a country.
9 The most compelling evidence is provided by the monthly *Politbarometer* surveys conducted by the *Forschungsgruppe Wahlen*, Mannheim, West Germany. In these surveys, all respondents rate all parties on a ten-point sympathy scale; among respondents intending to vote for the Greens the SPD is consistently ranked second and, on average, is the only other party rated positively.
10 The 1983 state election in Hesse left both the Social Democrats and the alliance between Christian Democrats and Liberals short of a majority of parliamentary seats. After nine months of negotiations the Greens agreed to informally support an SPD government and helped to elect Börner as Prime Minister. A formal coalition was formed 18 months later giving the Greens

their first ever cabinet post. Joschka Fischer, a veteran of the student movement of the 1970s, became Minister of Environmental Affairs.

11 The Republicans – so successful in Berlin – were organizationally not prepared to contest the municipal elections in Hesse. Instead, the NPD (which is officially classified as an extreme party and which is banned in Berlin under the allied statutes) gained 6.6 percent in the city of Frankfurt.

12 Quite recently though, at their 18th Congress in March of 1989, the Italian Communists have formally adopted a document proclaiming a new face of Communism – with much emphasis on ecology and feminism – promoted by the party's new leader, Achille Occhetto (according to a report in the *New York Times* of March 25, 1989, p. 8).

13 For example, the Social Democrat/Greens coalition in Berlin – mentioned above – has produced the first ever state-level cabinet in West Germany with a female majority, though still led by a male mayor.

14 In the words of a Greens leader (and veteran of both the French and the German student movement of the late 1960s), Daniel Cohn-Bendit: "All over Europe it is becoming clear that the problems of modernization mean that we need the ecological movement to offer solutions" (as quoted in a story by James Markham in the *New York Times* of April 12, 1989, p. A10).

15 In the first round of the municipal elections in March 1989 the environmentalists won about 15 percent of the vote in several major French cities, somewhat slipping in the second round due to intense efforts of the Socialists to woo voters with environmental concerns. In the elections to the European Parliament (for which a proportional system is used) in June 1989, the French ecologists won 10.6 percent of the vote and 9 seats in the European parliament. The virtually unknown British Greens also shocked political observers by garnering 14.5 of the vote in the European elections.

References

Abramson, Paul. 1983. *Political Attitudes in America: Formation and Change* San Francisco, CA: Freeman.
Accornero, Aris. 1967. *I comunisti in fabbrica: Documenti delle lotte operaie*, no. 1. Florence: Centro G. Francovich.
Accornero, Aris, Renato Mannheimer, and Chiara Saraceno, eds. 1983. *L'identita comunista: I militanti, le strutture, la cultura del PCI.* Rome: Editori Riuniti.
Adams, Carolyn, and Katherine Winston. 1980. *Mothers at Work.* New York: Longman.
Aktion Sühnezeichen/Friedensdienste. 1982. *Regionale atomwaffenfreie Zonen.* Berlin: AS/F. September.
Alber, Jens. 1985. Modernisierung, neue Spannungslinien und die politischen Chancen der Grünen. *Politische Vierteljahresschrift*, 26: 211–26.
Alber, Jens. 1986. Der Wohlfahrtsstaat in der Wirtschaftskrise: Eine Bilanz der Sozialpolitik in der Bundesrepublik seit den frühen siebziger Jahren. *Politische Vierteljahresschrift*, 27: 28–60.
Alberoni, Francesco. 1968. *Statu nascenti.* Bologna: Mulino.
Alberoni, Francesco. 1979. Movimenti e instituzioni nell'Italia tra il 1960 e il 1970. In Luigi Graziano and Sidney Tarrow, eds, *La crisi italiana*, vol. 1. Turin: Einaudi.
Alberoni, Francesco. 1984. *Movement and Institution.* New York: Columbia University Press.
Alemann, Ulrich von, and Rolf Heinze, eds. 1979. *Verbände und Staat.* Opladen: Westdeutscher Verlag.
Alinsky, Saul. 1971. *Rules for Radicals.* New York: Random House.
Almond, Gabriel, and Sidney Verba. 1963. *The Civic Culture.* Princeton, NJ: Princeton University Press.
Apter, David, and Nagayo Sawa. 1984. *Against the State.* Cambridge, MA: Harvard University Press.
Bahro, Rudolf. 1986. *Building the Green Movement.* Philadelphia: New Society Publishers.
Baker, Kendall, Russell Dalton, and Kai Hildebrandt. 1981. *Germany Transformed.* Cambridge, MA: Harvard University Press.
Banks, Olive. 1981. *Faces of Feminism: A Study of Feminism as a Social Movement.* New York: St Martin's Press.

Barbagli, Marzio, and Piergiorgio Corbetta. 1978. Partito e movimento: Aspetti del rinnovamento del PCI. *Inchiesta*, 8: 3–46.

Barkan, S. E. 1979. Strategic, Tactical and Organizational Dilemmas of the Protest Movement Against Nuclear Power. *Sociological Problems*, 27: 19–37.

Barnes, Samuel, Max Kaase, et al. 1979. *Political Action: Mass Participation in Five Western Democracies*. Beverly Hills, CA: Sage Publications.

Bartolini, Stefano. 1983. The Membership of Mass Parties: The Social Democratic Experience, 1889–1978. In Hans Daalder and Peter Mair, eds, *Western European Party Systems*. Beverly Hills, CA: Sage Publications.

Bassnett, Susan. 1986. *Feminist Experiences: The Women's Movement in Four Cultures*. London: Allen and Unwin.

Baude, Annika. 1979. *Public Policy and Changing Family Patterns in Sweden 1930–1977*. Stockholm: Swedish Center for Working Life, National Board of Welfare.

Baumgarten, Jürgen, ed. 1982. *Linkssozialisten in Europa*. Hamburg: Junius Verlag.

Beccalli, Bianca. 1971: Scioperi e organizazzione sindacale: Milano 1950–1970. *Rassegna Italiana di sociologia*, 12: 83–120.

Becchelloni, Giovanni, ed. 1973. *Cultura e ideologia della nuova sinistra: Materiali per un inventorio della cultura politica della riviste del dissenso marxista*. Milan: Comunita.

Beck, Ulrich. 1983. Jenseits von Stand und Klasse? Soziale Ungleichheiten, gesellschaftliche Individualisierungsprozesse und die Entstehung neuer sozialer Formationen und Identitäten. *Soziale Welt*, 34: 35–74.

Becker, Horst, and Bodo Hombach. 1983. *Die SPD von Innen: Bestandsaufnahme an der Basis der Partei*. Bonn: Verlag Neue Gesellschaft.

Bedürfnisanstalt Blockade. 1983. *Graswurzelrevolution*. December.

Beer, Samuel H. 1982. *Britain Against Herself: The Political Contradictions of Collectivism*. New York: W. W. Norton.

Bell, Daniel. 1976. *The Cultural Contradictions of Capitalism*. New York: Basic Books.

Benford, Robert. 1984. The Interorganizational Dynamics of the Austin Peace Movement. Unpublished thesis, the Univesity of Texas at Austin.

Berger, Peter L., B. Berger, and H. Kellner. 1973. *The Homeless Mind: Modernization and Consciousness*. New York: Random House.

Berger, Suzanne. 1979. Politics and Anti-politics in Western Europe in the Seventies. *Daedalus*, 108: 27–50.

Berger, Suzanne D., ed. 1981. *Organizing Interests in Western Europe*. Cambridge: Cambridge University Press.

Beyme, Klaus von. 1986. Neue soziale Bewegungen und politische Parteien. *Das Parlament – Beilage: Aus Politik und Zeitgeschichte*. October 10.

Bianchi, Suzanne, and Daphne Spahn. 1986. *American Women in Transition*. New York: Russell Sage Foundation.

Billiet, J. 1984. On Belgian Pillarization. *Acta Politica*, 19: 117–28.

Blackmer, Donald L. M., and Sidney Tarrow. 1975. *Communism in Italy and France*. Princeton, NJ: Princeton University Press.

Block, J. 1981. Some Enduring and Consequential Structures of Personality. In Albert Rabin et al., *Further Explorations in Personality*. New York: Wiley.

Bobbio, N. 1986. *The Future of Democracy*. Oxford: Polity Press.

Böltken, Ferdinand, and Wolfgang Jagodzinski. 1985. Postmaterialism in the European Community, 1970–1980: Insecure Value Orientations in an Environment of Insecurity. *Comparative Political Studies*, 17: 453–84.

Boggs, Carl. 1986. *Social Movements and Political Power*. Philadelphia: Temple University Press.

Bolton, Charles. 1972. Alienation and Action: A Survey of Peace-Group Members. *American Journal of Sociology*, 78: 537–61.

Bornschier, Volker. 1988. *Westliche Gesellschaft im Wandel*. Frankfurt: Campus.

Bouchier, David. 1984. *The Feminist Challenge*. New York: Schocken Books.

Boudon, Raymond. 1977. *Effects pervers et ordre social*. Paris: Presses Universitaires de France.

Boy, Daniel. 1981. Le vote ecologiste en 1978. *Revue francaise de science politique*, 31: 394–416.

Brand, Karl-Werner. 1982. *Neue soziale Bewegungen: Entstehung, Funktion und Perspektive neuer Protestpotentiale*. Opladen: Westdeutscher Verlag.

Brand, Karl-Werner, ed. 1985. *Neue soziale Bewegungen in Westeuropa und in den USA: Ein internationaler Vergleich*. Frankfurt/New York: Campus.

Brand, Karl-Werner, Detlef Büsser, and Dieter Rucht. 1986. *Aufbruch in eine neue Gesellschaft: Neue soziale Bewegungen in der Bundesrepublik Deutschland*, revised edn. Frankfurt/New York: Campus.

Brim, Orville. 1966. Socialization through the Life Cycle. In Orville Brim and Stanton Wheeler, eds, *Socialization after Childhood*. New York: Wiley.

Brim, Orville, and Jerome Kagan, eds. 1980. *Constancy and Change in Human Development*. Cambridge, MA: Harvard University Press.

Brock, Peter. 1968. *Pacifism in the United States: From the Colonial Era to the First World War*. Princeton, NJ: Princeton University Press.

Brock, Peter, 1970. *Twentieth Century Pacifism*. New York: Van Nostrand.

Brock, Peter, 1972. *Pacifism in Europe to 1914*. Princeton, NJ: Princeton University Press.

Budge, Ian, David Robertson, and Derek Hearl, eds. 1987. *Ideology, Strategy and Party Change: Spatial Analyses of Post-War Election Programmes in 19 Democracies*. Cambridge: Cambridge University Press.

Bürklin, Wilhelm. 1984. *Grüne Politik: Ideologische Zyklen, Wähler und Parteiensystem*. Opladen: Westdeutscher Verlag.

Bürklin, Wilhelm. 1985. The Greens: Ecology and the New Left. In George Romoser and Peter Wallach, eds, *West German Politics in the Mid-Eighties*. New York: Praeger.

Bürklin, Wilhelm P. 1987a. Governing Left Parties Frustrating the Radical Non-established Left: The Rise and Inevitable Decline of the Greens. *European Sociological Review*, 3: 109–26.

Bürklin, Wilhelm P. 1987b. Why study Political Cycles? *European Journal of Political Research*, 15: 131–26.

Bürklin, Wilhelm P. 1988a. A Politico-Economic Model Instead of a Sour Grapes Logic: A Reply to Herbert Kitschelt's Critique: *European Sociological Review*, 4: 161–6.

Bürklin, Wilhelm. 1988b. Wertwandel oder zyklische Wertaktualisierung. In Heiner Meulemann and H. Luthe, eds, *Wertwandel: Faktum oder Fiktion*? Frankfurt: Campus.

Byrd, Peter. 1985. The Development of the Peace Movement in Britain. In Werner Kaltefleiter and Robert Pfaltzgraff, eds, *The Peace Movements in Europe and the United States*. London: Croom Helm.

Byrne, Paul. 1981. *The Campaign for Nuclear Disarmament*. London: Croom Helm.

Cambridge Women's Peace Collective. 1984. *My Country is the Whole World*. London: Pandora Press.

Capra, Fritjof, and Charlene Spretnak. 1984. *Green Politics*. New York: Dutton.

Cattaneo Institute. 1968. *La presenza sociale del PCI e della DC*. Bologna: Mulino.

Cawson, Alan. 1986. *Corporatism and Political Theory*. Oxford: Basil Blackwell.

Cazzaniga, Gian Mario. 1967. Cronache e documenti del movimento studentesco. *Nuovo Impegno*, 8: 19–37.

Ceulers, Jan (directing debate). 1981. Evaluatie van de Partiecratie. *Res Publica*, 23: 155–77.

Chandler, William M., and Alan Siaroff. 1986. Postindustrial Politics in Germany and the Origins of the Greens. *Comparative Politics*, 18: 303–25.

Clark, Kitson. 1955. The Romantic Element – 1830 to 1850. In John Plumb, ed., *Studies in Social History: A Tribute to G. M. Trevelyan*. London: Longmans, Green and Co.

Converse, Philip E. 1970. Attitudes and Nonattitudes: Continuation of a Dialogue. In Edward R. Tufte, ed., *The Quantitative Analysis of Social Problems*. Reading, MA: Addison-Wesley.

Costa, Paul, and Robert McCrae. 1980. Still Stable After All These Years. In Paul Baltes and Orville Brim, eds, *Life-Span Development and Behavior*, vol. 3. New York: Academic Press.

Costain, Anne N., and W. Douglas Costain. 1986. The Decline of Political Parties and the Rise of Social Movements in America. Paper presented at the 81st Annual Meeting of the American Sociological Association, New York.

Cotgrove, Stephan. 1982. *Catastrophe or Cornucopia*. New York: Wiley.

Cotrove, Stephan, and Andrew Duff. 1980. Environmentalism, Middle-Class Radicalism and Politics. *Sociological Review*, 28: 333–51.

Cotgrove, Stephan, and Andrew Duff. 1981. Environmentalism, Values and Social Change. *British Journal of Sociology*, 32: 92–110.

Coultas, Valerie. 1981. Feminists Must Face the Future. *Feminist Review*, 7: 30–40.

Crewe, Ivor, and David Denver. 1985. *Electoral Change in Western Democracies: Patterns and Sources of Electoral Volatility*. London: Croom Helm.

Crouch, Colin. 1978. The Changing Role of the State in Industrial Relations in Western Europe. In Colin Crouch and Alessandro Pizzorno, eds, *The Resurgence of Class Conflict in Western Europe since 1968*, vol. 2. New York: Holmes and Meier/ London: Macmillan.

Crouch, Colin. 1983. Pluralism and the New Corporatism: A Rejoinder. *Political Studies*, 31: 452–60.

Crozier, Michel et al. 1975. *The Crisis of Democracy*. New York: New York University Press.

Curtis, Russell L., and Louis A. Zurcher. 1973. Stable Resources of Protest Movements: The Multi-Organizational Field. *Social Forces*, 52: 53–61.

Curtis, Russell L., and Louis A. Zurcher. 1974. Social Movements. *Social Problems*, 21: 236–70.

Dalton, Russell. 1988a. *Citizen Politics in Western Democracies*. Chatham, NJ: Chatham Publishers.

Dalton, Russell. 1988b. The Environmental Movement and West European Party Systems. Paper presented at the annual meeting of the Midwest Political Science Association.

Dalton, Russell, Scott C. Flangan, and Paul A. Beck, eds. 1984. *Electoral Change in Advanced Industrial Democracies: Realignment or Dealignment?* Princeton, NJ: Princeton University Press.

Dalton, Russell et al. 1986. Environmental Action in Western Europe. Unpublished report, Tallahassee: Florida State University.

Daniele, Piergiorgio. 1978. L'organizazzione dell-utenza casa a Milano: Rapporto tra sviluppo dei sindacati inquilini e territorio urbano. BA thesis, Faculty of Philosophy, State University of Milan.

Della Porta, Donatella. 1988. Recruitment Processes in Clandestine Political Organizations: Italian Left-Wing Terrorism. In Bert Klandermans, Hanspeter Kriesi, and Sidney Tarrow, eds, *From Structure to Action: Comparing Social Movement Research across Cultures*. Greenwich, CT: JAI Press.

Della Porta, Donatella, and Sidney Tarrow. 1986. Unwanted Children: Political Violence and the Cycle of Protest in Italy, 1966–1978. *European Journal of Political Research*, 14: 607–32.

Deschouwer, Kris, and Patrick Stouthuysen. 1984. *L'electorat d'Agalev*. Brussels: Centre de Recherche et d'Information Socio-Politiques.

Diani, Mario, and Giovanni Lodi. 1988. Three in One: Currents in the Ecological Movement in Milan. In Bert Klandermans, Hanspeter Kriesi and Sidney Tarrow, eds, *From Structure to Action: Comparing Movement Paticipation Across Cultures*. Greenwich, CT: JAI Press.

Donati, Paolo R. 1984. Organization between Movement and Institution. *Social Science Information*, 23: 837–59.

Downing, Paul, and Gordon Brady. 1974. The Role of Citizen Interest Groups in Environmental Policy Formation. In M. White, ed., *Nonprofit Firms in a Three Sector Economy*. Washington: Urban Institute.

Dreier, H. 1987. Staatliche Legitimität, Grundgesetz und neue soziale Bewegungen. In J. Marko and A. Stolz, eds, *Demokratie und Wirtschaft*. Wien: Böhlau.

Dubois, Pierre. 1978. New Forms of Industrial Conflict. In Colin Crouch and Alessandro Pizzorno, eds, *The Resurgence of Class Conflict in Western Europe Since 1968*, vol. 2 New York: Holmes and Meier/London: Macmillan.

Dunlap, R., and Karl van Liere. 1978. The "New Environmental Paradigm." *Journal of Environmental Education*, 9: 10–19.

Duverger, Maurice. 1954. *Political Parties*. London: Methuen.

Eduards, Maud. 1981. Sweden. In Jill Hills and Joni Lovenduski, eds, *The Politics of the Second Electorate*. London: Routledge and Kegan Paul.

Eduards, Maud, Beatrice Halsaa, and Gege Skjeie. 1985. Equality: How Equal? In Elina Haavio-Mannila et al., *Unfinished Democracy*. Elmsford, NY: Pergamon Press.

Eduards, Maud, Beatrice Halsaa, and Gege Skjeie. 1986. The Participation of Women in the Political Process in Sweden. Swedish Report to the Council of Europe.

Eisinger, Peter K. 1973. The Conditions of Protest Behavior in American Cities. *American Political Science Review*, 67: 11–28.

Eldersveld, Samuel J. 1964. *Political Parties: A Behavioral Analysis*. Chicago: Rand McNally.

Epstein, Leon D. 1967. *Political Parties in Western Democracies*. New York: Praeger.

Equal Opportunity Commission (EOC). 1983. *Eighth Annual Report*. Manchester.

Ergas, Yasmine. 1982. 1968–79: Feminism and the Italian Party System. *Comparative Politics*, 14: 253–80.

Eulau, Heinz. 1989. Crossroads of Social Science. In Heinz Eulau, ed., *Crossroads of Social Science: The ICPSR 25th Anniversary Papers*. New York: Agathon Press.

Evans, Richard J. 1977. *The Feminists: Women's Emancipation Movements in Europe, America and Australasia 1840–1920*. London: Croom Helm.

Everts, Philip. 1982. The Mood of the Country. *Acta Politica*, 17: 497–553.

Faber, Mient Jan. 1985. *Min x Min = Plus*. Weesp: Da Haan.

Falke, Wolfgang. 1982. *Die Mitglieder der CDU: Eine empirische Studie zum Verhaltnis von Mitglieder- und Organisationsstruktur der CDU 1971–1977.* Berlin: Duncker and Humblot.

Faslane Peace Camp. 1984. *Faslane: Diary of a Peace Camp.* Edinburgh: Polygon Books.

Feist, Ursula, Manfred Güllner, and Klaus Liepelt. 1978. Structural Assimilation versus Ideological Polarization. In Max Kaase and Klaus von Beyme, eds, *Elections and Parties.* Beverly Hills, CA: Sage Publications.

Ferree, Myra Marx. 1987a. Strategy and Tactics of the Women's Movement: The Role of Political Parties. In Mary Katzenstein and Carol Mueller, eds, *The Women's Movements of the United States and Western Europe.* Philadelphia: Temple University Press.

Ferree, Myra Marx. 1987b. Feminist politics in the U.S. and West Germany. In Mary Katzenstein and Carol Mueller, eds, *The Women's Movements of the United States and Western Europe.* Philadelphia: Temple University Press.

Ferree, Myra Marx, and Beth Hess. 1985. *Controversy and Coalition: The New Feminist Movement.* Boston: Twayne.

Fiorina, Morris. 1981. *Retrospective Voting in America.* New Haven, CT: Yale University Press.

Fireman, Bruce, and William Gamson. 1979. Utilitarian Logic in the Resource Mobilization Perspective. In Mayer Zald and John McCarthy, eds, *The Dynamics of Social Movements.* Cambridge, MA: Winthrop.

Flanagan, Scott. 1987. Value Change in Industrial Society. *American Political Science Review,* 81: 1303–19.

Flannery, Kate, and Sara Roelof. 1984 Local Government Women's Comittees. In Joy Holland, ed., *Feminist Action,* vol. 1. London: Battle Axe Books.

Flexner, Eleanor. 1975. *Century of Struggle: The Women's Rights Movement in the United States.* Revised edn, Cambridge: Harvard University Press.

Fogt, Helmut. 1984. Basisdemokratie oder Herrschaft der Aktivisten? Zum Politikverständnis der Grünen. *Politische Vierteljahresschrift,* 25: 97–114.

Fogt, Helmut. 1987. Die Grünen und die Neue Linke: Zum innerparteilichen Einfluss des organisierten Linksradikalismus. In M. Langner, ed., *Die Grünen auf dem Prüfstand.* Bergisch-Gladbach: Lübbe Verlag.

Foss, Daniel A. and Ralph Larkin. 1986. *Beyond Revolution: A New Theory of Social Movements.* South Hadley, MA: Bergin and Garvey.

Freeman, Christopher, ed. 1983. *Long Waves and the World Economy.* London: Butterworth.

Freeman, Jo. 1975. *The Politics of Women's Liberation.* New York: Longman.

Freeman, Jo. 1979. Resource Mobilization and Strategy. In Mayer Zald and John McCarthy, eds, *The Dynamics of Social Movements.* Cambridge, MA: Winthrop.

Freeman, Jo. 1983. A Model for Analyzing the Strategic Options of Social Movement Organizations. In Jo Freeman, ed., *Social Movements of the Sixties and Seventies.* New York and London: Longman.

Freeman, Jo. 1985–6. The Political Culture of the Democratic and Republican Parties. *Political Science Quarterly,* 101: 327–356.

Freeman, Jo. 1987. Whom you Know vs. Whom you Represent. In Mary Katzenstein and Carol Mueller, eds, *The Women's Movements of the United States and Western Europe.* Philadelphia: Temple University Press.

Freeman, Jo. 1988. Women at the 1988 Democratic Convention. *PS,* 21: 875–81.

Frohlich, Norman, et al. 1971. *Political Leadership and Collective Goods.* Princeton, NJ: Princeton University Press.

Fuchs, Dieter. 1984. Die Aktionsformen der neuen sozialen Bewegungen. In Jürgen Falter, Christian Fenner, and Michael Greven, eds, *Politische Willensbildung und Interessenvermittlung*. Opladen: Westdeutscher Verlag.

Fuchs, Dieter. 1989. *Die Unterstützung des politischen Systems der Bundesrepublik Deutschland*. Opladen: Westdeutscher Verlag.

Gallagher, Orvoell R. 1957. Voluntary Associations in France. *Social Forces*, 38: 153–60.

Galli, Giorgio, and Alfonso Prandi. 1970. *Patterns of Political Participation in Italy*. New Haven, CT: Yale University Press.

Gamson, William A. 1968. *Power and Discontent*. Homewood, IL: Dorsey Press.

Gamson, William A. 1975. *The Strategy of Social Protest*. Homewood, IL: Dorsey Press.

Gamson, William A. 1987. Introduction. In Mayer Zald and John McCarthy, eds, *Social Movements in an Organizational Society*. New Brunswick, NY/Oxford: Transaction Books.

Gelb, Joyce. 1989. *Feminism and Politics: A Comparative Perspective*. Berkeley: University of California Press.

Gelb, Joyce, and Marian Palley. 1987. *Women and Public Policies*. 2nd edn, Princeton, NJ: Princeton University Press.

Gerlach, Luther, and Virginia Hines. 1970. *People, Power, Change*. Indianapolis: Bobbs-Merrill.

Gerretson, Rob, and Marcel van der Linden. 1982 Die Pazifistisch-Sozialistische Partei der Niederlande, In Jürgen Baumgarten, ed., *Linkssozialisten in Europa: Alternativen zu Sozialdemokratie und kommunistischen Parteien*. Hamburg: Junius.

Giddens, Anthony. 1976. *The New Rules of Sociological Method*. London: Hutchinson and Company.

Giddens, Anthony. 1986. Modernity, Ecology and Social Transformation. Unpublished manuscript.

Gitlin, Todd. 1980. *The Whole World Is Watching*. Berkeley: University of California Press.

Glenn, Norval. 1974. Aging and Conservatism. *Annals of the American Academy of Political and Social Science*, 415: 176–86.

Glenn, Norval. 1980. Values, Attitudes and Beliefs. In Orville Brim and Jerome Kagan, eds, *Constancy and Change in Human Development*. Cambridge, MA: Harvard University Press.

Gordon, David, R. Edwards, and M. Reich. 1982. *Segmented Work, Divided Workers: The Historical Transformation of Labor in the United States*. Cambridge: Cambridge University Press.

Gordon, C. Wayne, and Nicholas Babchuk. 1959. A Typology of Voluntary Associations. *American Sociological Review*, 24: 1049–81.

Goss, Sue. 1984. Women's Initiatives in Local Government. In D. Boddy and C. Fudge, eds, *Local Socialism*. London: Macmillan.

Gramsci Institute. 1962. *Tendenze nel capitalismo italiano*. Rome: Riuniti.

Grant, Wyn, ed. 1985. *The Political Economy of Corporatism*. London: Macmillan.

Graziolo, Marco. 1979. Il movimento studentesco in Italia nell/anno accademico 1967–68: Ricostruzione ed analisi. BA thesis, Faculty of Political Science, State University of Milan.

Griffin Larry J., Michael E. Wallace, and Beth A. Rubin. 1986. Capitalist Resistance to the Organization of Labor before the New Deal: Why? How? Success? *American Sociological Review*, 51: 147–67.

Groeneveld, Frits. 1984. Het IKV moet ophouden het CDA te vermanen. *NRC Handelsblad*. November 29.

Grofman, Bernard, and Edward Muller. 1973. The Strange Case of Relative Gratification and Protest Potential. *American Political Science Review*, 67: 514–15.

Guggenberger, Bernd. 1980. *Bürgerinitiativen in der Parteiendemokratie*. Stuttgart: Kohlhammer.

Guggenberger, Bernd, and Udo Kempf, eds. 1984. *Bürgerinitiativen und repräsentatives System*. 2nd edn, Opladen: Westdeutscher Verlag.

Gundelach, Peter. 1984. Social Transformation and New Forms of Voluntary Associations. *Social Science Information*, 23: 1049–81.

Gurney, Joan, and Kathleen Tierney. 1982. Relative Deprivation and Social Movements: A Critical Look at Twenty Years of Theory and Research. *Sociological Quarterly*, 23: 33–47.

Gurr, Ted R. 1970. *Why Men Rebel*. Princeton, NJ: Princeton University Press.

Gusfield, Joseph. 1981. Social Movements and Social Change: Perspectives of Linearity and Fluidity. In Louis Kriesberg, ed., *Research in Social Movements, Conflicts and Change*, vol. 4. Greenwich, CT: JAI Press.

Habermas, Jürgen. 1973. *Legitimation Crisis*. Boston: Beacon Press.

Habermas, Jürgen. 1981. *Theorie des kommunikativen Handelns*, vol. 2. Frankfurt: Suhrkamp.

Habermas, Jürgen. 1984. *Observations on the "Spiritual Situation of the Age."* Cambridge, MA: MIT Press.

Hammans, P. 1987. *Das politische Denken der neueren deutschen Staatslehre in der Bundesrepublik*. Opladen: Westdeutscher Verlag.

Handler, Joel. 1978. *Social Movements and the Legal System*. New York: Academic Press.

Hanmer, Julia. 1977. Community Action, Women's Aid and the Women's Liberation Movement. In Marjorie Mayo, ed., *Women in the Community*. London: Routledge and Kegan Paul.

Hardin, Russell. 1982. *Collective Action*. Baltimore, MD: Johns Hopkins University Press.

Hardy, Dennis. 1979. *Alternative Communities in Nineteenth Century England*. London and New York: Longman.

Harrison, Reginald J. 1980. *Pluralism and Corporatism: The Political Evolution of Modern Democracies*. London: George Allen and Unwin.

Hastings, Elizabeth Hann, and Philip K. Hastings, eds. 1982. *Index to International Public Opinion, 1978–1980*. Westport, CT: Greenwood Press.

Hastings, Elizabeth Hann, and Philip K. Hastings, eds. 1984. *Index to International Public Opinion, 1982–1983*. Westport, CT: Greenwood Press.

Hastings, Elizabeth Hann, and Philip K. Hastings, eds. 1985. *Index to International Public Opinion, 1983–1984*. Westport, CT: Greenwood Press.

Hauss, Charles. 1978. *The New Left in France: The Unified Socialist Party*. Westport, CT: Greenwood Press.

Hays, Samuel P. 1958. *Conservation and the Gospel of Efficiency: The Progressive Conservation Movement, 1890–1920*. Cambridge, MA: Harvard University Press.

Hayward, Jack. 1976. Institutional Inertia and Political Impetus in France and Britain. *European Journal of Political Research*, 4.

Heberle, Rudolf. 1967. *Hauptprobleme der Politischen Soziologie*. Stuttgart: Enke (published in English in 1951 as *Social Movements: An Introduction to Political Sociology*. New York: Appleton Century).

Heirich, Max. 1968. *The Spiral of Conflict: Berkeley, 1964.* New York: Columbia University Press.
Hellman, Judith Adler. 1987. *Journeys among Women: Feminism in Five Italian Cities.* New York and London: Oxford University Press.
Hellman, Stephen. 1975. The PCI's Alliance Strategy and the Case of the Middle Classes. In Donald Blackmer and Sidney Tarrow, eds, *Communism in Italy and France.* Princeton, NJ: Princeton University Press.
Hellman, Stephen. 1980. Il PCI e la eredita ambigua dell-autunno caldo: Evidenza del caso Torinese. *Il Mulino,* Autumn.
Helm, Jutta. 1980. Citizen Lobbies in West Germany. In Peter Merkl, ed., *Western European Party Systems.* New York: Free Press.
Henig, Jeffrey R. 1982. *Neighborhood Mobilization, Redevelopment and Response.* New Brunswick, NY: Rutgers University Press.
Hernes, Helga. 1982. The Role of Women in Voluntary Associations. Part II of the preliminary study submitted to the Steering Committee of Human Rights (CDDH), Council of Europe.
Hernes, Helga. 1983. Women and the Welfare State. Paper presented at the conference on "The Transformation of the Welfare State," Bellagio, Italy.
Hernes, Helga. 1984. Women and the Welfare State. In Harriet Holter, ed., *Patriarchy in a Welfare Society.* Oslo: Universitetsforlaget.
Hernes, Helga. 1987. *Welfare State and Women Power.* Oslo: Universitetsforlaget.
Hewlett, Sylvia. 1986. *A Lesser Life: The Myth of Women's Liberation in America.* New York: Morrow.
Hibbs, Douglas A. 1976. On the Political Economy of Long-Run Trends in Strike Activity. *British Journal of Political Science,* 8: 153–75.
Higham, John. 1970. The Reorientation of American Culture in the 1890s. In J. Highman, *Writing American History: Essays on Modern Scholarship.* Bloomington, IN/London: Indiana University Press.
Hildebrandt, Kai, and Russell Dalton. 1978. The New Politics: Political Change or Sunshine Politics? In Max Kaase, ed., *Election and Parties.* London and Beverly Hills, CA: Sage Publications.
Himmelweit, Hilde, Patrick Humphreys, and Marianne Jaeger. 1985. *How Voters Decide.* 2nd edn, Milton Keynes, UK: Open University Press.
Hine, David. 1986. Leaders and Followers. In William E. Paterson and Alastair H. Thomas, eds, *The Future of Social Democracy.* Oxford: Clarendon Press.
Hirsch, Fred. 1976. *Social Limits to Growth.* Cambridge, MA: Harvard University Press.
Hirsch, Joachim, and Roland Roth. 1986. *Das neue Gesicht des Kapitalismus: Vom Fordismus zum Post-Fordismus.* Hamburg: VSA.
Hirschman, Albert. 1982. *Shifting Involvements: Private Interests and Public Action.* Princeton, NJ: Princeton University Press.
Hobsbawm, Eric. 1975. *The Age of Capital, 1848–1875.* New York: Scribner.
Hofstadter, Richard. 1962. *The Age of Reform: From Bryan to F.D.R.* London: Jonathan Cape.
Homberg, Soren. 1986. Gender/Party vs. Questions Dealing with Women in Sweden. SAS Institute, University of Gothenburg.
Huntington, Samuel. 1974. Postindustrial Politics: How Benign Will It Be? *Comparative Politics,* 6: 147–77.
Huntington, Samuel. 1975. The Democratic Distemper. *Public Interest,* 41.

Huntington, Samuel. 1981. *American Politics: The Promise of Disharmony*. Cambridge, MA: Harvard University Press.

Huysen, Andreas, and Klaus R. Scherpe, eds. 1986. *Die Postmoderne: Zeichen eines kulturellen Wandels*. Hamburg: Rowohlt.

Hynes, Samuel. 1968. *The Edwardian Turn of Mind*. Princeton, NJ: Princeton University Press.

Imhof, Kurt, and G. Romano. 1988. Krise und sozialer Wandel. In Ota Weinberger, ed., *Internationales Jahrbuch für Rechtsphilosophie und Gesetzgebung*. Wien: Verband der Wissenschaftlichen Gesellschaften Österreichs.

Inglehart, Ronald. 1971. The Silent Revolution in Europe: Intergenerational Change in Post-Industrial Societies. *American Political Science Review*, 65: 991–1017.

Inglehart, Ronald. 1977. *The Silent Revolution: Changing Values and Political Styles among Western Publics*. Princeton, NJ: Princeton University Press.

Inglehart, Ronald. 1979. Political Action: The Impact of Values, Cognitive Level, and Social Background. In Samuel Barnes, Max Kaase et al., *Political Action*. Beverly Hills, CA: Sage Publications.

Inglehart, Ronald. 1981. Post-Materialism in an Environment of Insecurity. *American Political Science Review*, 75: 880–990.

Inglehart, Ronald. 1984. The Changing Structure of Political Cleavages in Western Society. In Russell Dalton et al., eds, *Electoral Change in Advanced Industrial Democracies*. Princeton, NJ: Princeton University Press.

Inglehart, Ronald. 1985a. New Perspectives on Value Change: Responses to Lafferty and Knutsen, Savage, Boeltken and Jagodzinski. *Comparative Political Studies*, 17: 485–532.

Inglehart, Ronald. 1985b. Aggregate Stability and Individual-Level Change in Mass Belief Systems: The Level of Analysis Paradox. *American Political Science Review*, 79: 97–117.

Inglehart, Ronald. 1987a. Value Change in Industrial Societies. *American Political Science Review*, 81: 1289–303.

Inglehart, Ronald. 1987b. The Renaissance of Political Culture: Central Values, Political Economy and Stable Democracy. Paper presented at annual meeting of American Political Science Association, Chicago.

Inglehart, Ronald. 1990. *Culture Shift in Advanced Industrial Society*. Princeton, NJ: Princeton University Press.

Inglehart, Ronald, and Hans-Dieter Klingemann. 1976. Party Identification, Ideological Preference, and the Left-Right Dimension among Western Mass Publics. In Ian Budge et al., eds, *Party Identification and Beyond*. London and New York: Wiley.

Jacobs, Danny, and Joop Roebroek. 1983. *Nieuwe Sociale Bewegingen in Vlaanderen en Nederland*. Antwerpen: Uitgeverij Leon Lesoil.

Jagodzinski, Wolfgang. 1984. Wie transformiert man Labile In Stabile Relationen? Zur Persistenz postmaterialistischer Wertorientierungen. *Zeitschrift für Soziologie*, 13: 225–42.

Japp, Klaus P. 1984. Selbsterzeugung oder Fremdverschulden: Thesen zum Rationalismus in den Theorien sozialer Bewegungen. *Soziale Welt*, 35: 313–29.

Jenkins, J. Craig. 1981. Sociopolitical Movements. In Samuel Long, ed., *Handbook of Political Behavior*, vol. 4. New York: Plenum.

Jenkins, J. Craig. 1983. Resource Mobilization Theory and the Study of Social Movements. *Annual Review of Sociology*, 9: 527–53.

Jenkins, J. Craig. 1987. Interpreting the Stormy Sixties: Three Theories in Search of a Political Age. In R. G. Braungart, ed., *Research in Political Sociology*, vol. 3. Greenwich, CT: JAI Press.

Jennings, M. Kent, and Jan van Deth, eds. 1990. *Continuities in Political Action: A Longitudinal Study of Political Orientations in Three Western Democracies*. Berlin: de Gruyter.

Jennings, M. Kent, and Gregory Markus. 1984. Partisan Orientations over the Long Haul. *American Political Science Review*, 78: 1000–18.

Jennings, M. Kent, and Richard Niemi. 1981. *Generations and Politics*. Princeton, NJ: Princeton University Press.

Jensen, Jane. 1982. The Modern Women's Movement in Italy, France, and Great Britain. In Richard F. Thomasson, ed., *Comparative Social Research*, vol. 5. Greenwich, CT: JAI Press.

Jensen, Jane. 1983. Success Without Struggle: The Modern Women's Movement in France. Paper presented at the Cornell University Workshop on the Women's Movement in Comparative Perspective.

Kaase, Max. 1984a. Politische Beteiligung in den 80er Jahren: Strukturen und Idiosynkrasien. In Jürgen W. Falter, Christian Fenner, and Michael T. Greven, eds, *Politische Willensbildung und Interessenvermittlung*. Opladen: Westdeutscher Verlag.

Kaase, Max. 1984b. The Challenge of the 'Participatory Revolution' in Pluralist Democracies. *International Political Science Review*, 5: 299–318.

Kaase, Max. 1986a. Massenkommunikation und politischer Prozess. In Max Kaase, ed., *Politische Wissenschaft und politische Ordnung. Analysen zu Theorie und Empirie demokratischer Regierungsweise*. Opladen: Westdeutscher Verlag.

Kaase, Max. 1986b. Zur Legitimität des politischen Systems in den westlichen Demokratien. In Albrecht Randelzhofer and Werner Süss, eds, *Konsens und Konflikt: 35 Jahre Grundgesetz*. Berlin: de Gruyter.

Kaase, Max. 1986c. Die Entwicklung des Umweltbewusstseins in der Bundesrepublik Deutschland. In Rudolf Wildenmann, ed., *Umwelt-Wirtschaft-Gesellschaft: Wege zu einem neuen Grundverständnis*. Stuttgart: Staatsministerium Baden-Württemberg.

Kaase, Max. 1990. Mass Participation. In M. Kent Jennings and Jan van Deth, eds, *Continuities in Political Action*. Berlin: de Gruyter.

Kaase, Max, and Alan Marsh. 1979. Distribution of Political Action. In Samuel Barnes, Max Kaase, et al., *Political Action*. Beverly Hills, CA: Sage Publications.

Kaltefleiter, Werner, and Robert Pfaltzgraff, eds. 1985. *The Peace Movements in Europe and the United States*. London: Croom Helm.

Kanter, Rosabeth. 1972. *Commitment and Community*. Cambridge, MA: Harvard University Press.

Katzenstein, Mary Fainsod, and Carol McClurg Mueller, eds. 1987. *The Women's Movements of the United States and Western Europe*. Philadelphia: Temple University Press.

Kelly, Petra. 1984. *Fighting for Hope*. Boston, MA: South End Press.

Kelman, Steven. 1984. Party Strength and Governability in the Face of New Political Demands. Unpublished manuscript, Cambride, MA: J. F. Kennedy School of Government.

Kergoat, Jacques. 1982. Die Parti Socialiste Unifie in Frankreich. In Jürgen Baumgarten, ed., *Linkssozialisten in Europa*. Hamburg: Junius.

Kesselman, Mark, and Joel Krieger. 1987. *Eureopean Politics in Transition*. Lexington, MA: Heath.

Kielbowicz, Richard, and Clifford Scherer. 1986. The role of the Press in the

Dynamics of Social Movements. In Louis Kriesberg, ed., *Research in Social Movements*, vol. 9. Greenwich: JAI Press.

Kirchheimer, Otto. 1966. The Transformation of the European Party Systems. In Joseph LaPalombara and Myron Weiner, eds, *Political Parties and Political Development*. Princeton, NJ: Princeton University Press.

Kirkpatrick, Jeanne. 1976. *The New Presidential Elite*. New York: Russell Sage Foundation.

Kitschelt, Herbert. 1980. *Kernenergiepolitik*. Frankfurt: Campus.

Kitschelt, Herbert. 1984. *Der ökologische Diskurs*. Frankfurt: Campus.

Kitschelt, Herbert. 1985. New Social Movements in West Germany and the United States. In Maurice Zeitlin, ed., *Political Power and Social Theory*, vol. 5. Greenwich, CT: JAI Press.

Kitschelt, Herbert. 1986. Political Opportunity Structures and Political Protest: Anti-Nuclear Movements in Four Democracies. *British Journal of Political Science*, 16: 58–95.

Kitschelt, Herbert. 1988a. Left-libertarian Parties: Explaining Innovation in Competitive Party Systems. *World Politics*, 40: 194–234.

Kitschelt, Herbert. 1988b. Organization and Strategy in Belgian and West German Ecology Parties. *Comparative Politics*, 20: 127–54.

Kitschelt, Herbert. 1988c. The Life Expectancy of Left-libertarian Parties: Does Structural Transformation or Economic Decline Explain Party Innovation? *European Sociological Review*, 4: 155–60.

Kitschelt, Herbert. 1989. *The Logics of Party Formation: Structure and Strategy of Belgian and West German Ecology Parties*. Ithaca, NY: Cornell University Press.

Kitschelt, Herbert, and Staf Hellemans. 1990. *Beyond the European Left: Political Action in Left-libertarian Parties*. Durham, NC: Duke University Press.

Klandermans, Bert. 1986. New Social Movements and Resource Mobilization: The European and American Approaches. *International Journal of Mass Emergencies and Disasters*, 4: 13–39.

Klandermans, Bert. 1988. The Formation and Mobilization of Consensus. In Bert Klandermans, Hanspeter Kriesi, and Sidney Tarrow, eds, *From Structure to Action: Comparing Movement Participation Across Cultures*. Greenwich, CT: JAI Press.

Klandermans, Bert, ed. 1989a. *Organizing for Change: Social Movement Organizations Across Cultures*. Greenwich, CT: JAI Press.

Klandermans, Bert. 1989b. Interorganizational Networks. In Bert Klandermans, ed., *Organizing for Change: Social Movement Organizations Across Cultures*. Greenwich, CT: JAI Press.

Klandermans, Bert, and Dirk Oegema. 1987. Campaigning for a Nuclear Freeze: Grassroots Strategies and Local Governments in the Netherlands. In Richard Braungart, ed., *Research in Political Sociology*, vol. 3. Greenwich, CT: JAI Press.

Klandermans, Bert, Hanspeter Kriesi and Sidney Tarrow, eds. 1988. *From Structure to Action: Comparing Movement Participation Across Cultures*. Greenwich, CT: JAI Press.

Klandermans, Bert, and Sidney Tarrow. 1988. Introduction. In Bert Klandermans, Hanspeter Kriesi, and Sidney Tarrow, eds, *From Structure to Action*. Greenwich, CT: JAI Press.

Knafla, Leonore, and Christine Kulke. 1987. 15 Jahre neue Frauenbewegung. In Roland Roth and Dieter Rucht, eds, *Neue soziale Bewegungen in der Bundesrepublik Deutschland*. Frankfurt/New York: Campus.

Krabbe, Wolfgang R. 1974. *Gesellschaftsveränderung durch Lebensreform.* Göttingen: Vandenhoeck and Ruprecht.

Kretschmer, Winfried. 1988. Wackersdorf: Wiederaufbereitung im Widerstreit. In Ulrich Linse et al., *Von der Bittschrift zur Platzbesetzung.* Bonn: Dietz.

Kretschmer, Winfried, and Dieter Rucht. 1987. Beispiel Wackersdorf. In Roland Roth and Dieter Rucht, eds, *Neue soziale Bewegungen in der Bundesrepublik Deutschland.* Frankfurt/New York: Campus.

Kriesi, Hanspeter. 1982. *AKW-Gegner in der Schweiz.* Diessenhofen: Ruegger.

Kriesi, Hanspeter. 1985. *Bewegung in der Schweizer Politik, Fallstudien zu politischen Mobilierungsprozessen in der Schweiz.* Frankfurt/New York: Campus.

Kriesi, Hanspeter, and Philip van Praag. 1987. Old and New Politics: The Dutch Peace Movement and the Traditional Political Organizations. *European Journal of Political Research,* 15.

Kuechler, Manfred. 1984. Die Friedensbewegung in der BRD: Alter Pazifismus oder neue soziale Bewegung. In Jürgen W. Falter, Christian Fenner, and Michael T. Greven, eds, *Politische Willensbildung und Interessenvermittlung.* Opladen: Westdeutscher Verlag.

Kuechler, Manfred. 1986. Maximizing Utility at the Polls? A Replication of Himmelweit's 'Consumer Model of Voting' with German Election Data of '83. *European Journal of Political Research,* 14: 81–95.

Kuechler, Manfred. 1990. Public Perceptions of the Parties' Economic Competence. In Helmut Norpoth, Michael Lewis-Beck and Jean-Dominique Lafay, eds, *Economics and Politics: The Calculus of Support.* Ann Arbor: University of Michigan Press.

Kuhnle, Stein, Kaare Strom, and Lars Svasand. 1986. The Norwegian Conservative Party: Setback in an Era of Strength. *West European Politics,* 9: 448–71.

Kuisel, Richard F. 1981. *Capitalism and the State in Modern France.* Berkeley, CA: University of California Press.

Lang, Kurt, and Gladys Lang. 1961. *Collective Dynamics.* New York: Crowell.

Lange, Peter. 1983. *Union Democracy and Liberal Corporatism: Exit, Voice and Wage Regulation in Postwar Europe.* Ithaca, NY: Cornell Studies in International Affairs.

Lange, Peter, Sidney Tarrow, and Cynthia Irvin. 1989. Phases of Mobilization: Social Movements and Political Party Recruitment. *British Journal of Political Science.*

Langguth, Gerd. 1986. *The Green Factor in German Politics.* Boulder, CO: Westview Press. (published in German in 1984 as *Der grüne Faktor.* Zürich: Interfrom.)

Langner, M., ed. 1987. *Die Grünen auf dem Prüfstand.* Bergisch-Gladbach: Lübbe.

LaPalombara, Joseph. 1964. *Interest Groups in Italian Politics.* Princeton, NJ: Princeton University Press.

LaPalombara, Joseph. 1987. *Democracy, Italian Style.* New Haven, CT: Yale University Press.

Lavelle, Jini. 1983. Children Need Smiles, Not Missiles: Planning a Walk. In Lynne Jones, ed., *Keeping the peace.* London: Women's Press.

Lawson, R. 1983. Origins and Evolution of a Social Movement Strategy. *Urban Affairs Quarterly,* 18: 371–95.

Lears, Jackson T. J. 1981. *No Place of Grace: Antimodernism and the Transformation of American Culture 1880–1920.* New York: Pantheon Books.

Lehmbruch, Gerhard. 1977. Liberal Corporatism and Party Government. *Comparative Political Studies,* 10: 91–126.

Lehmbruch, Gerhard. 1984. Concertation and the Structure of Corporatist Networks. In John H. Goldthorpe, ed., *Order and Conflict in Contemporary Capitalism.* Oxford: Clarendon Press.

Lehmbruch, Gerhard. 1985. Neocorporatism in Western Europe: A Reassessment of the Concept in Cross-National Perspective. Paper presented to the International Political Science Association, XIII World Congress, Paris.

Lehmbruch, Gerhard, and Philippe Schmitter, eds. 1982. *Patterns of Neo-Corporatist Policy-Making.* Beverly Hills, CA, and London: Sage Publications.

Les français et leur défense. 1983. *Alternatives non violentes*, 49: 44–57.

Levinson, Daniel. 1978. *The Seasons of a Man's Life.* New York: Knopf.

Lijphart, Arend. 1977. Political Theories and the Explanation of Ethnic Conflict in the Western World. In Milton J. Esman, ed., *Ethnic Conflict in the Western World.* Ithaca: Cornell University Press.

Linse, Ulrich. 1983. *Zurück o Mensch zur Mutter Erde: Landkommunen in Deutschland 1890–1933.* München: DTV.

Linse, Ulrich. 1986. *Ökopax und Anarchie: Eine Geschichte der ökologischen Bewegungen in Deutschland.* München: DTV.

Linse, Ulrich, et al. 1988. *Von Bittschrift zur Platzbesetzung.* Bonn: Dietz.

Lipset, Seymour. 1981. Revolt against modernity. In P. Torsvik, ed., *Mobilization, Center–Periphery Structures and Nation-building.* Bergen: Universitetsforlaget.

Lipset, Seymour Martin, and Stein Rokkan. 1967. Cleavage Structures, Party Systems and Voter Alignments: An Introduction. In S. M. Lipset and Stein Rokkan, eds, *Party Systems and Voter Alignments.* New York: Free Press.

Lipset, Seymour Martin, and William Schneider. 1983. *The Confidence Gap: Business, Labor and Government in the Public Mind.* New York: Free Press.

Lipsky, Michael. 1968. Protest as a Political Resource. *American Political Science Review*, 62: 1144–58.

Livingston, Kenneth. 1984. Local Socialism. In D. Boddy and C. Fudge, eds, *Local Socialism.* London: Macmillan.

Logue, John. 1982. *Socialism and Abundance: Radical Socialism in the Danish Welfare State.* Minneapolis: University of Minnesota Press.

Loo van der, Hans, Erik Snel, and Bart van Steenbergen. 1984. *Een wenkend perspectief?* Amersfoort: De Horstink.

Lorenz, Einhart. 1982. Linkssozialismus in Norwegen. In Jürgern Baumgarten, ed., *Linkssozialisten in Europa.* Hamburg: Junius.

Lovenduski, Joni. 1986. *Women and European Politics: Contemporary Feminism and Public Policy.* Amherst: University of Massachusetts Press.

Lowe, Philip D. 1983. Values and Institutions in the History of British Nature Conservation. In A. Warren and F. B. Goldsmith, eds, *Conservation in Perspective.* Chichester: Wiley.

Lowe, Philip, and Jane Goyder. 1983. *Environmental Groups in Politics.* London: George Allen and Unwin.

Lowe, Philip, and Wolfgang Rüdig. 1986. Political Ecology and the Social Sciences. *British Journal of Political Science*, 16: 513–50.

Lowi, Theodore. 1971. *The Politics of Disorder.* New York: Basic Books.

Lüdtke, Alf. 1984. Protest – oder: Die Faszination des Spektakulären. In Heinrich Volkmann and Jürgen Bergmann, eds, *Sozialer Protest.* Opladen: Westdeutscher Verlag.

Luhmann, Niklas. 1986. *Ökologische Kommunikation.* Opladen: Westdeutscher Verlag.

Lumley, Robert. 1983. Social Movements in Italy, 1968–78. Ph.D. thesis, Centre for Contemporary Cultural Studies, University of Birmingham.

Lund, Basstrup. 1982. Sozialistische Volkspartei (SF) und Linkssozialisten (VS). In Jürgen Baumgarten ed., *Linkssozialisten in Europa*. Hamburg: Junius.

McAdam, Doug. 1982. *Political Process and the Development of Black Insurgency*. Chicago, IL: University of Chicago Press.

McAdam, Doug. 1983. Tactical Innovation and the Pace of Insurgency. *American Sociological Review*, 48: 735–54.

McArthur, John R., and Bruce R. Scott. 1969. *Industrial Planning in France*. Cambridge, MA: Harvard Graduate School of Business Administration.

McCarthy, John, D. and Mayer N. Zald. 1973. *The Trends of Social Movements in America*. Morristown, PA: General Learning Press.

McCarthy, John D., and Mayer N. Zald. 1977. Resource Mobilization and Social Movements: A Partial Theory. *American Journal of Sociology*, 82: 1212–41.

McCarthy, John D., and Mayer N. Zald. 1987. Resource Mobilization and Social Movements: A Partial Theory. In Mayer N. Zald and John D. McCarthy, eds, *Social Movements in an Organizational Society*. New Brunswick, NY/Oxford: Transaction Books.

McKenzie, Robert. 1955. *British Political Parties*. London: Heinemann.

Magna, Nino. 1978. Per una storia dell'operaismo in Italia: Il trentennio postbellico. In Fabrizio D'Agostini, ed., *Operaismo e centralita' operaia*. Rome: Editori Riuniti.

Maguire, Diarmuid. 1988. Leftwing Parties and Peace Movements in Western Europe: New Wine in Old Bottles. Paper presented to the American Political Science Association Annual Meeting, Washington DC, September.

Mansbridge, Jane L. 1986. *Why We Lost the ERA*. Chicago, IL: University of Chicago Press.

Marchant, Neil. 1983. Sanity's Guide to the New Parliament. *Sanity*, August.

Marsh, Jan. 1982. *Back to the Land: The Pastoral Impulse in Victorian England, 1880–1914*. London: Quartet Books.

Marshall, T. H. 1950. *Citizenship and Social Class*. Cambridge. Cambridge University Press.

Marshall, T. H. 1964. *Class, Citizenship and Social Development*. Garden City, NY: Doubleday.

Martin, Ross. 1983. Pluralism and the New Corporatism. *Political Studies*, 31: 86–102.

Marx, Gary T. 1979. External Efforts to Damage or Facilitate Social Movements: Some Patterns, Explanations, Outcomes and Complications. In Mayer Zald and John McCarthy, eds, *The Dynamics of Social Movements*. Cambridge, MA: Winthrop.

Marx, Gary, T. and James Wood. 1975. Strands of Theory and Research in Collective Behavior. In Alex Inkeles, ed., *Annual Review of Sociology*. Palo Alto, CA: Annual Reviews.

Melucci, Alberto. 1980. The New Social Movements: A Theoretical Approach. *Social Science Information*, 19: 199–226.

Melucci, Alberto. 1981. Ten Hypotheses for the Analysis of New Movements. In D. Pinto, ed., *Contemporary Italian Sociology*. Cambridge: Cambridge University Press.

Melucci, Alberto. 1984a. An End to Social Movements? *Social Science Information*, 23: 819–35.

Melucci, Alberto, ed., 1984b. *Altri codici*. Aree di movimen to nella metropili, Bologna.

Melucci, Alberto. 1985. The Symbolic Challenge of Contemporary Movements. *Social Research*, 52: 789–815.

Melucci, Alberto. 1988. Getting Involved: Identity and Mobilization in Social Movements. In Bert Klandermans, Hanspeter Kriesi, and Sidney Tarrow, eds, *From Structure to Action: Comparing Social Movement Research Across Cultures*. Greenwich, CT: JAI Press.

Merkl, Peter, and Kay Lawson, eds. 1988. *When Parties Fail*. Princeton, NJ: Princeton University Press.

Milbrath, Lester. 1984. *Environmentalists: Vanguard for a New Society*. Buffalo, NY: State University of New York Press.

Milbrath, Lester. 1989. *Envisioning a Sustainable Society: Learning Our Way Out*. Buffalo, NY: State University of New York Press.

Minnion, John, and Philip Bolsover, eds. 1983. *The CND Story*. London: Allison and Busby.

Mintzel, Alf. 1984. *Die Volkspartei: Typus und Wirklichkeit*. Opladen: Westdeutscher Verlag.

Mitchell, Robert. 1979. National Environmental Lobbies and the Apparent Illogic of Collective Action. In C. Russell, ed., *Collective Decision Making*. Baltimore, MD: Johns Hopkins University Press.

Moe, Terry. 1980. *The Organization of Interests*. Chicago, IL: University of Chicago Press.

Mohler, Peter. 1986. Mustertreue Abbidung: Ein Weg zur Lösung des Stabilitäts-Fluktuationsproblems in Panelumfragen. *ZUMA Nachrichten*, December: 31–44.

Morris, Aldon. 1984. *The Origins of the Civil Rights Movement: Black Communities Organizing for Change*. New York: Free Press.

Mortimer, Jeylan, and Roberta Simmons. 1978. Adult Socialization. *Annual Review of Sociology*, 4: 421–54.

Mueller, Carol. 1978. Riot Violence and Protest Outcomes. *Journal of Political and Military Sociology*, 6: 49–65.

Mueller, Carol. 1987. Collective Consciousness, Identity Transformation and the Rise of Women in Public Office in the United States. In Mary Katzenstein and Carol Mueller, eds, *The Women's Movements of the United States and Western Europe*. Philadelphia: Temple University Press.

Muller, Edward. 1972. A Test of a Partial Theory of Potential for Political Violence. *American Political Science Review*, 66: 928–59.

Muller, Edward. 1979. *Aggressive Political Participation*. Princeton, NJ: Princeton University Press.

Muller, Edward, and Karl-Dieter Opp. 1986. Rational Choice and Rebellious Collective Action. *American Political Science Review*, 80: 471–88.

Müller-Rommel, Ferdinand. 1982. Ecology Parties in Western Europe. *West European Studies*, 5: 68–74.

Müller-Rommel, Ferdinand. 1984. Die Grünen aus Sicht der Wahl und Elitenforschung. In Thomas Kluge, ed., *Die Grünen*. Frankfurt: Fischer.

Müller-Rommel, Ferdinand. 1985a. New Social Movements and Smaller Parties: A Comparative Perspective. *West European Studies*, 8: 41–54.

Müller-Rommel, Ferdinand. 1985b. Social Movements and the Greens: New Internal Politics in Germany. *European Journal of Political Research*, 13: 53–67.

Müller-Rommel, Ferdinand, ed. 1989. *New Politics in Western Europe: The Rise and the Success of Green Parties and Alternative Lists*. Boulder, CO: Westview Press.

Müller-Rommel, Ferdinand, and Thomas Poguntke. 1989. The Unharmonious Family: Green Parties in Western Europe. In Eva Kolinsky, ed., *The Greens in West Germany*. Oxford: Berg Publishers.

Mushaben, Joyce 1989. The Struggle Within. In Bert Klandermans, ed., *Organizing for Change*. Greenwich, CT: JAI Press.

Namenwirth, Zvi J. and Robert P. Weber. 1987. *Dynamics of Culture*. Boston, MA: Allen and Unwin.

Nash, Roderick. 1967. *Wilderness and the American Mind*. New Haven, CT, and London: Yale University Press.

Nedelmann, Birgitta. 1984. New Political Movements and Changes in Processes of Intermediation. *Social Science Information*, 23: 1029–48.

Neidhardt, Friedhelm. 1985. Einige Ideen zu einer allgemeinen Theorie sozialer Bewegungen. In Stefan Hradil, ed., *Sozialstruktur im Umbruch: Karl Martin Bolte zum 60. Geburtstag*. Opladen: Westdeutscher Verlag.

Nelkin, Dorothy, and Michael Pollak. 1981. *The Atom Besieged: Extraparliamentary Dissent in France and Germany*. Cambridge, MA: MIT Press.

Oberschall, Anthony. 1973. *Social Conflict and Social Movements*. Englewood Cliffs, NJ: Prentice-Hall.

Oberschall, Anthony. 1980. Loosely Structured Collective Conflict: A Theory and an Application. In Louis Kriesberg, ed., *Research in Social Movements, Conflict, and Change*. Greenwich, CN: JAI Press.

Offe, Claus, 1981. The Attribution of Public Status to Interest Groups: Observations on the West German Case. In Suzanne Berger, ed., *Organizing Interests in Western Europe*. Cambridge: Cambridge University Press.

Offe, Claus. 1983. Competitive Party Democracy and the Keynesian Welfare State. *Policy Science*, 15: 225–46.

Offe, Claus. 1984. *Contradictions of the Welfare State*. Cambridge, MA: MIT Press.

Offe, Claus. 1985. New Social Movements: Changing Boundaries of the Political. *Social Research*, 52: 817–68.

Offe, Claus. 1987a. Challenging the Boundaries of Institutional Politics: Social Movements since the Sixties. In Charles S. Maier, ed., *Changing Boundaries of the Political*. Cambridge: Cambridge University Press.

Offe, Claus. 1987b. The Utopia of the Zero-Option: Modernity and Modernization as Normative Political Criteria. *Praxis International*, 7.

Oliver, P., and G. Marwell. 1985. A Theory of the Critical Mass. *American Journal of Sociology*, 91.

Olofsson, Gunnar. 1988. After the Working Class Movement? An Essay on What's "New" and What's "Social" in the New Social Movements. *Acta Sociologica*, 31: 15–34.

Olsen, Johan. 1983. *Organized Democracy: Political Institutions in a Welfare State*. Bergen: Universitetsforlaget.

Olson, Mancur. 1965. *The Logic of Collective Action*. Cambridge, MA: Harvard University Press.

Opp, Karl-Dieter. 1986. Soft Incentives and Collective Action: Parties in the Anti-Nuclear Movement. *British Journal of Political Science*, 16: 87–112.

Opp, Karl-Dieter et al. 1984. *Soziale Probleme und Protestverhalten*. Opladen: Westdeutscher Verlag.

O'Riordan, Timothy. 1971. The Third American Conservation Movement: New Implications for Public Policy. *Journal of American Studies*, 5: 155–71.

Oudsten, Eymert den. 1984. Public Opinion and Nuclear Weapons. In *SIPRI Yearbook, 1984*. London: Taylor and Francis.

Panitch, Leo. 1979. The Development of Corporatism in Liberal Democracies. In

Philippe C. Schmitter and Gerhard Lehmbruch, eds, *Trends Toward Corporatist Intermediation.* Beverly Hills, CA, and London: Sage Publicaitons.

Papadakis, Elim. 1984. *The Green Movement in West Germany.* New York: St Martin's Press.

Pappi, Franz-Urban. 1988. *Neue soziale Bewegungen und Wahlverhalten in der Bundesrepublik.* Unpublished manuscript, University of Kiel, West Germany.

Parisi, Arturo, and Gianfranco Pasquino. 1980. Changes in Italian Electoral Behavior: The Relationships between Parties and Voters. In Peter Lange and Sidney Tarrow, eds, *Italy in Transition: Conflict and Consensus.* London: Cass.

Parkin, Francis. 1968. *Middle Class Radicalism.* New York: Praeger.

Paestella, Jukka. 1989. Finland: The Vihreät. In Ferdinand Müller-Rommel, ed., *New Politics in Western Europe.* Boulder, CO: Westview Press.

Paterson, William E., and Alistair H. Thomas, eds. 1986. *The Future of Social Democracy.* Oxford: Clarendon Press.

Perlmutter, Edward, 1988. Intellectuals and Urban Protest: Extraparliamentary Politics in Turin, Italy, 1968–1976. Ph.D. thesis, Harvard University.

Pero, Luciano. 1967. La crisi del Movimento studentesco: Indicazioni per una comprensione e una soluzione. *Questitalia,* 114–15: 53–69.

Pfenning, Uwe. 1987. Organistionsstruktur, Mitgliedschaft und parteipolitische Aktivitäten der Grünen Rheinland Pfalz: Eine politisch-soziologische Analyse einer Ortsverbandsbefragung im Jahr 1984. MA thesis, Faculty of Social Sciences at the University of Mannheim.

Piven, Francis Fox, and Richard Cloward. 1977. *Poor People's Movements.* New York: Vintage.

Pizzorno, Alessandro. 1978. Political Exchange and Collective Identity in Industrial Conflict. In Colin Crouch and Alessandro Pizzorno, eds, *The Resurgence of Class Conflict in Western Europe since 1968,* vol. 2. New York: Holmes and Meier/London: Macmillan.

Pizzorno, Alessandro. 1981. Interests and Parties in Pluralism. In Suzanne Berger, ed., *Organizing Interests in Western Europe.* Cambridge: Cambridge University Press.

Poguntke, Thomas, 1987a. New Politics and Party Systems. *West European Politics,* 10: 76–88.

Poguntke, Thomas. 1987b. The Organization of a Participatory Party: The German Greens. *European Journal of Political Research,* 15: 609–33.

Porritt, Jonathan. 1984. *Seeing Green.* London: Blackwell.

Pourquoi créer des zones dénucléarisées? 1983. *Alerte Atomique,* 87 (January–February): 1–11.

Przeworski, Adam. 1985. *Capitalism and Social Democracy.* Cambridge: Cambridge University Press.

Rabier, Jacques-Rene, Helene Riffault, and Ronald Inglehart. 1988. *Eurobarometer 25: April 1986.* Ann Arbor, MI: Inter-University Consortium for Political and Social Research.

Rammstedt, Otthein. 1978. *Soziale Bewegung.* Frankfurt am Main: Suhrkamp.

Rapoport, Ronald B., Alan I. Abramowitz, and John McGlennon, eds. 1986. *The Life of the Parties. Activists in Presidential Politics.* Lexington, KY: University of Kentucky Press.

Raschke, Joachim. 1980. Politik und Wertwandel in den westlichen Demokratien *Das Parlament – Beilage: Aus Politik und Zeitgeschichte.* September 6.

Raschke, Joachim, ed. 1982. *Bürger und Parteien.* Opladen: Westdeutscher Verlag.

Raschke, Joachim. 1983. Jenseits der Volkspartei. *Das Argument,* 25: 54–65.

Raschke, Joachim. 1985. *Soziale Bewegungen: Ein historisch systematischer Grundriß*. Frankfurt/New York: Campus.

Regalia, Ida, Mario Regini, and Emilio Reyneri. 1978. Labour Conflicts and Industrial Relations in Italy. In Colin Crouch and Alessandro Pizzorno, eds, *The Resurgence of Class Conflict in Western Europe since 1968*, vol. 1. New York: Holmes and Meier/London: Macmillan.

Rendall, Jane. 1984. *The Origins of Modern Feminism: Women in Britain, France and the United States, 1780–1860*. London: Macmillan.

Reuband, Karl Heinz. 1985. Politisches Selbstverständnis und Wertorientierungen von Anhängern und Gegnern der Friedensbewegung. *Zeitschrift für Parlamentsfragen*, 16: 25–45.

Reyneri, Emilio. 1978. 'Maggio strisciante': L'inizio della mobilitazione operaia. In Alessandro Pizzorno et al., *Lotte operaie e sindacato: Il ciclo 1968–1972 in Italia*. Bologna: Mulino.

Richardson, Jeremy John, and A. G. Jordan. 1979. *Governing Under Pressure: The Public Policy Process in a Post-Parliamentary Democracy*. Oxford: Martin Robertson.

Ridley, F. F. 1984. The Citizen Against Authority: British Approaches to the Redress of Grievances. *Parliamentary Affairs*, 37: 1–32.

Riesenberger, Dieter. 1985. *Geschichte der Friedensbewegung in Deutschland: Von den Anfängen bis 1933*. Göttingen: Vandenhoeck.

Riley, Matilda, and Kathleen Bond. 1983. Beyond Ageism. In Matilda Riley, Beth Hess, and Kathleen Bond, eds, *Aging in Society*. Hillsdale, NJ: Lawrence Erlbaum.

Robertson, David. 1976. *A Theory of Party Competition*. London: Wiley.

Rochon, Thomas. 1988. *Mobilizing for Peace: Antinculear Movements in Western Europe*. Princeton, NJ: Princeton University Press.

Rohrschneider, Robert. 1988. Citizen Attitudes Toward Environmental Issues: Selfish or Selfless? *Comparative Political Studies*, 21: 347–67.

Rokkan, Stein. 1966. Norway: Numerical Democracy and Corporate Pluralism. In Robert A. Dahl, ed., *Political Opposition in Western Democracies*. New Haven, CT: Yale University Press.

Roper Organization. 1980. *Virginia Slims 1980 Public Opinion Poll*. New York.

Rose, Anne C. 1981. *Transcendentalism as a Social Movement, 1830–1850*. New Haven, CT: Yale University Press.

Rose, Arnold M. 1954. Voluntary Associations in France. In Arnold M. Rose, ed., *Theory and Method in the Social Sciences*. Minneapolis: University of Minnesota Press.

Roth, Roland. 1987. Kommunikationsstrukturen und Vernetzungen in neuen sozialen Bewegungen. In Roland Roth and Dieter Rucht, eds, *Neue soziale Bewegungen in der Bundesrepublik Deutschland*. Frankfurt/New York: Campus.

Roth, Roland. 1988. In und gegen Institutionen: Anmerkungen zur paradoxen Situation neuer sozialer Bewegungen. In W. Luthardt and A. Waschkuhn, eds, *Politik und Repräsentation*. Marburg: SP-Verlag.

Roth, Roland, and Dieter Rucht. 1987a. Einleitung. In Roland Roth and Dieter Rucht, eds, *Neue soziale Bewegungen in der Bundesrepublik Deutschland*. Frankfurt/New York: Campus.

Roth, Roland, and Dieter Rucht, eds. 1987b. *Neue soziale Bewegungen in der Bundesrepublik Deutschland*. Frankfurt/New York: Campus.

Rowbotham, Sheila. 1973. *Hidden from History*. London: Pluto Press.

Rubart, Frauke. 1985. Neue soziale Bewegungen und alte Parteien in Schweden: Politischer Protest zwischen Autonomie und Integration. In Karl-Werner Brand, ed., *Neue soziale Bewegungen in Westeuropa und in den USA*. Frankfurt/New York: Campus.

Rucht, Dieter. 1980. *Von Wyhl nach Gorleben*. München: Beck.

Rucht, Dieter. 1984a. *Flughafenprojekte als Politikum*. Frankfurt: Campus.

Rucht, Dieter. 1984b. Zur Organisation der neuen sozialen Bewegungen. In Jürgen Falter, Christian Fenner, and Michael Greven, eds., *Politische Willenbildung und Interessenvermittlung*. Opladen: Westdeutscher Verlag.

Rucht, Dieter. 1985. Social Movements in France and West Germany since 1948. Paper delivered at the Council of European Studies Conference of Europeanists, Washington DC.

Rucht, Dieter. 1987a. Notizen zum Verhältnis von sozialen Bewegungen und politischen Parteien. *Zeitschrift für Sozialforschung*, 27: 297–314.

Rucht, Dieter. 1987b. Von der Ökologiebewegung zur Institution? In Roland Roth and Dieter Rucht, eds, *Neue soziale Bewegungen in der Bundesrepublik Deutschland*. Frankfurt/New York: Campus.

Rucht, Dieter. 1988. Themes, Logics, and Arenas of Social Movements: A Structural Approach. In Bert Klandermans, Hanspeter Kriesi, and Sidney Tarrow, eds, *From Structure to Action: Comparing Social Movement Research across Cultures*. Greenwich, CT: JAI.

Rucht, Dieter. 1989. Environmental Movement Organization in West Germany and France: Structure and Interorganizational Relations. In Bert Klandermans, ed., *Organizing for Change: Social Movement Organizations across Cultures*. Greenwich, CT: JAI Press.

Rüdig, Wolfgang, and Philip Lowe. 1986. The Withered "Greening" of British Politics. *Political Studies*, 34: 262–84.

Saarlvik, Bo, and Ivor Crewe. 1983. *Decade of Dealignment*. Cambridge/New York: Cambridge University Press.

Sabel, Charles F. 1981. The Internal Politics of Trade Unions. In Suzanne Berger, ed., *Organizing Interests in Western Europe*. Cambridge: Cambridge University Press.

Sabel, Charles. 1982. *Work and Politics*. Cambridge: Cambridge University Press.

Sainsbury, Diane. 1985. Women's Routes to National Legislatures. Paper presented at the annual meeting of the Swedish Political Science Association.

Sartori, Giovanni. 1966. European Political Parties: The Case of Polarized Pluralism. In Joseph LaPalombara and Myron Weiner, eds, *Political Parties and Political Development*. Princeton, NJ: Princeton University Press.

Schenk, Herrad. 1981. *Die feministische Herausforderung: 150 Jahre Frauenbewegung in Deutschland*. Munich: Beck Verlag.

Schenk, Herrad. 1982. Pazifismus in der ersten Frauenbewegung. In Ruth Esther Geiger and Anna Johannesson, eds, *Nicht Friedlich und nicht Still*. Munich: Frauenbuchverlag.

Schennink, Ben. 1988. Dynamics of the Peace Movement in the Netherlands. In Bert Klandermans, Hanspeter Kriesi, and Sidney Tarrow, eds, *From Structure to Action: Comparing Movement Participation Across Cultures*. Greenwich, CT: JAI Press.

Schennink, Ben, Ton Bertrand, and Hans Fun. 1982. *De 21 november demonstranten: wie zijn ze en wat willen ze?* Amsterdam: Jan Mets.

Schlesinger, Joseph A. 1984. On the Theory of Party Organization. *Journal of Politics*, 46: 369–400.

Schlesinger, Arthur M. jr. 1986. *The Cycles of American History*. Boston, MA: Houghton Mifflin Company.

Schmid, Günter. 1982. Zur Soziologie der Friedensbewegung und der Jugendprotesters. *Das Parlament – Beilage: Aus Politik und Zeitgeschichte.*

Schmidt, Manfred G. 1982. Does Corporatism Matter? Economic Crisis, Politics and Rates of Unemployment in Capitalist Democracies in the 1970s. In Gerhard Lehmbruch and Philippe Schmitter, eds, *Patterns of Neo-Corporatist Policy-Making.* Beverly Hills, CA, and London: Sage Publications.

Schmidt, Manfred G. 1984. Konkurrenzdemokratie, Wohlfahrtsstaat und neue soziale Bewegungen. *Das Parlament – Beilage: Aus Politik und Zeitgeschichte.* March 17.

Schmitt, Hermann. 1987. *Neue Politik in alten Parteien.* Opladen: Westdeutscher Verlag.

Schmitt, Rüdiger. 1987. Was bewegt die Friedensbewegung zum sicherheitspolitischen Protest der achtziger Jahre? *Zeitschrift für Parlamentsfragen*, 18: 110–36.

Schmitter, Philippe. 1979. Still the Century of Corporatism? In Philippe Schmitter and Gerhard Lehmbruch, eds, *Trends toward Corporatist Intermediation.* London: Sage Publications.

Schmitter, Philippe C. 1981. Interest Intermediation and Regime Governability in Contemporary Western Europe and North America. In Suzanne Berger, ed., *Organizing Interests in Western Europe.* Cambridge: Cambridge University Press.

Schmitter, Philippe C. 1983. Democratic Theory and Neo-Corporatist Practice. *Social Research*, 50: 885–928.

Schmitter, Philippe C. 1984. Still the Century of Corporatism? *Review of Politics*, 36: 85–131.

Schmitter, Philippe C., and Gerhard Lehmbruch, eds. 1979. *Trends Toward Corporatist Intermediation.* Beverly Hills, CA, and London: Sage Publications.

Schnapp, Alain, and Pierre Vidal-Naquet. 1971 *The French Student Uprising: November 1967–June 1968: An Analytical Record.* Boston, MA: Beacon.

Schönbohm, Wulf. 1985. *Die CDU wird moderne Volkspartei.* Stuttgart: Klett-Cotta.

Sciubba, Roberto, and Rossana Sciubba Pace. 1976. *Le comunita di base in Italia*, 2 vols. Rome: Coines.

Scott, Hilda. 1982. *Sweden's Right to be Human.* London: Allison and Busby.

Sars, David. 1983. On the Persistence of Early Political Predispositions. In L. Wheeler, ed., *Review of Personality and Social Psychology*, vol. 4. Beverly Hills, CA: Sage Publications.

Sevill, Lydia. 1985. Twinning: A Network of Support. *Sanity*, December.

Sheail, John. 1976. *Nature in Trust: The History of Conservation in Britain.* London: Blackie.

Sieferle, Rolf P. 1984. *Fortschrittsfeinde? Opposition gegen Technik und Industrie von der Romantik bis zur Gegenwart.* München: C. H. Beck.

Simon, Herbert. 1979. Rational Decision Making in Business Organizations. *American Economic Review*, 69: 493–513.

Sjoeblom, Gunnar. 1983. Political Change and Political Accountability. In Hans Daalder and Peter Mair, eds, *Western European Party Systems.* Beverly Hills, CA: Sage Publications.

Smelser, Neil. 1962. *Theory of Collective Behavior.* New York: Free Press.

Smith, Gordon, 1984a. *Politics in Western Europe.* 4th edn, London: Heinemann.

Smith, Gordon, 1984b. Social Movements and Party Systems in Western Europe. In M. Kolinsky and William Patterson, eds, *Social and Political Movements in Western Europe.* New York: St Martin's Press.

Sofri, Ariano. 1985 Intervento. In Democrazia Proletaria, ed., *1968/76: Le vere ragioni.* Milan: Mazzotta.

Spriano, Paolo. 1964. *The Occupation of the Factories.* London: Pluto.

Stacey, Margaret, and Marion Price. 1980. *Women, Power and Politics.* London: Tavistock.

Staggenborg, Susan. 1986. Coalition Work in the Pro-Choice Movement: Organizational and Environmental Opportunities and Obstacles. *Social Problems*, 33: 374–90.

Stephenson, S. M., and C. J. Brotherton. 1979. *Industrial Relations: A Social Psychological Approach.* Chichester: Wiley.

Sylos-Labini, Paolo. 1975. *Saggio sulle classi sociali.* Bari: Laterza.

Szabo, Stephen. 1983. *The Successor Generation.* London: Butterworth.

Tarrow, Sidney. 1967. *Peasant Communism in Southern Italy.* New Haven, CT: Yale University Press.

Tarrow, Sidney. 1974. Partisanship and Political Exchange in Italian Local Politics. *Sage Professional Papers in Contemporary Political Sociology*, vol. 1. Beverly Hills, CA: Sage Publications.

Tarrow, Sidney. 1983. *Struggling to Reform: Social Movements and Policy Change During Cycles of Protest.* Ithaca, NY: Cornell University (Western Societies Paper No. 15).

Tarrow, Sidney. 1986. Comparing Social Movements. *International Journal of Mass Emergencies and Disasters*, 4: 39–54.

Tarrow, Sidney. 1988a. National Politics and Collective Action: A Review of Recent Research in Western Europe and the United States. *Annual Review of Sociology*, 14: 421–40.

Tarrow, Sidney. 1988b. Old Movements in New Cycles: The Career of a Neighborhood Religious Movement in Italy. In Bert Klandermans, Hanspeter Kriesi, and Sidney Tarrow, eds, *From Structure to Action: Comparing Movement Participation Across Cultures.* Greenwich, CT: JAI Press.

Tarrow, Sidney. 1989. *Democracy and Disorder: Social Conflict, Protest and Politics in Italy, 1965–1975.* Oxford: Oxford University Press.

Taylor, Richard, and Colin Pritchard. 1980. *The Protest Makers: The British Nuclear Disarmament Movement of 1958 to 1965. Twenty Years On.* Oxford: Pergamon.

Taylor, Verta. 1988. Social Movement Continuity: The Women's Movement in Abeyance. Unpublished manuscript, Department of Sociology, Ohio State University.

Terchek, Ronald J. 1974. Protest and Bargaining. *Journal of Political Research*, 11: 133–45.

Tierney, Karen J. 1982. The Battered Women Movement and the Creation of the Wife Beating Problem. *Social Problems*, 29: 207–20.

Tilly, Charles. 1977. Hauptformen kollektiver Aktion in Westeuropa, 1500–1975. *Geschichte und Gesellschaft*, 3: 153–63.

Tilly, Charles. 1978. *From Mobilization to Revolution.* Reading: Addison-Wesley.

Tilly, Charles. 1984. *Big Structures, Large Processes, Huge Comparisons.* New York: Russell Sage Foundation.

Tilly, Charles. 1985. European Violence and Collective Action since 1700. *Social Research*, 52: 714–47.

Tilly, Charles. 1986. *The Contentious French.* Cambridge, MA: Harvard University Press.

Touraine, Alain. 1971. *The May Movement: Revolt and Reform.* New York: Random House.

Touraine, Alain. 1976. Crisis or Transformation? In Norman Birnbaum, ed., *Beyond the Crisis.* New York: Oxford University Press.

Touraine, Alain. 1977. *The Self-Production of Society*. Chicago, IL: University of Chicago Press.
Touraine, Alain. 1981. *The Voice and the Eye. An Analysis of Social Movements*. Cambridge: Cambridge University Press.
Touraine, Alain. 1983. *Anti-Nuclear Protest*. Cambridge: Cambridge University Press.
Tullock, Gordon. 1971. The Paradox of Revolution. *Public Choice*, 11: 425–47.
Turner, Ralph. 1970. Determinants of Social Movement Strategies. In T. Shibutani, ed., *Collective Behavior*. Englewood Cliffs, NJ: Prentice-Hall.
Turner, Ralph, and Lewis Killian. 1972. *Collective Behavior*. 2nd edn, Englewood Cliffs, NJ: Prentice-Hall.
Turner, Ralph, and Lewis Killian. 1987. *Collective Behavior*. 3rd edn, Englewood Cliffs, NJ: Prentice-Hall.
Van der Gaag, Nikki. 1985. Women Organizing. In Gorgina Ashworth and Lucy Bonnerjea, ed., *The Invisible Decade*. Aldershot, Hants: Gower.
Versteylen, Luc. 1981. *Wat nu met Agalev*? Borgerhout: Uitgeverij Stil Leven.
Vester, Heinz-Günter. 1985. Modernismus und Postmodernismus: Intellektuelle Spielereien? *Soziale Welt*, 36: 3–26.
Walsh, Edward. 1981. Resource Mobilization and Citizen Protest in Communities Around Three Mile Island. *Social Problems*, 29: 1–21.
Walsh, Edward. 1989. *Democracy in the Shadows: Citizen Mobilization in the Wake of the Accident at Three Mile Island*. New York: Greenwood.
Walsh, Edward, and Rex Warland. 1983. Social Movements in the Wake of a Nuclear Accident. *American Sociological Review*, 48: 764–80.
Wasmuht, Ulrike C. 1987. Die Entstehung und Entwicklung der Friedensbewegungen der achtziger Jahre. In Roland Roth and Dieter Rucht, eds, *Neue soziale Bewegungen in der Bundesrepublik Deutschland*. Frankfurt/New York: Campus.
Wassenberg, Arthur F. P. 1982. Neo-Corporatism and the Quest for Control: the Cuckoo Game. In Gerhard Lehmbruch and Philippe C. Schmitter, eds, *Patterns of Neo-Corporatist Policy-Making*. Beverly Hills, CA, and London: Sage Publications.
Watts, Nicholas S. J. 1987. Mobilisierungspotential und gesellschaftspolitische Bedeutung der neuen sozialen Bewegungen: Ein Vergleich der Länder der Europäischen Gemeinschaft. In Roland Roth and Dieter Rucht, eds, *Neue soziale Bewegungen in der Bundesrepublik Deutschland*. Frankfurt/New York: Campus.
Webb, Sidney, and Beatrice Webb. 1920. *Industrial Democracy*. London: Longmans, Green and Co.
Weber, Max. 1978. *Economy and Society*. 2 vols, Berkeley: University of California Press.
Wellhofer, Spencer E. 1985. The Electoral Effectiveness of Party Organization: Norway 1945–1977, *Scandinavian Political Studies*, 8: 171–85.
Wellhofer, Spencer E., and Timothy M. Hennessey. 1974. Models of Political Party Organization and Strategy. In Ivor Crewe, ed., *Elites in Western Democracy*, British Political Sociology Yearbook. London: Croom Helm.
Westle, Bettina. 1989. *Politische Legitimität*. Baden-Baden: Nomos.
Wiener, Martin. 1981. *English Culture and the Decline of the Industrial Spirit, 1850–1950*. Cambridge: Cambridge University Press.
Wiesendahl, Elmar. 1987. Neue soziale Bewegungen und moderne Demokratietheorie. In Roland Roth and Dieter Rucht, eds, *Neue soziale Bewegungen in der Bundesrepublik Deutschland*. Frankfurt/New York: Campus.
Wiesenthal, Helmut. 1987. Sieben Thesen zur Interaktion zwischen Grünen und neuen sozialen Bewegungen. Paper prepared for the meeting of the Work Group

"New Social Movements" in the German Political Science Association, July 10–12, in Meckenheim-Merl.

Wiesenthal, Helmut. 1988. Rehe am Weltmarkt. Unpublished manuscript.

Wilensky, Harold L. 1976. *The 'New Corporatism,' Centralization, and the Welfare State.* Beverly Hills, CA, and London: Sage Pubications.

Wilson, Frank L. 1979. The Revitalization of French Parties. *Comparative Political Studies,* 12: 82–103.

Wilson, Frank L. 1983. Interest Groups and Politics in Western Europe: The Neo-Corporatist Approach. *Comparative Politics,* 16: 105–23.

Wilson, Frank L. 1988. When Parties refuse to Fail: The Case of France. In Kay Lawson and Peter Merkl, eds, *When Parties Fail* Princeton, NJ: Princeton University Press.

Wilson, James Q. 1962. *The Amateur Democrat.* Chicago, IL: University of Chicago Press.

Wilson, James Q. 1973. *Political Organizations.* New York: Basic Books.

Winter, Lieven De. 1981. De parteipolitisering als instrument van particratie: Een overzicht van de ontwikkeling sinds de Tweede Wereldoorlog. *Res Publica,* 23: 53–107.

Wolsfeld, Gadi. 1986. Political Action Repertoires: The Role of Efficacy. *Comparative Political Studies.* 19: 104–29.

Women against the Bomb (Dorothy Thompson, ed.). 1983. *Over Our Dead Bodies.* London: Virago Press.

Zablocki, Benjamin. 1980. *Alienation and Charisma: A Study in Contemporary American Communes.* New York: Free Press.

Zajonc, Robert B. 1980. Feeling and Thinking: Preferences Need No Inferences. *American Psychologist,* 35: 151–75.

Zald, Mayer N. 1987. The Future of Social Movements. In Mayer Zald and John McCarthy, eds, *Social Movements in an Organizational Society.* New Brunswick/Oxford: Transaction Books.

Zald, Mayer N., and Roberta Ash. 1966. Social Movement Organizations. *Social Forces,* 44: 327–41.

Zald, Mayer, and Michael Berger. 1987. Social Movements in Organizations: Coup d'Etat, Bureaucratic Insurgency and Mass Movement. In Mayer Zald and John McCarthy, eds, *Social Movements in an Organizational Society.* New Brunswick/Oxford: Transaction Books.

Zald, Mayer N., and John D. McCarthy. 1980. Social Movement Industries: Competition and Cooperation Among Movement Organizations. In Louis Kriesberg, ed., *Research in Social Movements, Conflicts and Change,* vol. 3. Greenwich, CT: JAI Press.

Zald, Mayer N., and Bert Useem. 1982. Movement and Countermovement: Loosely Coupled Conflict. Paper presented at the 77th Annual Meeting of the American Sociological Association, San Francisco.

Zald, Mayer N., and John D. McCarthy, eds. 1987. *Social Movements in an Organizational Society: Collected Essays.* New Brunswick/Oxford: Transaction Books.

Zald, Mayer, and Bert Useem. 1987. Movement and Countermovement Interaction. In Mayer N. Zald and John D. McCarthy, eds, *Social Movements in an Organizational Society.* New Brunswick, NJ: Transaction Books.

Zentralarchiv. 1987. *Wahlstudie 1987* (panel study). Cologne: Zentralarchiv für empirische Sozialforschung, University of Cologne.

Zeuner, B. 1984. Parlamentarisierung der Grünen. *Prokla,* 15.

Zolberg, Aristide. 1972. Moments of Madness. *Politics and Society,* 2: 183–207.

Index